Satoyama Initiative Thematic Review

Series Editor

Tsunao Watanabe, United Nations University Institute for the Advanced Study of Sustainability (UNU-IAS), Tokyo, Japan

This Open Access book series aims to make timely and targeted contributions for decision-makers and on-the-ground practitioners by producing knowledge concerning "socio-ecological production landscapes and seascapes" (SEPLS) – areas where production activities help maintain biodiversity and ecosystem services in various forms while sustainably supporting the livelihoods and well-being of local communities. Each volume is designed as a compilation of case studies providing useful knowledge and lessons focusing on a specific theme that is important for SEPLS, accompanied by a synthesis chapter that extracts lessons learned through the case studies to present them for policy-relevant academic discussions. The series is also intended to contribute to efforts being made by researchers and scientists to strengthen the evidence base on socio-ecological dynamics and resilience, including those under the Intergovernmental Science-Policy Platform on Biodiversity and Ecosystem Services (IPBES) and the Convention on Biological Diversity (CBD).

The promotion and conservation of SEPLS have been the focus of the Satoyama Initiative, a global effort to realise societies in harmony with nature. In 2010, the International Partnership for the Satoyama Initiative (IPSI) was established to implement the concept of the Satoyama Initiative and promote various activities by enhancing awareness and creating synergies among those working with SEPLS. As a unique platform joined by governmental, intergovernmental, nongovernmental, private-sector, academic, and indigenous peoples' organisations, IPSI has been collecting and sharing the information, lessons, and experiences on SEPLS to accumulate a diverse range of knowledge. Case studies to be published in this series will be solicited from practitioners, researchers, and others working on the ground, who are affiliated with IPSI member organizations and closely involved and familiar with the activities related to SEPLS management.

Maiko Nishi • Suneetha M. Subramanian •
Philip Varghese •
Juliano Sènanmi Hermann Houndonougbo
Editors

Ensuring Ecological Connectivity in Socio-Ecological Production Landscapes and Seascapes (SEPLS)

Springer

Editors
Maiko Nishi
United Nations University Institute for the Advanced Study of Sustainability (UNU-IAS)
Tokyo, Japan

College of Bioresource Sciences
Nihon University
Fujisawa, Japan

Philip Varghese
Akita International University
Akita, Japan

United Nations University Institute for the Advanced Study of Sustainability (UNU-IAS)
Tokyo, Japan

CHRIST University
Bangalore, India

Suneetha M. Subramanian
United Nations University Institute for the Advanced Study of Sustainability (UNU-IAS)
Tokyo, Japan

Juliano Sènanmi Hermann Houndonougbo
Integrated Science and Engineering for Sustainable Societies
Chuo University
Tokyo, Japan

United Nations University Institute for the Advanced Study of Sustainability (UNU-IAS)
Tokyo, Japan

ISSN 2731-5169 ISSN 2731-5177 (electronic)
Satoyama Initiative Thematic Review
ISBN 978-981-95-1473-1 ISBN 978-981-95-1474-8 (eBook)
https://doi.org/10.1007/978-981-95-1474-8

This Springer imprint is published by the registered company Springer Nature Singapore Pte Ltd.
The registered company address is: 152 Beach Road, #21-01/04 Gateway East, Singapore 189721, Singapore

Foreword

Ecological connectivity is a fundamental component of both biodiversity conservation and climate resilience. It supports pollination, enhances genetic diversity and promotes ecosystem restoration and overall health. Conventional approaches to conservation, focusing on listed species and protected areas, have proven insufficient—failing to arrest ecological fragmentation and aggravating the global biodiversity crisis under increasing anthropogenic pressures. This underscores the urgency of rethinking conservation strategies.

From the Aichi Biodiversity Targets to the Kunming–Montreal Global Biodiversity Framework, international agreements have increasingly emphasized the need for well-connected landscapes and seascapes. Yet efforts to implement these ambitious agendas have been faced with myriad conceptual, methodological and practical challenges. Despite some recent improvements, only 7% of the world's landmass is both protected and connected, far less than the 17% pursued under Aichi Biodiversity Target 11 (UNEP-WCMC and IUCN 2021).

This book discusses the growing relevance of ecological connectivity in global efforts on biodiversity conservation and sustainable development. It takes a similar approach to previous volumes in the Satoyama Initiative Thematic Review (SITR) series, using diverse case studies from partner organizations of the International Partnership for the Satoyama Initiative (IPSI) to gather insights from socio-ecological production landscapes and seascapes (SEPLS). By synthesizing local and regional perspectives, the book reaffirms the crucial importance of balancing ecological and socio-economic priorities while aiming for ecological connectivity. It also showcases innovative spatial planning approaches and examples of stakeholder collaboration, and provides potential practical pathways to achieve equitable, sustainable and climate-resilient systems.

The theme of this edition of the SITR is both critical and timely, addressing the urgent need for ecological connectivity while keeping the sustainable use of natural resources at the core of the discussion. The diverse case studies drawn from four continents provide an engaging, insightful and inspiring read. I sincerely hope that readers find the volume a valuable resource that informs and supports your research, policy development and other activities.

Shinobu Yume Yamaguchi
Director, United Nations University
Institute for the Advanced Study of Sustainability
Tokyo, Japan

Foreword

Since the last decade, the United Nations University Institute for the Advanced Study of Sustainability (UNU-IAS) has been working with the partners of the International Partnership for the Satoyama Initiative (IPSI) on the Satoyama Initiative Thematic Review (SITR) publication series. This series aims to curate experiences from diverse socio-ecological production landscapes and seascapes (SEPLS) that speak to important topics of policy relevance, including the mandates of the Convention on Biological Diversity (CBD), the Intergovernmental Science-Policy Platform on Biodiversity and Ecosystem Services (IPBES), and broader topics related to the sustainable development goals (SDGs), among others.

In this volume, the focus is on ecological connectivity in the context of SEPLS. Across the different chapters, we have sought to illustrate how ecological connectivity is conceptualized and fostered at the SEPLS scale, what challenges are encountered by the actors in sustaining such connectivity, what drivers affect it, and what opportunities we can leverage towards ensuring the flow of natural processes and their benefits for human well-being.

This topic is highly relevant within the CBD context and is a subject of an upcoming IPBES assessment. Besides these, it is of high significance within the Convention on Migratory Species (CMS) and the UN Convention to Combat Desertification (UNCCD) that place emphasis on equitable and sustainable spatial planning and ecological connectivity.

As in previous volumes, in this volume too I am pleased to note the diversity of experiences across the selected case studies. Again, I am struck by the fact that despite the different contexts, several common messages recur that include similar challenges and contextually driven approaches to arrive at outcomes that are supportive of flourishing ecosystems and biodiversity, their connectedness, and a good quality of human life through innovative practices.

As is our practice, this volume too was developed in the spirit of providing credible, well-analysed information through a process of co-design and discursive analysis amongst IPSI partners, and in line with the new Strategy and Plan of Action of IPSI.

I hope you will enjoy the experiences highlighted in this volume and are inspired to uptake the lessons in your own contexts and work.

Tsunao Watanabe
Series Editor, Satoyama Initiative Thematic Review,
Visiting Research Fellow of Biodiversity and Society,
United Nations University Institute for the Advanced Study of Sustainability
Director, The Secretariat of the International Partnership for the Satoyama Initiative
Tokyo, Japan

Preface

The Satoyama Initiative is "a global effort to realize societies in harmony with nature," started through a collaboration between the United Nations University (UNU) and the Ministry of the Environment of Japan. The initiative focuses on the revitalization and management of "socio-ecological production landscapes and seascapes" (SEPLS), areas where production activities help maintain biodiversity and ecosystem services in various forms while sustainably supporting the livelihoods and well-being of local communities. In 2010, the International Partnership for the Satoyama Initiative (IPSI) was established to implement the concept of the Satoyama Initiative and promote various activities by enhancing awareness and creating synergies among those working with SEPLS. IPSI provides a unique platform for organizations to exchange views and experiences and to find partners for collaboration. As of April 2025, 337 members have joined the partnership, including governmental, intergovernmental, non-governmental, private-sector, academic, and Indigenous peoples' organizations.

The Satoyama Initiative promotes the concept of SEPLS through a threefold approach that argues for connection of land and seascapes holistically for management of SEPLS (see Fig. 1). This often means involvement of multiple sectors at the landscape or seascape scale, under which it seeks to: (1) consolidate wisdom in securing diverse ecosystem services and values, (2) integrate traditional ecosystem knowledge and modern science, and (3) explore new forms of co-management systems. Furthermore, activities for SEPLS management cover multiple dimensions, such as equity, poverty alleviation, forest restoration, and incorporation of traditional knowledge for sustainable management practices in primary production processes such as agriculture, fisheries, and forestry (UNU-IAS & IGES 2015).

As one of its core functions, IPSI serves as a knowledge-sharing platform through the collection and sharing of information and experiences on SEPLS, providing a place for discussion among members and beyond. Over 320 case studies have been collected and are shared on the IPSI website, providing a wide range of knowledge covering diverse issues related to SEPLS. Discussions have also been held to further strengthen IPSI's knowledge-facilitation functions, with members suggesting that efforts should be made to produce knowledge on specific issues in SEPLS in order

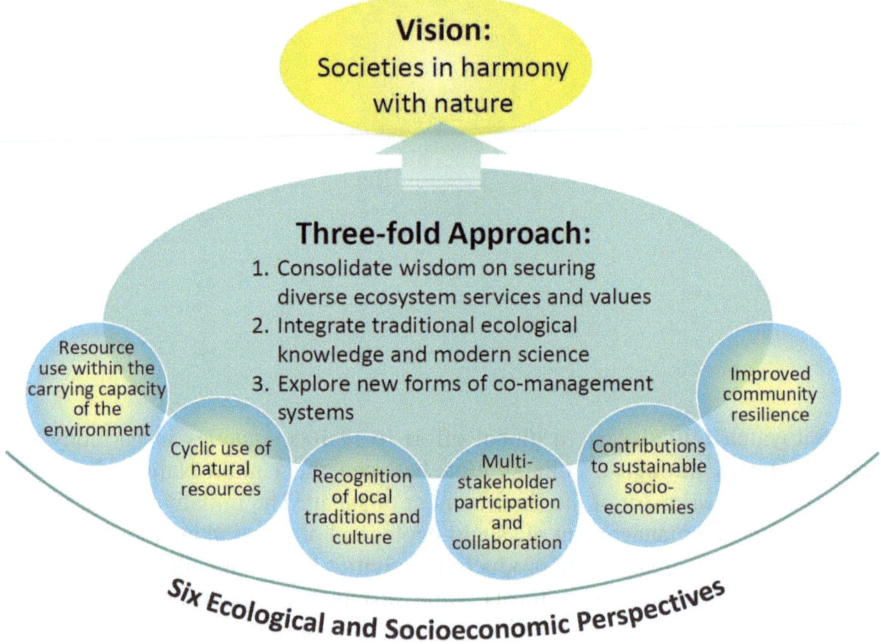

Fig. 1 The conceptual framework of the Satoyama Initiative

to make more targeted contributions to decision makers and on-the-ground practitioners.

It is in this context that a project to create a publication series titled the Satoyama Initiative Thematic Review (SITR)[1] was initiated in 2015. The SITR series was developed as a compilation of case studies providing useful knowledge and lessons focusing on a specific theme that is important for SEPLS. The overall aim of the SITR publications is to collect experiences and relevant knowledge, especially from practitioners working on the ground, considering their usefulness in providing concrete and practical knowledge and information as well as their potential to contribute to policy recommendations. Each volume is also accompanied by a synthesis chapter which extracts lessons learned through the case studies, presenting them for policy-relevant and academic discussions. This series also contributes to efforts being made by researchers to strengthen the evidence base for policymaking concerning social-ecological dynamics and resilience, including those under the Intergovernmental Science-Policy Platform on Biodiversity and Ecosystem Services (IPBES) and the Convention on Biological Diversity (CBD). Nine volumes have been published since 2015 for this series on various timely topics such as knowledge enhancement, biodiversity mainstreaming, sustainable livelihoods, effective

[1]The previous volumes of the SITR series are available at: https:/satoyamainitiative.org/featured_activities/sitr/

area-based conservation, multiple values associated with sustainable use of biodiversity, transformative change, biodiversity-health-sustainability nexus, ecosystem restoration, and business and biodiversity.

Building on the earlier volumes that were in-house publications of UNU, the sixth edition and onward have been published by Springer Nature to reach out to a broader range of readers. In particular, the SITR has become a Springer book series from the seventh volume, with the aim of ensuring consistent and coherent contributions to science-policy-practice interfaces while enhancing the publications' impact and reach a wide audience. Furthermore, a review committee has been established since the seventh volume by inviting experienced and knowledgeable experts in SEPLS management from the IPSI community to engage in the review process for publication. This expert group has helped to ensure credibility and reinforce the quality of the contents and at the same time to facilitate sharing of expertise and collaboration among IPSI members. The SITR series currently serves as one of the flagship projects underpinning the effort to attain "Knowledge Co-Production, Management, and Uptake," one of the five strategic objectives for the Strategy and Plan of Action 2023–2030 that was adopted at the Ninth IPSI Global Conference (IPSI-9) in 2023.

Similar to the previous volumes, this tenth edition was developed through a multi-stage process, including both peer review and discussions among the authors and reviewers at a workshop. The authors had several opportunities to receive feedback, which helped them to improve their manuscripts in substance, quality, and relevance. First, each manuscript received comments from the editors, the review committee members, and authors of two other chapters. Throughout the peer review process, each author received feedback from all these reviewers who were requested to comment on whether the manuscript was easy to understand and informative, addressed key questions of the volume's focus, and provided useful lessons as contributions to the theme of the volume. The aforementioned workshop was then held in person in Tokyo on June 23–25, 2024 to enable the exchange of feedback between authors and reviewers. The basic ideas contained in the synthesis of the concluding chapter were developed from presentations and discussions during the workshop, and the chapter was made available for two rounds of review by authors and reviewers before finalization.

The above process offers an opportunity for authors from both academic and non-academic organizations to contribute to generating knowledge in an accessible and interactive way, as well as to provide high-quality papers written in simple language for academics and a broader audience alike. It is our hope that this publication will be useful in providing information and insights to practitioners, researchers, and policymakers on the importance of long-term collaborative management of SEPLS that helps to promote and enhance sustainable business activities through which biodiversity can thrive. This, we hope, will prompt policymaking that strengthens such integrated and holistic management approaches.

We would like to thank all the authors who contributed their case studies. We also appreciate the continued commitment and support for the thorough review process by the four experts in the review committee: Guido Gualandi, Stefan Hotes,

Uma Khumairoh, and Mordecai Owidi Ogada. We are also grateful to the other participants in the case study workshop who provided insightful remarks and valuable inputs into the discussions. These individuals include Wataru Suzuki and Jennifer Kelleher. We also thank the colleagues of UNU-IAS and Prime International Co., Ltd. who were supportive and instrumental in organizing the workshop and facilitating the publication process: Saori Aizawa, Alexandra Franco Guajardo, Himangana Gupta, Ikuyo Kaseda, Miles Lambert-Peck, Rina Miyake, Tomonobu Tosaki, Nicholas Turner, Tomoko Yamamoto, and Madoka Yoshino. Our gratitude extends to the UNU administration, especially Francesco Foghetti, and to colleagues from the United Nations Geospatial for all the instrumental support through the publication process. Furthermore, we acknowledge Susan Yoshimura who skilfully proofread the manuscripts.

Publication of this volume as a new Springer Book would not have been possible without the helpful guidance of Mei Hann Lee and Momoko Asawa from Springer and the institutional support and leadership of Shinobu Yume Yamaguchi and Tsunao Watanabe from UNU-IAS. Our grateful thanks are also due to the Ministry of the Environment, Japan, for supporting the activities of IPSI and its secretariat hosted by UNU-IAS.

Fujisawa, Japan — Maiko Nishi
Tokyo, Japan — Suneetha M. Subramanian
Akita, Japan — Philip Varghese
Tokyo, Japan — Juliano Sènanmi Hermann Houndonougbo

Reference

UNU-IAS & IGES (eds.) (2015) Enhancing knowledge for better management of socio-ecological production landscapes and seascapes (SEPLS). Satoyama Initiative Thematic Review, vol. 1, United Nations University Institute for the Advanced Study of Sustainability, Tokyo

Acknowledgement

Review Committee

Guido Gualandi
Stefan Hotes
Uma Khumairoh
Mordecai Ogada

Editorial Support

Madoka Yoshino
Tomoko Yamamoto

English Proofreading

Susan Yoshimura

Contents

Editors and Contributors

About the Editors

Maiko Nishi is an Adjunct Professor at United Nations University Institute for the Advanced Study of Sustainability, and Professor at Department of Global Coexistence Studies, College of Bioresource Sciences, Nihon University. Her research interests include social-ecological system governance, regional planning, and agricultural land policy.

Suneetha M. Subramanian is a Research Fellow at United Nations University Institute for the Advanced Study of Sustainability. Her research interests include biodiversity and human well-being with a focus on equity, traditional knowledge, community well-being, and social-ecological resilience.

Philip Varghese is a JSPS-UNU Postdoctoral Fellow at Akita International University, Japan, and United Nations University Institute for the Advanced Study of Sustainability during 2022–2024. He currently teaches in the Department of Political Science and International Studies at CHRIST University in India. His research interests include tourism and development politics, Indigenous communities, and sustainability.

Juliano Sènanmi Hermann Houndonougbo is a JSPS-UNU Postdoctoral Fellow at Chuo University, Japan, and United Nations University Institute for the Advanced Study of Sustainability. His research interests include applied landscape ecology, environmental change, and rapid urbanization.

About the Contributors

Chapter 2

Abdulai Sulemana is a conservation ecologist working as a Programme Manager at Advocates for Biodiversity Conservation, Ghana, focusing on promoting sustainable agriculture practices and environmental protection in local communities.

Daniel Enti is an agriculturist working as a Programme Officer at Advocates for Biodiversity Conservation in Ghana. His work focuses on promoting agroforestry, supporting farmers, local communities, and government agencies in adopting climate-smart technologies.

Vincent Awotwe-Pratt is an Ecologist and works with data to influence natural resource and environmental decision-making. He has been working with communities and farmer groups by providing them with tools and approaches for conservation decisions.

Chapter 3

Chemuku Wekesa is a Landscape Ecologist with Kenya Forestry Research Institute (KEFRI). He leads action research on ecosystem functions, diversity, and dynamics with special focus on biodiversity conservation, agroecology, Indigenous knowledge systems, and restoration of degraded ecosystems.

Leila Ndalilo is an Environmental Social Scientist at Kenya Forestry Research Institute (KEFRI). Her main research focus is in the area of Socioeconomics, Policy and Governance and largely entails integration of community perspectives into forestry research.

Krystyna Swiderska is a Principal Researcher and team leader for traditional knowledge and biocultural heritage at the International Institute for Environment and Development (IIED). Her work focuses on protecting the interlinked biodiversity and cultural heritage.

Gladys Rutto is a Forest Ecologist with Kenya Forestry Research Institute (KEFRI). Her research focuses on conducting assessments and evaluation of forest ecosystems to guide management and conservation initiatives.

Chapter 4

Yu-Chun Chan was a Project Manager at the Tse-Xin Organic Agriculture Foundation. He works on projects with local residents and Indigenous stakeholders to promote the transformation of local agriculture and industry to live in harmony with nature.

Jing-Juan Hsu is a Coordinator at the Tse-Xin Organic Agriculture Foundation, responsible for the promotion of organic agriculture and international exchanges.

Jia-Xin Yang is a Coordinator at the Tse-Xin Organic Agriculture Foundation, responsible for communicating with Indigenous farmers in the region and supporting their efforts to transform traditional agriculture into a modern, sustainable business.

Chun-En Xiao works as a part-time staff member in the Tse-Xin Organic Agriculture Foundation, focusing on the cultural heritage and revitalization of the Nan'an Bunun community.

Chapter 5

Jeeranuch Sakkhamduang works as an expert at the Thailand Environment Institute, Thailand, in projects focused on applying nature-based solutions for climate change adaptation, forest governance, sustainable agriculture, and water resource management.

Rattanaporn Charoenchad is a Research Assistant at Thailand Environment Institute, Thailand, and works in projects focused on forest resource management, urban forest, and the Thailand State of Environment Report.

Chapter 6

Emilio R. Díaz-Varela is an Associate Professor at the University of Santiago de Compostela, Spain. His research interests are social-ecological systems analysis and management, spatial analysis in complex patterns, and sustainable development and resilience in rural and urban systems.

Pedro Álvarez-Álvarez is an Associate Professor in the Department of Organisms and Systems Biology, University of Oviedo, Spain. He is interested in sustainable forest management, complex forests, and traditional chestnut (*Castanea sativa* Mill.) stands.

José C. Pérez-Girón is a Postdoctoral Researcher at the Botany Department of the University of Granada, Spain. He is interested in primary production indicators and the potential effects of climate change on plant and animal species.

José V. López-Bao is a Conservation Scientist at the Spanish National Research Council (CSIC). He is interested in the integration of quantitative, interdisciplinary, and transdisciplinary approaches to face conservation challenges.

Fernando Ballesteros is a biologist and nature documentary filmmaker. He was Coordinator of EU LIFE projects at the Brown Bear Foundation. He passed away in August 2023 during the completion of this study.

María Gómez Manzanedo is a forestry engineer specializing in brown bear habitat management and conservation and ecosystem restoration.

María Párraga holds a PhD in Animal Sciences (University of Padova, Italy), with a specialty in radio tagging and wildlife tracking. Currently, she is a LIFE Project Coordinator at the Brown Bear Foundation, Spain.

Guillermo Palomero is President of the Brown Bear Foundation, Spain. He is a naturalist specializing in the conservation of large carnivores in the Cantabrian Mountains and Pyrenees.

Chapter 7

Andrés Quintero-Ángel is a PhD candidate in Environmental Sciences and Executive Director of Corporación Ambiental y Forestal del Pacifico (CORFOPAL), who majored in conservation and use of biodiversity with ethnic communities.

Sebastian Orjuela-Salazar a biologist, Coordinator of the "Conserva Aves Initiative" at Fondo Acción, who majored in planning and declaration of protected areas and other effective area-based conservation measures.

Mauricio Quintero-Ángel a Doctor in Environmental Sciences and Full Professor at Universidad del Valle at Palmira Campus. He is interested in research of social-ecological systems, landscape planning, and rural development.

Leonor Valenzuela is a Doctor of Ecology and Coordinator of Analysis and Synthesis for Wildlife Conservation Society (WCS) Colombia, specializing in monitoring the effectiveness of biodiversity conservation and management measures and understanding the basic principles underlying observed patterns of diversity.

Diana Saavedr-Zuñiga a social worker specializing in social management, with 14 years of experience as a consultant and coordinator of disaster risk management projects, natural protected areas, and watershed management plans.

Daniel Osorio-Domínguez is a Doctor of Ecology and Evolutionary Biology, Scientific and Research Director of Corporación Ambiental y Forestal del Pacifico (CORFOPAL), and Professor at Universidad Javeriana-Cali, interested in the implementation/research of conservation in productive systems, spatial modelling, and climate change.

Chapter 8

Yoji Natori is an Associate Professor at Akita International University, teaching environmental science, conservation, and sustainable development. He is also an Honorary Advisor to Conservation International Japan and Vice Chair of the Japan Committee for IUCN.

Hamelda Francisca Majit is an Assistant Research Officer at Sabah Parks. She is involved in various programs related to environment, forestry, conservation, community engagement, management plans, and establishment of new parks. Previously, she was involved in the Kinabalu Ecolinc Project.

Remmy Alfie Awang is an Assistant Park Manager at Sabah Parks. He is responsible for matters related to enforcement and community on eight islands. Previously, he worked as a Community Engagement Officer for the Kinabalu Ecolinc Project.

Alfred Marinus is a Senior Park Ranger at Sabah Parks. He is responsible for fulfilling assigned duties, focusing on Crocker Range Biosphere programs across 10 districts.

Fazrullah Rizally Abdul Razak is a Senior Principal Assistant Director at Sabah Parks. He leads the Management Division, coordinating various programs/activities related to park management plans, environment, privatization matters, community engagement, and establishment of new parks, with a focus on long-term conservation and sustainability within the park.

Gerald Jetony (see chapter 9)

Alessandra Markos (see chapter 9)

Chapter 9

Gerald Jetony is Director of the Sabah Biodiversity Centre (SaBC), Malaysia, a state government department. He is experienced in policy formulation and natural resource management, particularly in the field related to community-based conservation as well as biodiversity conservation.

Isabelle Lackman is a Co-founder and Director of HUTAN, Malaysia. HUTAN is a non-governmental organization whose work focuses on empowering local communities to manage and protect wildlife in Kinabatangan, Sabah, Malaysia.

Alessandra Markos is an Assistant Director at Sabah Biodiversity Centre, Malaysia, a state government department. Alessandra is directly involved in the management, conservation, protection, and sustainable utilization of Sabah's biological resources.

Vivi Sandra Victor is an Assistant Administrative Officer at Sabah Biodiversity Centre (SaBC), Malaysia, a state government department under the Chief Minister's Department.

Chapter 10

Paulina G. Karim is a Part-time Assistant Professor and Postdoctoral Researcher at @ScapesLab, the College of Environmental Studies and Oceanography at National Dong Hwa University (NDHU). Her research interests include environmental management, integrated landscape-seascape approach (ILSA), IPLC engagement, and marine conservation.

Hwa-Ching Lin is the Director General of the Forestry and Nature Conservation Agency (FANCA). He has a longstanding interest in wildlife-human interactions and is currently promoting the integration of SEPLs and IPLC issues into forestry and biodiversity policies.

Chium-Tse Huang is the Division Director of FANCA-Hualien. He has been involved in development of various national nature conservation policies and currently leads the team of FANCA-Hualien in promoting the Taiwan Ecological Network in Hualien County.

Mei-Yin Chen is a Technical Specialist at FANCA-Hualien. She is FANCA's contact person for implementation of integrated landscape-seascape approaches (ILSA) and assessment of resilience at the Xinshe, Gangkou, Luoshan, and Fengnan SEPLS, Hualien County.

Yuan-Jyun Wang is a Technical Specialist at FANCA-Hualien. He specializes in habitat conservation and wildlife management and is responsible for implementation of the Taiwan Ecological Network in Hualien County, including spatial planning and multi-stakeholder participation.

Yu-Chuan Lo is the Chief of Nature Conservation Management Division, FANCA. Her research interests include forest recreation area planning, wildlife conservation, and habitat protection.

Chia-Tzu Chen is a Specialist at FANCA. She majored in environmental economics and has been engaged in the Taiwan Ecological Network and the evaluation of ecosystem services value.

Kuang-Chung Lee is a Professor at @ScapesLab, the College of Environmental Studies and Oceanography at National Dong Hwa University (NDHU). He specializes in community-based approaches to landscape conservation, protected areas, and promotion of the Satoyama Initiative.

Chih-Chin Shih is a Senior Specialist at FANCA. She is committed to promoting the Taiwan Ecological Network and effectively using biodiversity data for policy-making and spatial planning.

Chapter 11

Chen-Yang Lee is the Director General of the Agency of Rural Development and Soil and Water Conservation (ARDSWC) of the Ministry of Agriculture (MOA), leading the governmental agency to implement policy and strategies for soil and water conservation, rural regeneration, slopes, and supervision and management of landslides risk and warning systems.

Jung-Chun Chen is the Director of the Taichung Branch, ARDSWC. He leads the whole branch to implement soil and water conservation and rural construction projects in the jurisdiction.

Ling-Tsen Chen is the Division Chief of the Rural Construction Division, ARDSWC. She leads the division to implement rural regeneration plans including overall environmental improvement, construction of public facilities, renovation of individual homes, industrial revitalization, cultural preservation and utilization, and ecological conservation.

Chen-Fa Wu is a Professor in the Department of Horticulture at National Chung Hsing University (NCHU). Dr Wu's research interests and expertise include rural development, horticultural therapy, land-use planning, landscape design, and landscape ecology.

Szu-Hung Chen is an Assistant Professor in the International Agriculture Master Program at NCHU. She is also well known as Dr. Vickey Chen. Her research interests and expertise include sustainable development, environmental modelling, ecosystem services, community resilience assessment, and applications of GIS and spatial analysis.

Zhao-Hong Zhang is the Section Chief of the Rural Regeneration Construction section at the Taichung Branch, ARDSWC. His duty is assisting rural communities in maintaining rural ecology and culture and improving quality of life.

Ching-Chien Huang is the Director General at Wufeng Farmers' Association. He promotes a new agriculture value chain in the southern area of Taichung city.

Hao-Yun Chuang is a Senior Engineer at the Rural Construction Division, ARDSWC. He is in charge of implementing rural regeneration plans, including overall environmental improvement, construction of public facilities, renovation of individual homes, industrial revitalization, cultural preservation and utilization, and ecological conservation.

Chun-Hsien Lai is the Executive Director of the Taiwan Landscape Environment Association (TLEA). She serves as a project leader for assisting rural communities in implementing rural regeneration plans, guiding the operation of local organizations, and promoting sustainable rural development.

Luu Van Thong Trac is a PhD student in the Department of Horticulture at National Chung Hsing University. His research interests and expertise are agriculture, landscape ecology, and ecological conservation.

Hao-Wei Hsu is a Project Manager at TLEA. He focuses on promoting sustainable rural development and rural regeneration.

Yu-Chin Chen is an Assistant Engineer of the Rural Regeneration Construction section at the Taichung Branch, ARDSWC. She mainly incorporates the Satoyama Initiative into promoting rural regeneration.

Da-Li Lin is an Associate Research Fellow of the Taiwan Biodiversity Research Institute since 2011, working on biodiversity conservation, population trends and indicators, wildlife-habitat relationships, citizen science, and biodiversity-friendly agriculture.

Chapter 12

Somajita Paul is a former DS Kothari Postdoctoral Fellow and is currently involved in research, teaching, and consultancy in nature conservation and socioeconomic development.

Krishna Gopal Saxena is a Fellow of National Academy of Agricultural Sciences, India, and a Professor (retired), currently involved in research, teaching, and consultancy in ecology, natural resource management, and sustainable development.

S. Sreekesh is a Professor of Centre for the Study of Regional Development, Jawaharlal Nehru University, and geoscientist by training. He is currently involved in teaching and research in geo-environmental issues and climate change in environments such as forests, wetlands, coastal regions, mountains, and mining areas.

Chapter 13

Chen-Yang Lee (as written above)

Chi-Fang Chiu is the Director of the Tainan Branch, ARDSWC. He leads the whole branch to implement soil and water conservation and rural construction projects in the jurisdiction.

Kuei-Lin Fu is the Director of the Nantou Branch, ARDSWC. He leads the whole branch to implement soil and water conservation and rural construction projects in the jurisdiction.

Chen-Fa Wu (as written above)

Szu-Hung Chen (as written above)

Luu Van Thong Trac (as written above)

Jui-Min Chuis the Deputy Director of the Tainan Branch, ARDSWC. She is in charge of coordinating soil and water conservation and rural development with regional agricultural development policies in the jurisdiction.

Chuang-Chin Cheng is the Secretary of the Tainan Branch, ARDSWC. He is responsible for the planning, construction, leadership, sustainable management, human resource development, and promotion of rural revitalization and regeneration in rural communities.

Cheng Hsiu Tsai is the Section Chief of the Rural Regeneration Construction section at the Tainan Branch, ARDSWC. His duty is to assist rural communities in maintaining rural ecology and culture, improving quality of life, and building prosperous new rural areas.

Chien-Hung Tsai is an Assistant Engineer of the Rural Regeneration Construction section at the Tainan Branch, ARDSWC. She mainly incorporates the concepts and practices of the Satoyama Initiative into promoting rural regeneration, focusing on improving the quality of rural life, industrial development, and the protection of natural ecology and the environment.

Hao-Yun Chuang (as written above)

Chun-Hsien Lai (as written above)

Ming-Ju Lee is a Professor in the Department of Tourism and Leisure at National Penghu University of Science and Technology (NPU). His research interests and expertise include marine tourism, marine culture, recreation area planning, and stone weir culture and restoration.

Chen-Ju Wang is the CEO of Gain In Culture Studio. Wang is dedicated to rural community regeneration in Penghu. She is also acknowledged as the "Queen of Offshore Island Community Development" and the "Penghu Grandma." Her long-term commitment and practical spirit make her a role model admired by the younger generation.

Chun-Chie Hu is a Professor in the Department of Marine Recreation at NPU. His research interests and expertise are marine recreation, leisure fisheries, and strategic management.

Hsi-Chun Chen is an Assistant Professor in the Department of Soil and Water Conservation at NCHU. His research interests and expertise include rural development, landscape conservation, social-ecological innovation, and actor-network analysis.

Hao-Wei Hsu is a Project Manager at TLEA. He focuses on promoting sustainable rural development and rural regeneration.

Ting-Yu Linis a full-time research assistant at TLEA. He specializes in ecological surveys and the compilation of biodiversity data.

Chapter 1
Introduction

Maiko Nishi, Suneetha M. Subramanian, Philip Varghese, and Juliano Sènanmi Hermann Houndonougbo

Abstract This book documents 12 case studies providing useful knowledge and lessons concerning socio-ecological production landscapes and seascapes (SEPLS) with a focus on the roles of SEPLS in ensuring and enhancing ecological connectivity. With the understanding that landscape approaches underpinning SEPLS management can facilitate optimal spatial patterns for ecological connectivity toward more socially and ecologically sound outcomes, it explores how to minimize trade-offs and maximize synergies between multiple factors associated with SEPLS by ensuring the quality of ecological connectivity on a landscape or seascape scale. This book is relevant to the achievement of multiple sustainable development goals (SDGs), particularly SDGs 15 (Life on Land), 14 (Life below Water), and 17 (Partnerships for the Goals), whereas it offers practical and methodological insights

The original version of the chapter has been revised. A correction to this chapter can be found at https://doi.org/10.1007/978-981-95-1474-8_15

M. Nishi (✉)
United Nations University Institute for the Advanced Study of Sustainability (UNU-IAS), Tokyo, Japan

College of Bioresource Sciences, Nihon University, Fujisawa, Japan
e-mail: nishi.maiko@nihon-u.ac.jp

S. M. Subramanian
United Nations University Institute for the Advanced Study of Sustainability (UNU-IAS), Tokyo, Japan

P. Varghese
Akita International University, Akita, Japan

United Nations University Institute for the Advanced Study of Sustainability (UNU-IAS), Tokyo, Japan

CHRIST University, Bangalore, India

J. S. H. Houndonougbo
Integrated Science and Engineering for Sustainable Societies, Chuo University, Tokyo, Japan

United Nations University Institute for the Advanced Study of Sustainability (UNU-IAS), Tokyo, Japan

M. Nishi et al. (eds.), *Ensuring Ecological Connectivity in Socio-Ecological Production Landscapes and Seascapes (SEPLS)*, Satoyama Initiative Thematic Review, https://doi.org/10.1007/978-981-95-1474-8_1

into policy-making and implementation to help achieve multiple goals and targets of the Kunming–Montreal Global Biodiversity Framework (GBF), including 2050 Goal A (integrity, connectivity, and resilience of all ecosystems) and 2030 Targets 1 (spatial planning and effective management), 2 (ecosystem restoration), 3 (protected and conserved areas: 30x30), and 11 (ecosystem services and functions). This chapter provides a brief background, including the global importance of ensuring ecological connectivity and related challenges and opportunities, international processes as efforts to ensure and enhance ecological connectivity, and the relevance of SEPLS management to challenges and opportunities for ensuring ecological connectivity. It also clarifies the scope and objectives of this book and presents an overview of case studies and the organization of this book.

Keywords Socio-ecological production landscapes and seascapes · Ecological connectivity · Spatial planning · Biodiversity · Sustainability · Landscape approaches · Case studies · Science–policy–practice interface

1.1 Ecological Connectivity

Ecological connectivity underpins crucial ecological processes, retaining the functionality of ecosystems across biomes, habitats, and spatial scales (Lemieux et al. 2022; IPBES 2022). Key ecological features derived from connectivity include the maintenance of genetic diversity, pollination, seed dispersal, energy flow, nutrient cycling, and disease control. As such, ecological connectivity is essential for the survival of wild animals and plant species and, importantly, their migration. It also supports climate change mitigation and adaptation by allowing for the persistence and recovery of flora and fauna through the dispersal and movement of organisms and materials in response to a variety of environmental changes (Lemieux et al. 2022; Berkström et al. 2022).

With the increased understanding that conventional policies based on the preservation of listed species and protected areas are insufficient to curb the current trends of biodiversity loss, the concept of ecological connectivity has gained attention in the domain of conservation policies (Perrin et al. 2022). The traditional provisions for nature protection (e.g., nature reserves, Birds and Habitats Directives) can be effective in protecting large spaces of remarkable nature and preserving heritage species but are often limited in conserving ordinary biodiversity subjective to anthropogenic pressure, leading to landscape fragmentation.

In this context, in 2010, the Parties to the Convention on Biological Diversity (CBD) adopted the Strategic Plan for Biodiversity 2011–2020, where the importance of ecological connectivity was enshrined in its Aichi Target 11. This target calls for the conservation of at least 17% of terrestrial and inland water areas as well as 10% of coastal and marine areas of particular importance for biodiversity and ecosystem services by 2020 through "effectively and equitably managed, ecologically representative and well-connected systems" of protected areas and other effective area-based conservation measures (OECMs or conserved areas) to be "integrated into the wider landscape and seascape" (CBD 2010). It thus articulated the need to not only expand the areal extent of protected and conserved areas but also to ensure and enhance the quality of their management as systems for the achievement of the 2050 Vision for Biodiversity: "Living in harmony with nature".

However, progress made on the connectivity element of Aichi Target 11 has been limited (Gannon et al. 2019; Maxwell et al. 2020). Despite the achievement of protecting 17% of the terrestrial surface globally, the protected area considered connected accounts for only 7% (UNEP-WCMC and IUCN 2021). In particular, the areas that are most globally important for mammal movement continue to be unprotected, whereas 71% of them overlap with global priority areas for biodiversity (Brennan et al. 2022). Furthermore, the planning process of marine protected areas (MPAs) often neglects or oversimplifies an assessment of connectivity as it has only recently been added to the objectives of many MPAs (Roberts et al. 2021). Among 746 MPAs examined as those located in six countries or regions with advanced marine spatial planning (i.e., Australia, California, Canada, France, Hawaii, and the UK), only 11% took into account connectivity as a criterion for selection and design but increasingly so after 2007 (Balbar and Metaxas 2019).

In view of ever worsening global environmental crises that require meeting emergencies for biodiversity and climate change jointly, the global target was expanded to protect and conserve at least 30% of the world's lands and waters by 2030 in the Kunming–Montreal Global Biodiversity Framework (GBF) that was adopted at the 15th meeting of the Conference of the Parties to the CBD in December 2022 (Oppler et al. 2021). Given the urgency to address the challenges, a so-called "30x30" target was set out to take a grand step to pursue nature-positive pathways by rapidly scaling up efforts in protecting and conserving land, sea, and freshwater ecosystems around the world, wherein the importance of ecological connectivity is reiterated. Furthermore, given the increasing need to prioritize ecological connectivity in order to facilitate larger-scale ecosystem functionality and fortify the safety net for biodiversity in the climate change era (Oppler et al. 2021), the profile of ecological connectivity has been uplifted in the GBF within which the notions associated with ecological connectivity (e.g., spatial planning, integrity, and connectivity) have been incorporated throughout the overarching 2050 Goals (e.g., Goal A) and multiple 2030 Targets (e.g., Targets 1, 2, 3, and 11) (Lemieux et al. 2022).

Although ambitious global goals and targets have been set out, how to ensure the quality of ecological connectivity continues to be a challenge. The challenge entails both conceptual and methodological dimensions. As agreed internationally, ecological connectivity is defined as "the unimpeded movement of species and the flow of natural processes that sustain life on Earth" (CMS 2020, p. 1). However, it is conceptualized in varied ways by different people across scales and times and used differently in various disciplines (Beger et al. 2022; Butler et al. 2022). For instance, some may be concerned with connectivity to enable or enhance the movement of charismatic species for the purpose of promotion of ecotourism, while others may value connectivity that allows for propagule dispersal and fish spawning aggregations to design marine protected areas through the conservation of linked reefs. They thus refer to different phenomena as connectivity, which could be distinctive or even conflicting with each other. Consequently, conservation decisions may give rise to trade-offs among stakeholders who may value different types or dimensions of connectivity, impeding a unified approach to operationalize it in spatial planning and management (Beger et al. 2022).

This broad, complex, and value-laden conceptualization has methodological implications for how to measure ecological connectivity, assess its ecological functions and benefits, and deal with trade-offs. Certain kinds of connectivity may allow or accelerate species movements that have negative impacts on native ecosystems through biological invasions (Berkström et al. 2022). Conversely, fragmentation of ecosystems can exacerbate the impacts of either abrupt or progressive natural disturbances (e.g., floods, wildfires, and heat waves) and may hamper socioeconomic and cultural resources from being renewed at a landscape or seascape level (Butler et al. 2022). These outcomes appear to be unevenly distributed across different demographics and geographies, causing or amplifying the inequities that those most vulnerable or disadvantaged often confront more severely (Butler et al. 2022).

In this connection, the tenth session of the Plenary of the Intergovernmental Science-Policy Platform on Biodiversity and Ecosystem Services (IPBES) held in early 2023 agreed to undertake a methodological assessment of integrated biodiversity-inclusive spatial planning and ecological connectivity (so-called "spatial planning and connectivity assessment"). This assessment has been designed as a fast-track 2-year process to timely provide information for the implementation of the GBF, particularly its Targets 1 (spatial planning), 2 (ecosystem restoration), and 3 (protected and conserved areas). The requests for the assessment from the plenary include addressing an information gap concerning spatial planning measures to support conservation, restoration, and ecological connectivity; assessing corridor connectivity, landscape mosaics, and spatial patterns and planning to support effective conservation of biodiversity and nature's contributions to people; and assessing knowledge gaps for planning and investing in climate-ready marine fisheries and marine protected areas, among others (IPBES 2023).

As an integrated way to promote both conservation and development in a given area, landscape approaches facilitate a process to optimize the spatial patterns of natural and socioeconomic factors by designing and enhancing ecological connectivity for socially and ecologically sound outcomes. Such approaches have been practiced through managing socio-ecological production landscapes and seascapes (SEPLS) with the involvement of a range of stakeholders who negotiate and collaborate on their valued connectivity at a landscape or seascape level. In this respect, many case studies from the partners of the International Partnership for the Satoyama Initiative (IPSI) demonstrate how to minimize trade-offs and maximize synergies between multiple associated factors by ensuring the quality of ecological connectivity on a landscape or seascape scale. Hinging on local and regional perspectives, they also showcase the roles of landscape and seascape stewardship in enhancing ecological connectivity, while providing practical insights into more effective, sustainable, and equitable processes to operationalize connectivity in spatial planning and management.

1.2 Relevance of Socio-ecological Production Landscapes and Seascapes to Ecological Connectivity

Socio-ecological production landscapes and seascapes (SEPLS) are areas that involve a mosaic of ecosystems under different types of governance regimes but are ideally interconnected to allow for flows of ecological functions, resource movements, and movements of people for various purposes. The underlying principle is ensuring that ecological integrity is maintained through seamless connectivity between different land and sea use types. It is pertinent to note that SEPLS are multifunctional sites and have a social–ecological systemic outlook toward the use and management of resources and ecosystems.

One of the challenges that environmental policy-makers are seeking to address is to enhance ecological connectivity, whether within a site or in a region, to ensure less conflicted coexistence between humans and wildlife and also foster conservation and sustainable use of biodiversity and its diverse contributions to people (e.g., cultural, economic, provisioning, and sustaining). Multiple initiatives, as mentioned above, aim to restore such connectivity.

In this context, it appears quite evident that actions for managing SEPLS could provide lessons on opportunities and challenges we may encounter in sustaining ecological connectivity, given the multiple drivers and pressures on land and sea uses—primarily led by anthropogenic, development-related actions.

In this volume, we have compiled experiences from IPSI members that demonstrate how to minimize trade-offs and maximize synergies between multiple associated factors (social and ecological) by ensuring the good quality of ecological connectivity on a landscape or seascape scale. Hinging on local and regional perspectives, they also showcase the roles of land and seascape stewardship in enhancing ecological connectivity while providing practical insights into more effective, sustainable, and equitable processes to operationalize connectivity in spatial planning and management.

1.3 Objectives and Structure of This Book

This book focuses on the relevance of SEPLS to the aspects of ecological connectivity and integrated biodiversity-inclusive spatial planning. It looks at the strategies and approaches by which multiple stakeholders collaboratively minimize trade-offs, maximize synergies, and ensure quality ecological connectivity to achieve more effective, sustainable, and equitable processes of landscape and seascape planning and management.

With a focus on ensuring ecological connectivity in the context of SEPLS management, this book brings together the case studies that address the following questions:

- What are the multiple benefits derived through SEPLS management? And how have these benefits helped ensure and enhance the aspects of ecological connectivity?
- Are there any trade-offs and synergies among efforts to attain quality ecological connectivity in managing SEPLS? If so, what are they, and who has been losing or winning?
- How can we measure ecological connectivity, assess its ecological functions and benefits, and examine related trade-offs and synergies through managing SEPLS?
- How has SEPLS management in the study areas helped in operationalizing ecological connectivity in spatial planning and management and making the processes more effective, sustainable, and equitable?
- Have local and traditional knowledge and cultural diversity helped to ensure and enhance the quality of ecological connectivity for socially and ecologically sound outcomes? If so, how?

This volume presents 12 case studies that combine practical experiences and subject matter expertise. This chapter is followed by two cases from Africa (Chaps. 2 and 3), one case each from Europe and South America (Chaps. 6 and 7, and 8) cases from Asia (Chaps. 4, 5, 8, 9, 10, 11, 12, and 13).

Across all the chapters, the SEPLS perspective underscores connectivity as a multidimensional process that integrates the ecological, social, cultural, and economic dimensions of changes within landscapes and seascapes. The vantage points highlighted are community rights (Chap. 2), biocultural heritage (Chap. 3), traditional wisdom (Chaps. 4 and 8), community forest (Chap. 5), integrated (SEPLS) management/multi-stakeholder platforms (Chaps. 6, 7, 8, 9, and 10), ecological networks (Chaps. 10.and 13), sustainable agricultural practices (Chap. 11), and periurban habitat connectivity (Chap. 12).

Lastly, Chap. 14 synthesizes key messages and lessons learned from the case studies in regard to the conceptual and methodological considerations and future courses of action. This concluding chapter would serve as a useful reference for the policy formulation and the implementation of projects and initiatives that aim to maintain, enhance, or expand ecological connectivity. While reconceptualizing ecological connectivity in the context of SEPLS, it also showcases the key benefits, common assumptions, paradoxes, and useful indicators, tools, and methods for measuring, monitoring, and evaluating ecological connectivity.

A summary overview of the case studies is given in Table 1.1, while the map indicating the locations of the case studies is shown in Fig. 1.1.

Table 1.1 Overview of the case studies

Dominant landscapes and regions/areas	Chapter (country/region)	Ecosystem types	Key connectors	Efforts to ensure ecological connectivity
Forest—Africa	Chapter 2 (Ghana)	Forest, agricultural	Traditional and modern agroforestry techniques; wildlife	Multi-stakeholder collaboration with a landscape governance framework; various capacity-building programs; development of management plans; rehabilitation of cocoa farms
Forest—Africa	Chapter 3 (Kenya)	Forest, agricultural, mountain, coastal	Biocultural heritage territory (e.g., traditional knowledge, cultural values and practices, ancestral food systems, biodiversity)	Establishment of biocultural enterprises and communal fund for conservation; indigenous tree planting; traditional crops diversification; eco-cultural tourism
Forest—Asia	Chapter 4 (Chinese Taipei—Nanan)	Forest, agricultural, inland water, mountain	Medaka fish species	Organic farming; education, mentorship; ecotourism; payments for ecosystem services
Forest—Asia	Chapter 5 (Thailand)	Forest, agricultural, mountain	Wildlife sanctuary; community forest; low-impact tourism	Integrated forest management; spatial planning; joint patrolling
Forest—Europe	Chapter 6 (Spain)	Forest, agricultural, mountain	Wildlife; human populations; mountain areas	Spatial analysis; stakeholder mapping exercise; spatial planning; integrated management
Extensive: Mountain–agro—South America	Chapter 7 (Colombia)	Agricultural, mountain, dry forest	SEPLS such as Atuncela and El Chilcal	Community empowerment; management of productive systems
Extensive: Mountain–agro—Asia	Chapter 8 (Malaysia—Mt. Kinabalu)	Forest, agricultural, mountain	Connectivity between two parks (e.g., Kinabalu Park/Crocker Range Park); community forests; local communities	Establishment of community conserved areas; Forest restoration; promotion of sustainable agriculture; spatial planning; promotion of ecotourism; documentation of traditional knowledge; multi stakeholder platform

(continued)

Table 1.1 (continued)

Dominant landscapes and regions/ areas	Chapter (country/ region)	Ecosystem types	Key connectors	Efforts to ensure ecological connectivity
Extensive: Mountain–agro—Asia	Chapter 9 (Malaysia—Kinabatangan)	Forest, agricultural, inland water, mountain	Kinabatangan River; wildlife; biosphere reserve	Ecological corridor; zonation; stakeholder engagements
Extensive: Mountain–coastal—Asia	Chapter 10 (Chinese Taipei—Hualien)	Mountain, Forest, grassland, agricultural, inland water, coastal	Corridors for aquatic migratory species and native habitats for carnivorous plants and land crabs; socio-cultural dimension of connectivity to promote community resilience	Resilience assessment workshops; conservation corridors of Taiwan ecological network—Biodiversity-focused spatial planning policy; cross-sectoral and multi-stakeholder partnerships; community-based coastal patrol and beach clean-up; community-based monitoring and conservation and restoration activities
Agro—Asia	Chapter 11 (Chinese Taipei—Wufeng)	Agricultural	Agroecosystems; local farmers; raptors (black-winged kite)	Promotion of organic farming and eco-friendly practices; adoption of innovative pest control methods (biological rodent control); sustainable agriculture; establishment of centres for organic agriculture promotion and product certification
Peri-urban—Asia	Chapter 12 (India)	Forest, grassland, agricultural, inland water, dryland, mountain, urban, Peri-urban	Peri-urban areas; connectivity between protected areas	Green spaces in and around urban areas; buffer zone maintenance; sustainable production and consumption practices
Coastal—Asia	Chapter 13 (Chinese Taipei—Penghu)	Coastal, inter-tidal	Fish species; underwater habitats	Conservation of stone weirs; youth awareness and involvement in conservation and sustainable use activities

Fig. 1.1 Locations of the case studies (regions, landscapes, and/or seascapes) (Map template: Geospatial Information, United Nations). Note: Details of the case study locations, including geographic coordinates, are described in each chapter.

References

Balbar AC, Metaxas A (2019) The current application of ecological connectivity in the design of marine protected areas. Glob Ecol Conserv 17:e00569

Beger M, Metaxas A, Balbar AC, McGowan JA, Daigle RM, Kuempel CD, Treml EA, Possingham HP (2022) Demystifying ecological connectivity for actionable spatial conservation planning. Trends Ecol Evol 37(12):1079

Berkström C, Wennerström L, Bergström U (2022) Ecological connectivity of the marine protected area network in the Baltic Sea, Kattegat and Skagerrak: current knowledge and management needs. Ambio 51(6):1485–1503

Brennan A, Naidoo R, Greenstreet L, Mehrabi Z, Ramankutty N, Kremen C (2022) Functional connectivity of the world's protected areas. Science 376(6597):1101–1104

Butler EP, Bliss-Ketchum LL, de Rivera CE, Dissanayake ST, Hardy CL, Horn DA, Huffine B, Temple AM, Vermeulen ME, Wallace H (2022) Habitat, geophysical, and eco-social connectivity: benefits of resilient socio–ecological landscapes. Landsc Ecol 37:1–29

CBD (2010). Strategic plan for biodiversity 2011–2020 and the Aichi Targets "Living in Harmony with Nature". cbd.int/sp/

Convention on Migratory Species (CMS) (2020) Improving ways of addressing connectivity in the conservation of migratory species, resolution 12.26 (REV.COP13), Gandhinagar, India (17–22 February 2020). https://www.cms.int/sites/default/files/document/cms_cop13_res.12.26_rev.cop13_e.pdf

Gannon P, Dubois G, Dudley N, Ervin J, Ferrier S, Gidda S, Mackinnon K, Richardson K, Schmidt M, Seyoum-Edjigu E, Shestakov A (2019) Editorial essay: an update on progress towards Aichi biodiversity target 11. Parks 25(2):7–18

IPBES (2022) External communication: proposed assessment on connectivity. IPBES/9/INF/27. https://www.ipbes.net/node/40122

IPBES (2023) Report on the prioritization of requests, input and suggestions for additional elements of the rolling work programme of the Platform up to 2030. IPBES/10/10. https://www.ipbes.net/document-library-catalogue/ipbes1010

Lemieux CJ, Beazley KF, MacKinnon D, Wright P, Kraus D, Pither R, Crawford L, Jacob AL, Hilty J (2022) Transformational changes for achieving the Post-2020 global biodiversity framework ecological connectivity goals. FACETS 7(1):1008–1027

Maxwell SL, Cazalis V, Dudley N, Hoffmann M, Rodrigues ASL, Stolton S, Visconti P, Woodley S, Maron M, Strassburg B, Wenger AS, Jonas H, Venter O, Watson JEM (2020) Area-based conservation in the twenty-first century. Nature 586:217–227. https://doi.org/10.1038/s41586-020-2773-z

Oppler G, Hilty JA, Laur AT, Tabor G (2021) Connectivity conservation: the time is now. Parks Stewardship Forum 37(3):488–504

Perrin M, Bertrand N, Vanpeene S (2022) Ecological connectivity in spatial planning: from the EU framework to its territorial implementation in the French context. Environ Sci Pol 129:118–125

Roberts KE, Cook CN, Beher J, Treml EA (2021) Assessing the current state of ecological connectivity in a large marine protected area system. Conserv Biol 35(2):699–710

UNEP-WCMC and IUCN (2021) Protected planet report 2020: Tracking progress towards global targets for protected and conserved areas. livereport.protectedplanet.net/

Chapter 2
Giving Communities the Right to Govern Their Natural Resources: A Case of the Asunafo-Asutifi Landscape in the Ahafo Region of Ghana, West Africa

Abdulai Sulemana, Daniel Enti, and Vincent Awotwe-Pratt

Abstract The Asunafo-Asutifi Landscape Initiative, located in Ghana's Ahafo Region, aims to balance cocoa production with environmental conservation. Recognizing that cocoa farming, while vital to Ghana's economy, often drives deforestation, the initiative integrates traditional and modern agroforestry techniques to create a deforestation-free, climate-resilient cocoa landscape. The initiative seeks to improve farmers' livelihoods while preserving biodiversity and critical forest areas through collaboration among government agencies, private sector entities, and local communities. The landscape management plan incorporates sustainable farming, forest protection, and proactive community participation. By promoting biodiversity conservation, addressing the cocoa swollen shoot virus, and fostering ecological connectivity, the initiative emphasizes sustainable resource management aligned with the principles of Socio-Ecological Production Landscapes and Seascapes (SEPLS). As an example of how to harmonize environmental stewardship with economic growth, this project sets a model for integrated landscape management in Ghana and beyond. The initiative demonstrates Ghana's commitment to sustainable cocoa production, aiming for long-term ecological and economic benefits through comprehensive and collaborative measures.

Keywords Asunafo-Asutifi Landscape · Biodiversity conservation · Cocoa production · Deforestation · Agroforestry

A. Sulemana (✉) · D. Enti · V. Awotwe-Pratt
Advocates for Biodiversity Conservation, Greater Accra Region, Ghana
e-mail: sulemana@biodiversityadvocates.org

M. Nishi et al. (eds.), *Ensuring Ecological Connectivity in Socio-Ecological Production Landscapes and Seascapes (SEPLS)*, Satoyama Initiative Thematic Review, https://doi.org/10.1007/978-981-95-1474-8_2

2.1 Introduction

2.1.1 Background

Cocoa production is an essential component of Ghana's economy and has a rich history. Ghana is the world's second-largest cocoa producer, accounting for around 20% of global cocoa production (Amponsah-Doku et al. 2022). Most of the cocoa is grown by smallholder farmers (Roldan et al. 2013) who manage relatively small plots of land within the country's forest regions. These farmers rely on traditional farming methods and often face challenges such as pests, diseases, and fluctuating global cocoa prices. The cocoa industry in Ghana is critical to the livelihoods of millions of people, including farmers, traders, and processors. It provides foreign exchange earnings (Tropical Forest Alliance et al. 2023).

However, the industry faces sustainability challenges, both environmental (deforestation) and social (child labor) in nature. The expansion of cocoa farming has been a significant driver of deforestation in Ghana. In some areas, forests are cleared to make way for cocoa plantations, reducing biodiversity and ecological services (Kalischek et al. 2023; Wainaina et al. 2021). Deforestation also contributes to climate change by releasing stored carbon dioxide into the atmosphere (Baccini et al. 2012).

The Asunafo-Asutifi Landscape in Ghana is one area where cocoa production results in deforestation. The landscape comprises four administrative districts: Asutifi North, Asutifi South, Asunafo North, and Asunafo South and is situated in the Ahafo Region (Fig. 2.1 and Table 2.1). The landscape has an area of approximately 383,000 ha, of which over 30% is protected forest reserve. The landscape is home to several ethnic groups, including the Ahafo and Asante people and a diverse habitat that includes tropical rainforests, shrublands, and wetlands. The area provides essential resources such as food, water, and other necessities for the surrounding communities (Tropical Forest Alliance et al. 2023).

These four districts are major cocoa-producing areas, and farmers often expand their farms into forest areas to maximize their production and income. Although cocoa productivity in the landscape is high, the prevalence of diseased farms, particularly with the cocoa swollen shoot virus disease (CSSVD), is increasing, as shown in Fig. 2.2. In 2019, the Ghana Cocoa Board (COCOBOD) designated the Asunafo South district (Sankore Cocoa District) as a hotspot for this disease. The traditional forest frontier that once acted as a natural barrier to the virus has been disrupted, leading to its rapid spread. Another key challenge is the abundance of moribund cocoa farms resulting from aging farmers who, due to being resource-poor and having small holdings, can no longer maintain their farms. Also, cocoa trees past their bearing ages (2–25 years) need to be replanted. Diminishing timber resources in reserved areas is another factor. Legal and illegal logging are shifting toward cocoa farms, which now hold the remnants of native timber species and biodiversity in the landscape (Tropical Forest Alliance et al. 2023).

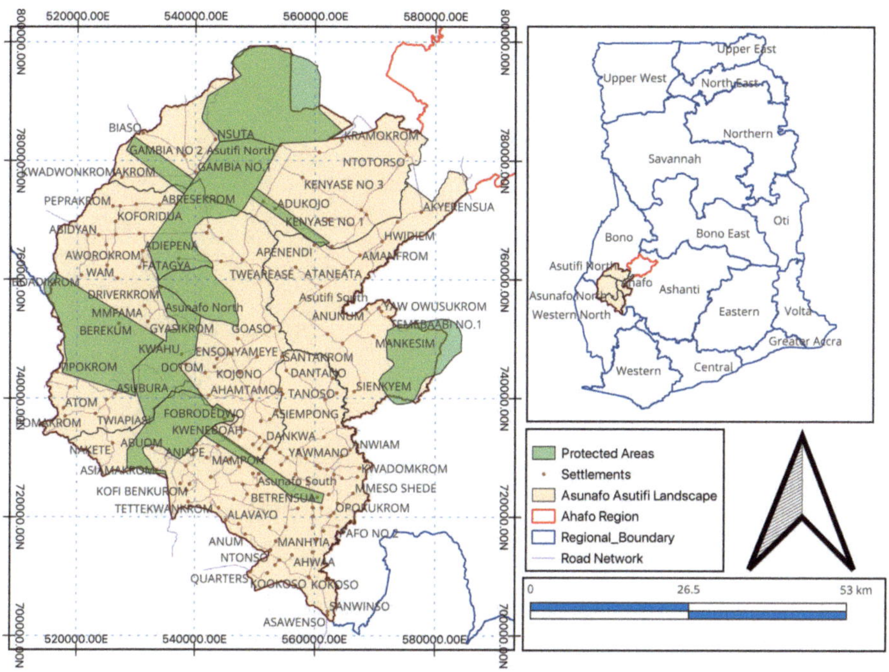

Fig. 2.1 Overview of the Asunafo-Asutifi cocoa landscape in the Ahafo Region of Ghana. (Map developed by Advocates for Biodiversity Conservation, 2024; Shapefiles sourced from the Forestry Commission of Ghana)

Table 2.1 Basic information of the study area

Country/Region	Ghana
Province	Southern Ghana
District	Asutifi North, Asutifi South, Asunafo North, and Asunafo South
Municipality (Region)	Ahafo
Size of geographical area (hectare)	383,000
Dominant ethnicity(ies), if appropriate	Akan
Size of case study/project area (hectare)	383,000
Dominant ethnicity in the project area	Akan
Number of direct beneficiaries (people)	470,400
Number of indirect beneficiaries (people)	705,600
Geographic coordinates (latitude, longitude)	6.5295° N, 2.5000° W, 7.0432° N, 2.5000° W, 6.9577° N, 2.4209° W

Source: Proforest (2020), based on the Report of the Baseline Socio-economic and Ecological Assessment of the Asunafo-Asutifi Hotspot Intervention Area, 2020

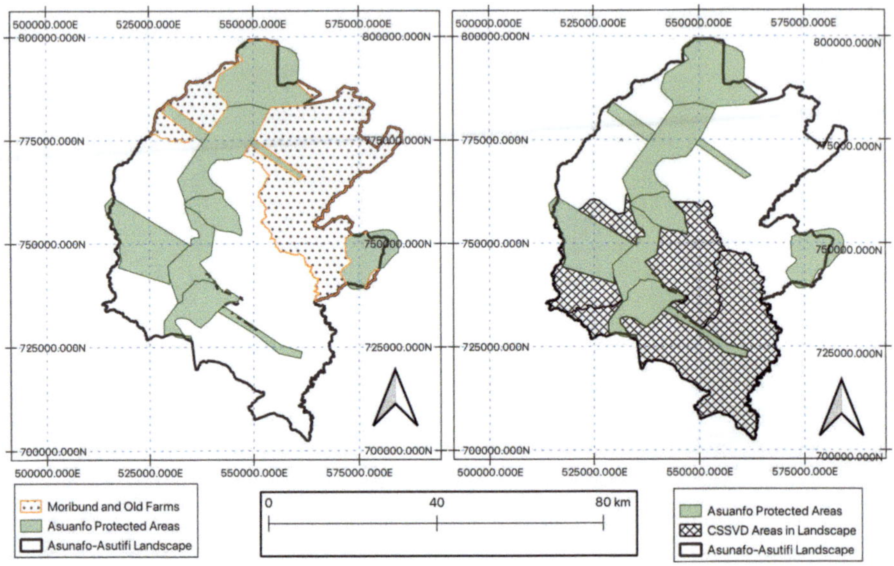

Fig. 2.2 Areas with cocoa swollen shoot disease and moribund cocoa farms in the Asunafo-Asutifi Landscape. (Map developed by Advocates for Biodiversity Conservation, 2024; Shapefiles sourced from the Forestry Commission of Ghana)

Stakeholders in the cocoa value chain (COCOBOD, cocoa purchasing companies, chocolate companies, local communities, civil society organizations (CSOs), and the government) have joined hands to tackle the issue of deforestation and promote sustainable cocoa production. Through a series of engagements (workshops and advocacy), they agreed on a common vision for the next 20 years (Tropical Forest Alliance et al. 2023). The vision was documented in the Asunafo-Asutifi Landscape management plan, which details a series of interventions to be implemented by stakeholders. One such intervention is the Asunafo-Asutifi Landscape Initiative, which was launched in 2023 by Ghana's Minister for Lands and Natural Resources, Mr. Samuel A. Jinapor. This initiative, led by the government and private sector, aims to create a deforestation-free, climate-resilient cocoa landscape. It focuses on sustainable production, forest protection, restoration (including planting native shade trees on cocoa farms), enhancing farmers' livelihoods, engaging with communities, and promoting social inclusion. Stakeholders are working to maintain and strengthen the connections between different ecosystems, habitats, and species within the human-influenced landscape while preserving the ecological integrity of protected areas. The initiative is intended to improve tree cover, provide safe passage for fauna between cocoa farms and nearby reserves, and help Ghana remain a leading producer of high-quality cocoa while ensuring the well-being of its cocoa farmers and preserving the environment.

The Asunafo-Asutifi Landscape Initiative is of utmost importance in the context of socio-ecological production landscape (SEPL) management. Its focus is on halting deforestation, preserving remnant forest patches, protecting biodiversity, and planting native tree species on cocoa farms to enhance the microclimatic conditions of the farms and improve cocoa yields. Traditional farming practices are integrated with modern agroforestry practices to maximize the cocoa yield and improve the biodiversity on farms and adjoining reserves. This, in essence, preserves the ecological connectivity of the landscape, allowing biodiversity to thrive, thus aligning seamlessly with SEPL principles (Nishi et al. 2023) and strategically addressing the challenges of extensive agriculture and deforestation.

Through this initiative, Ghana is not only taking proactive measures to mitigate environmental degradation but also setting the stage for a sustainable and resilient cocoa production landscape. The initiative incorporates forest protection, restoration, and community participation, making it a beacon for responsible and forward-thinking agricultural practices.

As Ghana moves toward a deforestation-free future, the Asunafo-Asutifi Landscape Initiative is a testament to the nation's commitment to balancing economic progress with environmental stewardship. By navigating the complexities of SEPL management, Ghana exemplifies a model that balances the needs of its people, the agricultural sector, and the ecosystem at large.

This chapter aims to provide an overview of the Asunafo-Asutifi Landscape Initiative's significance in the context of ecological connectivity and integrated biodiversity-inclusive spatial planning at the local level.

2.2 Approach and Methods

2.2.1 Stakeholder Identification

2.2.1.1 Landscape Governance Structure

The Asunafo-Asutifi Landscape Initiative relies on the collaboration of a diverse group of stakeholders working together to achieve the landscape restoration goals. To understand the various stakeholders present in the landscape, a mapping approach was used to categorize stakeholders into three categories based on their respective interests. These are the local communities and their leadership, private sector actors, and government agencies and their representatives. Traditional authorities and their governance structures were included in the local communities category because they hold land (especially) and benefits that are derived from the land in trust of the people.

2.2.1.2 Use of the Hotspot Intervention Area (HIA) Concept

The Hotspot Intervention Area (HIA) approach, a landscape governance framework evolved from the Community Resource Management Areas (CREMAs) model, integrates biodiversity conservation, land-use practices, and rural development into

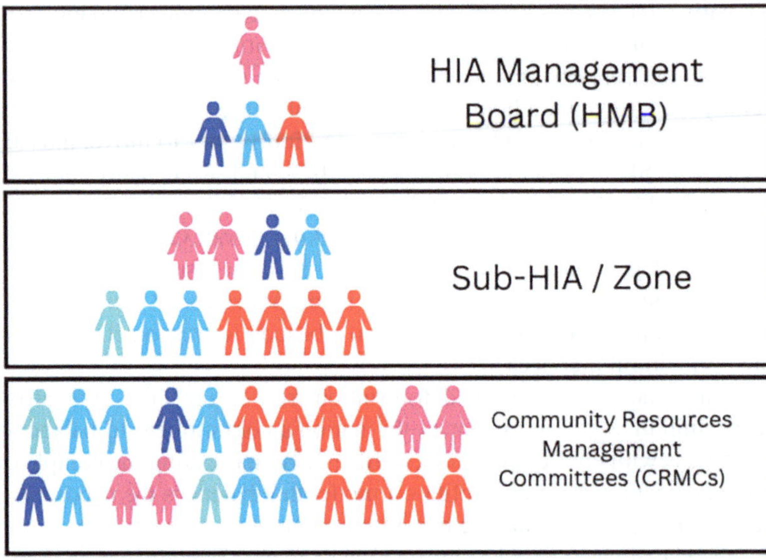

Fig. 2.3 Landscape governance structure in the Asunafo-Asutifi Landscape. (Source: Designed by authors)

the cocoa production landscape (Asare et al. 2013; Dugasseh et al. 2024; Foli et al. 2018). Initially focused on only wildlife connectivity and sustainable resource use outside protected areas, CREMAs have faced size limitations (25,000-30,000 ha), challenges in financial sustainability, and a lack of proper connectivity to other areas. The HIA concept aims to tackle this challenge by implementing a streamlined governance mechanism that maintains both the effectiveness and efficiency of the landscape governance structure. It establishes a platform for diverse stakeholders to interact, communicate, and share plans (Fig. 2.3). At the top (HMB), the governance structure represents the diverse groups of smaller governance structures along with leadership arrangements present in the landscape mimicking the cultural identity of the people and the players in the local economy. The Community Resource Management Committees (CRMCs) include representation from community structures, while the sub-HIAs or zones are bodies constituted with representation from these CRMCs. The HIA Management Board (HMB), the apex governance body in the landscape, includes 13 members elected from the various zones/sub-HIAs.

This governance structure collaborates with private sector representatives at the landscape level and the representatives of government agencies on a platform established to enhance sharing and collaboration called the Consortium. Admittance into the Consortium requires an entity to have interest in developing the landscape and a vision that aligns with the landscape vision. Their joint efforts focus on addressing deforestation, increasing productivity, preventing encroachment of farms into forest reserves, and improving the livelihoods of cocoa farmers. Regular capacity-building events are organized by the Consortium to assist stakeholders in implementing or monitoring various aspects of the Asunafo-Asutifi Hotspot Intervention Area (HIA) Management Plan, hereafter known as HIA Management Plan. This plan will guide the implementation of the Asunafo-Asutifi Landscape Initiative, which aims to reduce deforestation on and off-reserve, rehabilitate degraded land and forests, and improve livelihoods and local governance.

2.2.2 Methods

2.2.2.1 Data Collection

The information presented in this chapter was gathered using two basic approaches. These are conducting one-on-one interviews with key contacts within the Asunafo-Asutifi Landscape and holding meetings with various groups including the Consortium, HMB, sub-HIAs/zones, and selected CRMCs within the landscape, with most interviewees being farmers. A field visit was also conducted to gain first-hand experience of the activities in the field and interact with the farmers and members of the Consortium. Data were collected between January and June 2022. A total of 23 meetings were held, including 2 landscape governance meetings, 12 cocoa farm visits, 4 zonal meetings, 3 CRMC (out of 24) meetings, and 2 meetings with NGOs working in the landscape. The nonprobability (nonrandom) sampling technique was used to target CRMCs and active participants in the Asunafo-Asutifi Landscape Initiative. A total of 128 people were interviewed. Interviews were conducted in English and Twi (the most dominant language in the landscape) based on the interviewee's proficiency.

2.2.2.2 Analysis of Data

The study team used a simple yet iterative approach to analyze the information collected during the interaction to understand trends, similarities, diversions, and areas lacking information.

The interactions from the stakeholders' meetings were transcribed manually, and linkages were made with activities that promoted ecological connectivity in the Asunafo-Asutifi Landscape. These activities included setting up CREMAs and planting native shade trees on cocoa farms. Participants' responses and field observations were categorized according to these activities, leading to the creation of narratives for each one. In addition, other important information generated on the functioning of the landscape and its relation to the environmental, economic, and sociocultural development of the landscape was determined through these discussions.

2.2.3 *Asunafo-Asutifi Landscape*

2.2.3.1 General Description of the Area

Cocoa in Ghana is produced mainly in nearly six million hectares of agroforest mosaic landscape within the High Forest Zone (HFZ) (Tutu Benefoh 2018). According to Ghana's National Forest Reference Level, Ghana's deforestation rate has been approximately 2% per year (320,803 ha/year) since 2000 (Yeboah-Assiamah et al. 2023). Recent years have also seen a marked increase in the deforestation rate. From 2013 to 2015, the average annual deforestation rate in Ghana rose to 794,214 ha per annum (Damptey 2023). The major driver of deforestation is agricultural expansion. According to the Ghana Forest Plantation Strategy 2016–2040 (Forestry Commission 2016), over a quarter (27% of agriculture conversion) resulted from cocoa expansion, making it the single most important commodity driver of deforestation in the REDD+ program area.

To address the high deforestation rate in the HFZ of Ghana, the Government of Ghana led by the National REDD+ Secretariat (NRS) of the Forestry Commission (FC) has developed the Ghana Cocoa Forest REDD+ Program (GCFRP). The overall goal of the program is to improve land use and socioeconomic development in the cocoa-growing areas of Ghana by increasing cocoa yield on farmlands through intensification with climate-smart practices while preventing the expansion of cocoa farms into forest lands. The Asunafo-Asutifi Landscape is one of the priority HIAs that stakeholders intend to coalesce efforts and resources to address the issues of deforestation, low productivity, and farm encroachments into forest reserves while improving cocoa production and livelihood of cocoa farmers under the GCFRP.

2.2.3.2 Sociocultural Profile of the Landscape

The Asunafo-Asutifi HIA is mainly an agrarian landscape with a total land area of 386,532 ha and an estimated population size of 312,435 as of 2017, of which 50.88% are females. The settlement pattern is nucleated, with a majority of the population residing in rural areas. Inhabitants of the landscape are mainly indigenes, with a considerable migrant population from the northern regions of Ghana and Southern Côte d'Ivoire due to rich natural capital endowments, ranging from minerals to timber and arable lands for farming. Most of the lands within the Asunafo-Asutifi HIA are owned by families. There are also sharecropping and land lease systems, locally called *abunu*[1]and *abusa,*[2] with the *abunu* being the most

[1]*Abunu*: a sharecropping system where a tenant farmer cultivates a piece of land given by a landowner and then splits the farm yields equally with the landowner.

[2]*Abusa*: a sharecropping system where a tenant farmer takes care of an existing tree crop farm and shares the produce with the owner according to an agreed proportion, usually one-third or two-thirds of the produce, depending on the agreement between tenant and owner.

prevalent sharecropping system. Cocoa production within the landscape is very high, occupying about 70–80% of agricultural land.

There are eight Traditional Councils in the landscape, each of which comprises Divisional Chiefs and is headed by a Paramount Chief. The Traditional Councils are located at Goaso, Mim, and Akrodie in Asunafo North; Kukuom and Sankore in Asunafo South; and Kenyase #I, Kenyase #II, and Ntotroso in Asutifi North. Land ownership and access rights are governed by traditional systems, with about 80% of the lands owned by Stools.[3] The existing governance structure consists of a Hotspot Management Board (HMB), four sub-HIAs, and one CREMA, while identifying 30 zones that are yet to be developed into CREMAs.

2.2.3.3 Ecological Profile of the Landscape

In addition to the rich socioeconomic and cultural contexts, the Asunafo-Asutifi Landscape has significant ecological value. It is endowed with a high diversity of fauna and flora species as well as ecosystems that are significant to the conservation of the biodiversity of the area. Keystone megafauna such as elephants are no longer present in the Asunafo-Asutifi forest enclave, but there are large fauna and flora communities with a significant presence of species of global conservation concern in the landscape.

Land cover types identified in the study include closed and open forests, agricultural land, waterbodies, and human settlements. Notably, forests and cocoa farms comprise the majority of land use types in the landscape. There is also a significant presence of watersheds and riparian buffers in relatively good condition, particularly in forested areas. While there are no national parks, there are ten forest reserves which together make up 121,830 ha or about 37% of the landscape.

2.3 Results and Discussion

2.3.1 Operationalizing Ecological Connectivity in Spatial Planning and Management Processes

The strengthening of the landscape governance structure through various capacity-building programs and the creation of the Consortium enabled the development of management plans that take into consideration the various components of the landscape (e.g., off- and on-reserve areas, cocoa farms, and settlements) and how they are connected or otherwise. This platform also determined that managing each of these ecological zones as a single unit has many limitations and does not bring out

[3] Lands in southern Ghana are generally owned by traditional communities whose titular heads are the Stools.

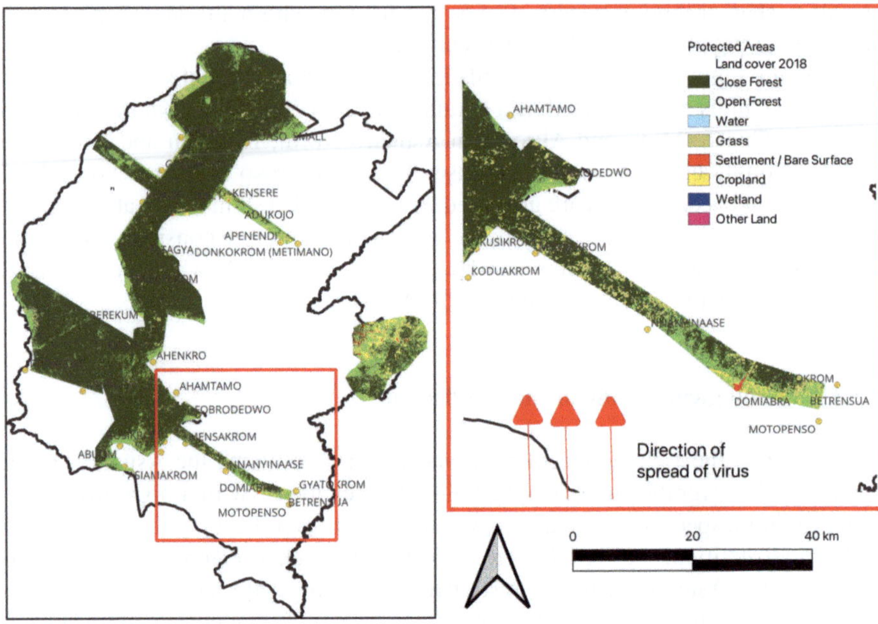

Fig. 2.4 Spread of CSSVD through the Asunafo-Asutifi Landscape. (Map developed by Advocates for Biodiversity Conservation, 2024; Shapefiles sourced from the Forestry Commission of Ghana)

their optimal value and use for local communities. For example, all protected areas are managed by the Forestry Commission under the Forestry Commission Act (ACT 571) (Forestry Commission 2016). Meanwhile, COCOBOD has the mandate to manage all cocoa trees and beans under the Ghana Cocoa Board Act, 1984 (PNDCL 81) (FAOLEX Database 2022). The platform realized that encroachment into the forest frontier in some parts of the landscape is allowing for the spread of CSSVD into new areas, therefore thwarting the efforts of COCOBOD and its partners in the rehabilitation of cocoa farms with over 300 million USD already invested in the program (COCOBOD 2020). Likewise, spatial planning revealed that the ongoing removal of forest in the Abonyire Forest Reserve breaks down the frontier and allows the disease to spread, as shown in Fig. 2.4.

The management planning process ensured the identification of critical ecological infrastructure and proposed specific interventions to maintain or improve it. Consideration was given to its function concerning other elements in the landscape to ensure optimization. Where trade-offs were identified, measures were proposed to mitigate their effects on agricultural productivity and human well-being. Over the years, chocolate and cocoa companies have promoted cocoa agroforestry and climate-smart cocoa production to manage pest and disease incidences within cocoa farms. However, these practices have also had unintended positive impacts, such as enhancing tree cover in the landscape and increasing the availability of timber species in off-reserve areas, thereby generating additional income for farmers. Furthermore, the enhanced connectivity of trees has facilitated the movement of species and genetic material between farms and natural forests. Farmers have realized and testified to the direct impact and benefit of keeping trees on their farms and the monetary and ecological benefits to them.

Through the partnerships between the chocolate and cocoa companies, stakeholders have been sensitized to the importance of cocoa farms as connectors to forest reserves. For example, CREMA, an innovative natural resource governance and landscape-level planning tool that authorizes communities to manage their natural resources for economic and livelihood benefits (Asare et al. 2013), have become an effective off-reserve management tool led by community members to mimic forest reserve conditions in off-reserve areas and ensure that communities are not overly reliant on forest reserves. The concept is being implemented in the Asunafo-Asutifi Landscape. Off-reserve areas that are not connected to on-reserve areas are being managed to provide ecosystem services and goods that natural forests provide so that communities benefit from harvesting non-timber forest products (NTFPs). An example is the harvesting of cola nuts, herbal medicine, and rattan from CREMA areas. These products are currently being harvested from the wild as there are no plantations in Ghana.

The above examples indicate that the HIA Management Plan has been developed taking into consideration the whole landscape and that each stakeholder group has brought into a vision for the entire landscape. There is a commitment from landscape actors (Fig. 2.5) to also update the management plan annually, with major revisions to be made after 5 years of implementation.

2.3.2 *Multiple Benefits from SEPL Management within the Landscape*

Analysis of data collected indicated that cocoa farmers appreciate tree retention on farms even if the trees are not planted for commercial purposes (Fig. 2.6). Based on the COCOBOD recommendation of 22 trees per hectare (Asare 2005), most cocoa

Fig. 2.5 Consortium of actors in the Asunafo-Asutifi landscape. (Photo credit: Advocates for Biodiversity Conservation 2022)

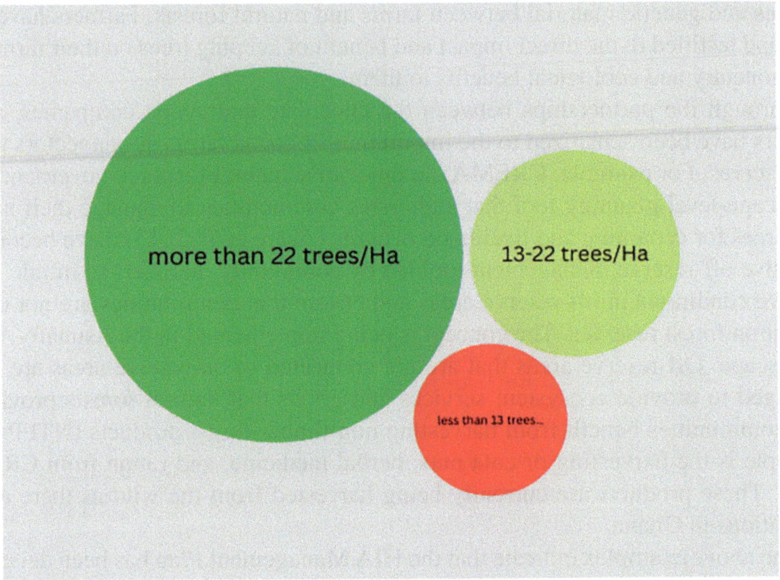

Fig. 2.6 Count of native trees on farms. (Source: Developed by Advocates for Biodiversity Conservation 2024)

farms had more than the recommended number and diversity of species. This could be attributed to the extensive education and awareness creation campaigns that have been undertaken by landscape Consortium members to extend the knowledge of the nontangible values of trees on farms coupled with the availability of preferred tree seedlings at the right time for farmers to plant. Farmers see the retention of trees as an "insurance policy" against future failure of the farm. This insurance policy includes the possibility of using the trees as a guarantee to receive credit either in the form of farm input or cash for school fees in off-seasons. In other cases, these trees serve as food security and supplement nutrition at home in the form of herbs, fruits, seeds, and medicine. Where there are larger quantities harvested, farmers can sell them for additional income for the family.

In addition, farmers are rewarded as part of the GCFRP with community projects, including the building of community schools, water projects, and monetary payments as incentives for trees on farms which contribute to increased carbon stocks. The landscape restoration program was officially launched in 2023, whereas the four administrative districts already contributed around 15% of the national cocoa output in 2022 (Tropical Forest Alliance et al. 2023). While a direct correlation with the landscape restoration initiative is not established, the influence of prior interventions by various civil society organizations in the districts may have contributed to the observed uptick in cocoa production for the year 2022.

The Asunafo-Asutifi Landscape Initiative incorporates smallholder livelihood improvement programs to complement the adoption of climate-smart cocoa and agroforestry. One such program is the Timber Tending Benefit (TTB) introduced by

Ghana's Ministry of Lands and Natural Resources as part of the Tree Tenure and Benefit Sharing policy reform. This initiative is designed to provide farmers with upfront payments for nurturing and protecting naturally occurring trees on their farmlands (IDH 2022). In Ghana, land ownership primarily follows customary practices, with approximately 80% held communally by families or tribes. Chiefs act as trustees, managing these lands for the collective benefit of the community members (Bugri and Yeboah 2017).

The TTB policy acts as an additional incentive for farmers aside from the benefit of trees as shade and the creation of a microclimate for cocoa to thrive. The Forestry Commission of Ghana by law manages the process and ensures that farmers who tend trees are paid as agreed by all stakeholders with social responsibility agreements (SRAs) for the local communities (Agyei and Adjei 2017). Through TTB, farmers who register, nurture, and protect the trees planted on their farms are incentivized to keep the trees on their farms. These trees on cocoa farms act as carbon sinks, helping to mitigate climate change and contributing to emission reduction goals. Additionally, these practices can potentially create a source of income for farmers through carbon credits (Afele et al. 2021). Furthermore, with the diversification of crops with food items like plantains, vegetables, and fruits, farmers can produce additional revenue for their families and communities (Djokoto et al. 2017). Timber and medicinal plants can further enhance their economic return.

2.3.3 Promoting Climate-Smart Cocoa (CSC) Production Practices

The promotion of climate-smart cocoa practices among cocoa farmers within the Asunafo-Asutifi Landscape is expected to yield a range of benefits, including enhancing farm resilience to climate change, boosting productivity and incomes, reducing environmental impacts, and elevating livelihoods (Bunn et al. 2019a). Shade tree retention, cover crop planting, compost application, and integrated pest management are all integral components of these modules.

Applying compost enhances soil fertility, structure, and water-holding capacity, reducing the reliance on chemical fertilizers and pesticides. Integrated pest management (IPM) is a holistic approach using biological, cultural, and chemical methods to control pests on cocoa trees, thereby reducing the need for chemical pesticides and promoting environmental and human health (Muhie 2022). These practices are encompassed in the climate-smart cocoa production principles that farmers are encouraged to implement on their farms. They do not only increase the resilience of the farm and farmers but also improve the general environment. Their adoption underscores a commitment to both economic prosperity and the enduring viability of cocoa farming operations.

Shade tree retention enhances temperature and humidity regulation, reduces wind speed, improves soil moisture, provides habitats for beneficial insects and birds, and mitigates the risk of pests and diseases (Olwig et al. 2024; Schooler et al. 2020). Planting cover crops improves soil fertility, reduces soil erosion, suppresses weeds, enhances water infiltration and retention, and lowers the risk of flooding (Olwig et al. 2024).

2.3.4 Community Ownership and Participation

The Asunafo-Asutifi Landscape initiative presents a unique opportunity to demonstrate the power of community-led ownership, management, and development of natural resources. By empowering local communities to co-manage resources, the Initiative has the potential to achieve lasting positive impacts on the environment and livelihoods and well-being of the people. Under the collaborative management policy of the Forestry Commission, CREMAs are established in off-reserve areas, with community members given the authority to manage them (Asare et al. 2013). This concept, which works at the community level, has been expanded to the district/zonal level through the Hotspot Intervention Areas (HIAs) of the GCFRP Strategy to give local communities representation at the regional level through the formation of the Hotspot Management Board (HMB). The HMB takes its strength from the local CREMA leadership (CRMCs) and represents communities and their interests at the landscape level. This structure interfaces with other landscape stakeholders including the private sector companies, NGOs/CSOs, and government representation at the local levels through the landscape Consortium platform that allows for the exchange of information, planning, and alignment of activities to the landscape management plan.

2.3.5 Trade-Offs, Synergies, and Incentives

2.3.5.1 Restoring Ecological Connectivity

The Asunafo-Asutifi Landscape Initiative places a strong emphasis on promoting trees on farms to bolster cocoa production, as outlined in the HIA Management Plan (unpublished). This endeavor has given rise to a microclimate on these farms that not only supports cocoa production but also fosters biodiversity, creating a habitat akin to a secondary forest and offering a secure environment for small mammals. Nevertheless, these efforts have sparked conflicting interests among stakeholders, particularly those in agriculture and other economically vital sectors such as timber extraction. Large-scale producers of annual crops reliant on inputs may experience reduced yields, while the timber industry is eager to harvest trees that are not readily accessible in forest reserves or are simply unavailable.

While increasing tree cover on farms is a beneficial practice, there may be unintended consequences, such as suboptimal farm yields and the destruction of farms by illegal timber merchants. Furthermore, creating a network of cocoa farms that connects with other habitats poses a risk to wildlife, as people may target these species if proper protection and education are not provided to keep people away.

2.3.5.2 Interconnected Ecosystems for Increased Climate Resilience

The Asunafo-Asutifi Landscape is situated in Ghana's HFZ, close to the transition zone (an area between savannah woodlands and forests). In particular, the northern parts of the landscape share a boundary with the transition zone of Ghana. As a result, certain areas are predicted to become unsuitable for cocoa production in the future (Bunn et al. 2019b) if interventions such as tree planting and the use of drought-resistant varieties are not implemented. As climate patterns shift, many species must adapt by relocating to areas with suitable conditions. Wildlife corridors, especially those created within the protected areas that connect isolated habitats, will facilitate migrations, allowing populations to find suitable habitats. It is expected that wildlife species will move toward the southern part of the Ahafo region further into the landscape.

Connected ecosystems are more likely to maintain essential ecological processes, such as nutrient cycling, pollination, and water regulation, under changing climatic conditions (Vignola et al. 2015). Connectedness in the Abonyire Forest Reserve (Fig. 2.4) has always been a natural barrier to the spread of diseases like the CSSVD. In recent times, increased logging in forest reserves (Oduro et al. 2015) has broken down this barrier, resulting in the spread of the disease further north in the landscape. Large, contiguous ecosystems can absorb the impacts of extreme weather events and other impacts of climate change more effectively than smaller, isolated areas (Tilman et al. 2014). Restoring the Abonyire Forest Reserve will restore the natural barrier and reduce the cost of cocoa rehabilitation.

2.3.5.3 Short- or Long-Term Gains? Balancing Trade-Offs and Improving Synergies

Promoting ecological connectivity often requires a balance between immediate and future benefits (Wiens 2009). While the long-term advantages of such investments are substantial, they may require short-term adjustments or sacrifices. In interviews, 78% of the farmers admitted to experiencing a reduction in short-term gains (labor-intensive and lower yields) but remained focused on sustained long-term benefits. These farmers are well-educated and fully grasp the journey they are on, and they also anticipate receiving incentives such as certification schemes, payment for ecosystem services, and REDD+ payments.

Generally, the Asunafo Asutifi Landscape Initiative has proven that it is possible to achieve a balance between conservation and cocoa production, leading to healthier ecosystems and improved livelihoods for local communities. This was evident in the participatory planning and monitoring platform that has been established for communities (landscape governance) and other stakeholders (Consortium) to discuss issues in the management plan. Planning is also done at the landscape level, which allows connectivity between the various ecological units within the landscape to be managed together. Farmers are encouraged to look at long-term benefits instead of short-term ones and engage in adaptive management of the landscape based on data contributed by each stakeholder.

2.4 Conclusion

The Asunafo-Asutifi Landscape Initiative is a testament to the transformative power of proactive SEPL management when local communities are empowered to participate in the management of natural resources. The introduction of the SEPL concept to guide the coordination of the landscape has offered a valuable framework for tackling the socio-ecological challenges present in the cocoa production landscape. Though at its initial stages with no real results currently in sight, the approach seeks to strike a balance between enhancing farm yields and safeguarding the natural processes that underpin cocoa production and the communities' well-being.

By acknowledging the unique nature and required management regimes of the various units that make up the landscape, such as forest reserves, farms, fallow areas, and waterbodies, while preserving the unique connectivity among them, the Initiative strategically tackles the pervasive challenges of extensive agriculture and deforestation. It aims for a more sustainable future for the farmers. The Initiative signifies Ghana's commitment to mitigate environmental degradation and facilitate a change in thinking toward a sustainable and resilient cocoa production landscape.

The Asunafo-Asutifi Landscape Initiative is characterized by the promotion of a repository of responsible and forward-thinking agricultural practices that are based on traditional farming practices and agroecological principles. Its success has the potential to serve as a model for the transformation of human-influenced landscapes into functional, sustainable ecosystems that balance ecological health with economic benefits. As the Asunafo-Asutifi Landscape Initiative unfolds, it will chart a course toward a future where environmental preservation and agricultural prosperity harmoniously coexist, setting a commendable standard for SEPL initiatives globally.

References

Afele JT, Dawoe E, Abunyewa AA, Afari-Sefa V, Asare R (2021) Carbon storage in cocoa growing systems across different agroecological zones in Ghana. Pelita Perkeb 37(1):32–49

Agyei FK, Adjei PO-W (2017) Representation without accountability in forestry: experiences from the social responsibility agreement in Ghana. Forest Policy Econ 80:34–43

Amponsah-Doku B, Daymond A, Robinson S, Atuah L, Sizmur T (2022) Improving soil health and closing the yield gap of cocoa production in Ghana—a review. Sci Afr 15:e01075

Asare R (2005) Cocoa agroforests in West Africa: a look at activities on preferred trees in the farming systems. Forest & Landscape Denmark (FLD), Copenhagen

Asare RA, Kyei A, Mason JJ (2013) The community resource management area mechanism: a strategy to manage African forest resources for REDD+. Philos Trans R Soc B: Biol Sci 368(1625):20120311

Baccini A, Goetz S, Walker W, Laporte N, Sun M, Sulla-Menashe D, Hackler J, Beck P, Dubayah R, Friedl M (2012) Estimated carbon dioxide emissions from tropical deforestation improved by carbon-density maps. Nat Clim Chang 2(3):182–185

Bugri J, Yeboah E (2017) A brief overview of land tenure arrangements in Ghana. In: Understanding changing land access and use by the rural poor in Ghana. International Institute for Environment and Development, pp 18–20

Bunn C, Fernández Kolb P, Asare R, Lundy MM (2019a) Climate smart cocoa in Ghana towards climate resilient production at scale, CCAFS info note. CGIAR Research Program on Climate Change, Agriculture and Food Security (CCAFS), Cali

Bunn C, Läderach P, Quaye A, Muilerman S, Noponen MR, Lundy M (2019b) Recommendation domains to scale out climate change adaptation in cocoa production in Ghana. Clim Ser 16:100123

COCOBOD (2020) President Akufo Adoo Launches National Cocoa Rehabilitation Programme, viewed 31 July 2024. Retrieved from https://cocobod.gh/news/president-akufo-addo-launches-national-cocoa-rehabilitation-programme

Damptey FG (2023) Effects of forest restoration on ecosystem attributes in a post-mining area in Ghana. Ph.D. thesis,. Brandenburg University of Technology Cottbus-Senftenberg

Djokoto JG, Afari-Sefa V, Addo-Quaye A (2017) Vegetable diversification in cocoa-based farming systems Ghana. Agric Food Secur 6:1–10

Dugasseh FA, Adams MA, Zandersen M (2024) Actor perceptions of the governance framework and non-carbon benefits from the Ghana cocoa forest REDD+ program: an extended Q-study of the Juabuso-Bia hotspot intervention area. Environ Manag 74(1):1–21

FAOLEX Database (2022) Ghana Cocoa Board Act 1984 (PNDCL 81), viewed 29 July 2024. Retrieved from https://www.fao.org/faolex/results/details/en/c/LEX-FAOC132746/

Foli S, Ros-Tonen MA, Reed J, Sunderland T (2018) Natural resource management schemes as entry points for integrated landscape approaches: evidence from Ghana and Burkina Faso. Environ Manag 62:82–97

Forestry Commission (2016) Ghana forest plantation strategy: 2016–2040

IDH (2022) Ghana cocoa & forests initiative 2022 annual report. IDH – the sustainable trade initiative, viewed 23 July 2024. Retrieved from https://www.idhsustainabletrade.com/publication/ghana-cocoa-forests-initiative-2022-annual-report/

Kalischek N, Lang N, Renier C, Daudt RC, Addoah T, Thompson W, Blaser-Hart WJ, Garrett R, Schindler K, Wegner JD (2023) Cocoa plantations are associated with deforestation in Côte d'Ivoire and Ghana. Nat Food 4(5):384–393

Muhie SH (2022) Novel approaches and practices to sustainable agriculture. J Agric Food Res 10:100446

Nishi M, Subramanian SM, Melaku A (2023) Synthesis: ecosystem restoration in the context of socio-ecological production landscapes and seascapes (SEPLS). In: Nishi M, Subramanian SM (eds) Ecosystem restoration through managing socio-ecological production landscapes and seascapes (SEPLS). Satoyama initiative thematic review, vol 8. Springer, pp 267–288

Oduro K, Mohren G, Pena-Claros M, Kyereh B, Arts B (2015) Tracing forest resource development in Ghana through forest transition pathways. Land Use Policy 48:63–72

Olwig MF, Skovmand Bosselmann A, Owusu K (2024) Agroforestry as climate change adaptation: the case of cocoa farming in Ghana. Springer

Roldan MB, Fromm I, Aidoo R (2013) From producers to export markets: the case of the cocoa value chain in Ghana. J Afr Dev 15(2):121–138

Schooler SL, Johnson MD, Njoroge P, Bean WT (2020) Shade trees preserve avian insectivore biodiversity on coffee farms in a warming climate. Ecol Evol 10(23):12960–12972

Tilman D, Isbell F, Cowles JM (2014) Biodiversity and ecosystem functioning. Annu Rev Ecol Evol Syst 45(1):471–493

Tropical Forest Alliance, Proforest, CDP (2023) Company landscape engagement for cocoa sustainability: progress and the path forward

Tutu Benefoh D (2018) Assessing land-use dynamics in a Ghanaian cocoa landscape. Ph.D. thesis, Universitäts-und Landesbibliothek, Bonn

Vignola R, Harvey CA, Bautista-Solis P, Avelino J, Rapidel B, Donatti C, Martinez R (2015) Ecosystem-based adaptation for smallholder farmers: definitions, opportunities and constraints. Agric Ecosyst Environ 211:126–132

Wainaina P, Minang PA, Duguma L, Muthee K (2021) A review of the trade-offs across different cocoa production systems in Ghana. Sustainability 13(19):10945

Wiens JA (2009) Landscape ecology as a foundation for sustainable conservation. Landsc Ecol 24:1053–1065

Yeboah-Assiamah E, Muller K, Domfeh KA (2023) Lessons from the past: a forest policy reform in Ghana through the feedback loop. Environ Dev Sustain 27:1–21

Chapter 3
Enhancing Ecological Connectivity in a Kaya Forest Landscape Through Biocultural Heritage Territory (BCHT) Management Model

Chemuku Wekesa, Leila Ndalilo, Krystyna Swiderska, and Gladys Rutto

Abstract Kaya forests are sacred biodiverse small forest patches within Kenya's Coast region, historically valued for their spiritual and cultural significance and protected by the Mijikenda community through customary laws and taboos. These forests host over half of the region's rare plants, alongside diverse fauna, providing essential ecosystem services crucial for local livelihoods. They supply timber, medicine, fruits, and water, the essential resources for the community. However, fragmentation and loss of Kaya forests threaten the biodiversity, culture, and livelihoods of the Mijikenda community. To address these challenges, a biocultural heritage territory (BCHT) approach was initiated. Led by Indigenous communities, this initiative aims to protect traditional knowledge, cultural values and practices, and ancestral food systems and enhance biodiversity conservation across four Kaya forests: Bomu, Fimboni, Mudzi Muvya, and Mudzi Mwiru in the Rabai cultural landscape. To guide the establishment of a BCHT management model, socioeconomic surveys were conducted through interviews involving 200 households in ten villages and five focus group discussions (FGDs) between September and October 2022. Findings show that most households earn under 200 USD monthly. Recognition of cultural values was highest among the elderly aged 65 years and above. Women and youth involvement in decision-making was 30% and 20%, respectively. Biodiversity assessments revealed 196 plant species, including 6 endangered endemic species, and 51 butterfly species across the Kaya forests. The generated information informed the establishment of biocultural enterprises and a communal fund for conservation. Furthermore, initiatives to promote agrobiodiversity through indigenous tree planting and traditional crops diversification enhance agrobiodiversity conservation and food security. The BCHT model is effective in

C. Wekesa (✉) · L. Ndalilo · G. Rutto
Kenya Forestry Research Institute (KEFRI), Nairobi, Kenya
e-mail: cwekesa@kefri.org

K. Swiderska
International Institute for Environment and Development (IIED), London, UK

M. Nishi et al. (eds.), *Ensuring Ecological Connectivity in Socio-Ecological Production Landscapes and Seascapes (SEPLS)*, Satoyama Initiative Thematic Review, https://doi.org/10.1007/978-981-95-1474-8_3

ensuring biodiversity-rich landscapes are managed in an integrated and sustainable manner that enhances ecological connectivity.

Keywords Biocultural heritage territory · Kaya forests · Micro-enterprises · Traditional knowledge · Biodiversity

3.1 Background

Biocultural heritage territories (BCHTs) have been defined as land-use mosaics encompassing Indigenous and traditional land tenure, production and exchange systems, cultural identity, and community organization with simultaneous goals of endogenous development and biodiversity conservation (Argumedo and Swiderska 2014). By linking "*heritage*" that is associated with the history and rights to "*bioculture,*" a holistic concept denoting the indivisibility of land and culture, BCHTs reflect land-use practices within Indigenous worldviews and cultural traditions (Swiderska et al. 2020). Born of Indigenous Peoples' memories and experiences, BCHTs embody and protect their world views, spiritual values, customary laws and institutions, and stewardship practices, form the backbone of local economies, and are home to critical genetic resources for food and agriculture. Globally, years of local adaptation and traditional knowledge (TK) creation and use have shaped BCHTs into highly diverse, productive, and resilient management systems that integrate sustainable development with biodiversity conservation (Swiderska et al. 2022).

BCHTs are a territorial self-governance and management approach that emerged from Indigenous Peoples' struggles for self-determination and a 20-year decolonizing participatory action–research (PAR) process in the Peruvian Potato Park (Swiderska et al. 2020). The main goal of BCHTs is holistic well-being, recognizing that "conservation" is a western concept, but holistic well-being encompasses the well-being of both people and nature and results in conservation as the outcome of an autonomous process (Swiderska et al. 2020). BCHTs are bottom-up, food-centered processes that strengthen Indigenous exchange systems, ritual culture and spirituality, and the continued co-evolution of land and culture from pre-colonial times (Swiderska et al. 2020). They arise from Indigenous traditions of landscape management expressed in holistic territorial concepts such as *Mudzini* of the Mijikenda in Rabai (Kenya) and *Ayllu* of Quechua communities in Peru (Swiderska et al. 2020; Wekesa et al. 2021). BCHTs aim to protect centers of origin and diversity of important foods (animals, plants, fish, and grasses), including native varieties and wild relatives that are important for resilience to climate change (Swiderska et al. 2020; Wekesa et al. 2021). Besides, they conserve rich wildlife and provide ecosystem services, while supporting a range of Kunming–Montreal Global Biodiversity Framework targets (CBD 2022) and Sustainable Development Goals (Swiderska et al. 2022).

Kaya forests and associated biodiverse landscapes represent a typical BCHT that is appropriated and valued, both symbolically and instrumentally, by Indigenous communities (Wekesa et al. 2021; Easdale et al. 2023). Within the agricultural landscapes associated with Kaya forests, traditional farming systems reflect the close relationship between human communities and local ecosystems. The transmission of knowledge, cultural practices, values, and beliefs, in which this relationship is materialized, or biocultural memory is key for these systems' conservation and adaptation to social–environmental changes (Ndalilo et al. 2020; Kaulen-Luks et al. 2022). Kaya forests and associated landscapes as a BCHT of the Indigenous Rabai community sustain rich biodiversity, cultivated and wild, through agroecological practices rooted in Indigenous cosmovision and cultural and spiritual values (Wekesa and Ndalilo 2018; Ndalilo et al. 2020; Wekesa et al. 2021; Swiderska et al. 2022).

Knowledge and understanding of existing community-led conservation models are critical in promoting integrated conservation and livelihood initiatives to sustainably manage biodiversity-rich landscapes and enhance ecological connectivity between the fragmented patches of Kaya forests for biodiversity conservation and support to local economies and livelihoods. Therefore, a study was conducted to document how Indigenous-led BCHT initiatives contribute to the protection of traditional knowledge, cultural values and practices, Indigenous food systems and territorialities of Indigenous Peoples, their livelihoods, conservation of biodiversity, and enhancement of ecological connectivity.

3.2 Materials and Methods

3.2.1 Study Sites and Communities

The study was undertaken in 2022 in Rabai sub-county in Kilifi County (Table 3.1). Rabai sub-county is mainly inhabited by the Rabai community, which is one of the nine Mijikenda community sub-tribes (Fig. 3.1).0F[1] The population of Rabai sub-county is 120,813, comprising of 48.5% males (58,571) and 51.5% females (62,242) (KNBS 2019). The sub-county covers an area of 207.8 km^2 (20,780 ha), with a population density of 581 persons per km^2. The number of households in the sub-county is 24,809 with an average of 4.9 persons per household (KNBS 2019). About 71% of people in Rabai live below the poverty line (Republic of Kenya 2013). The Rabai community occupies a hilly area that has moderate rainfall of 1300 mm per annum, and thus they practice mixed farming (both livestock keeping and crop farming). The rainfall is bimodal with two distinct wet seasons occurring in March–May and October–December. There have been considerable changes in rainfall patterns and amounts since 1982; this has led to negative effects on crop yields,

[1] The other sub-tribes are the Giriama, Chonyi, Duruma, Digo, Kambe, Jibana, Kauma, and Ribe.

Table 3.1 Region, area, and demographic information for the case study and BCHT intervention area (Source: Wekesa et al. 2021)

Region and demographic information for the case study	
Country	Kenya
Province/region	Coast
District/county	Kilifi
Size of geographical area	1,224,600 ha
Number of indirect beneficiaries	128,459 persons
Dominant ethnicity	Mijikenda
Area (ha), geographic location and number of people benefiting from BCHT intervention	
Size of case study/project area	580 ha
Number of direct beneficiaries	10,000 persons
Geographic coordinate (longitude and latitude)	S 03° 55′ 55” E 39° 35′ 46”
Dominant ethnicity	Rabai sub-tribe of the Mijikenda tribe

livestock production, water sources, and forest resources, increasing the likelihood of climate-related hazards like drought, flooding, incidences of pests and diseases, and food insecurity (Republic of Kenya 2018; Wekesa et al. 2017).

3.2.2 Research Methods

The study was conducted using decolonizing action research which involved members of the local community and community researchers in co-designing the research approach as well as data collection tools. Data collection was conducted between September and October 2022 by community researchers who were trained on the data collection process and actively participated in community dialogues to build consensus prior to the data collection exercise. A socioeconomic survey was conducted through the use of semi-structured household questionnaires and focus group discussions (FGDs). Household questionnaires were administered to 200 households living in 10 out of the 34 villages in Rabai-Kisurutini Ward, which is inhabited mainly by the Rabai community. The 10 villages are adjacent to four Kaya forests (Kaya Mudzi Muvya, Mudzi Mwiru, Kaya Bomu, and Fimboni). Five FGDs, each of which combined two adjacent villages, were held, and the target respondents composed of 15–20 participants, including elderly and middle-aged men and women, male and female youth, Indigenous farmers, forest resource users, and members of various micro-enterprise groups such as bee-keeping, briquette-making, broom-making, and coconut oil production. Thematic analysis was applied in organizing qualitative data into a series of patterns or themes, in this case, the contribution of various aspects (traditional governance systems, biocultural enterprises, and cultural values) to biodiversity conservation and ecological connectivity. Prior to data analysis, systematic coding of underlying ideas, assumptions, and

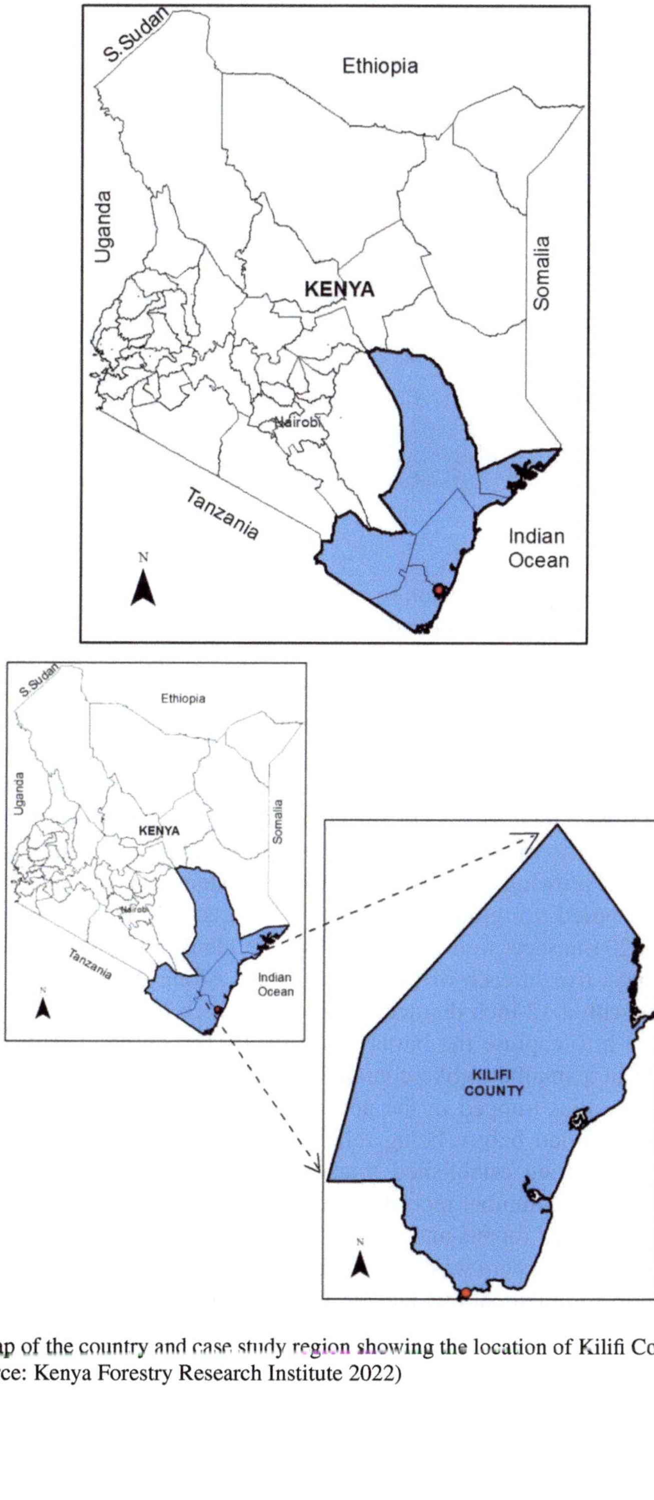

Fig. 3.1 Map of the country and case study region showing the location of Kilifi County, Kenyan Coast. (Source: Kenya Forestry Research Institute 2022)

Fig. 3.1 (continued)

conceptualizations was undertaken to make meaning of the various themes in relation to biodiversity conservation and ecological connectivity.

Ecological surveys were also conducted to identify tree species diversity, frequency of occurrence, and density of tree species using a combination of the following methods. First, 10 m radius circular plots were established for the assessment of population trends, species diversity, species frequency, and natural regeneration of the tree species, as well as the diversity of butterflies in the Kaya forests. Participatory transect walks were then used to identify and record the number of threatened and endangered species to triangulate information generated from the circular plots. The key tree species growing in association with threatened and endangered species were also recorded to guide conservation efforts in the four Kaya forests. Furthermore, 20 transects were established within the four Kaya forests and associated landscapes, five in each of the Kaya forests. Along the transects, insect and butterfly nets with a 12-inch diameter ring and 24-inch net depth were placed and left for 1 to 3 h to capture the butterflies. Locally tapped palm wine was placed inside the net in a small cup-like container to attract the butterflies with its strong scent. The butterflies trapped in the net were counted, and the species' scientific names were identified before being released back to the forests. Transect walks were undertaken along established transects whereby observed elephant shrews were counted and the number recorded. In addition, camera traps were mounted on trails in the four Kaya forests and used to capture small mammals. The number of animals captured on camera was used to estimate the population of elephant shrews in the forests to complement the data on the observed number of elephant shrews.

3.3 Results

3.3.1 Contribution of Traditional Governance Systems to Enhancing Ecological Connectivity and Biodiversity Conservation

The Kaya Council of Elders is an important traditional governance institution for the Mijikenda community that makes decisions in every aspect of community life. Their responsibilities include the regulation of natural resource use, resolution of disputes, and decision-making on trade with outsiders (Kibet and Nyamweru 2008). Results of the study established that traditional governance systems such as the Kaya Council of Elders play an important role in enhancing ecological connectivity and biodiversity conservation. This is done through arbitration of natural resource-based conflicts such as those concerning grazing areas, farm boundaries, and land encroachment, thus contributing to the social cohesion that is key to successful enforcement of rules and regulations for sustainable utilization of natural resources. The Kaya Council of Elders also deters illegal activities such as logging through enforcement of rules and regulations and day-to-day patrol and surveillance activities in the forests. In addition, the council provides guidance on suitable tree species for restoration of degraded areas both in the Kaya forests and farmlands based on their traditional ecological knowledge to ensure that the tree species in the forests and farmlands complement each other in the provision of ecosystem services, hence ensuring the landscape is ecologically connected. Studies have indicated that proper species-to-site matching and species association are critical for proper tree growth which contributes to enhancing ecological connectivity (Wekesa et al. 2018; Ndalilo et al. 2020). The priority indigenous tree species selected by Kaya elders for restoration of degraded areas in Rabai Kaya forests and associated landscapes are presented in Table 3.2.

3.3.2 Contribution of Biocultural Enterprises to Household Incomes, Ecological Connectivity, and Biodiversity Conservation

Farming is the main livelihood activity of the community. Previously, farming was the main source of food for the community, while palm wine tapping was a major source of income. However, this has changed since 2016 due to persistent droughts that have resulted in low agricultural productivity as well as the dwindling population of coconut trees—a significant number of coconut trees have dried out due to recurring droughts. The problem is further exacerbated by the harvesting of coconut trees due to the increased demand for timber and fuelwood. Thus, the contribution of wine tapping as an economic activity has significantly reduced. To fill the income

Table 3.2 Priority tree species for restoration of degraded areas in Kaya forests. (Source: Authors)

Name of tree	Uses
Parkia filicoidea	Fruit, timber
Ficus sycomorus	Firewood
Flueggea virosa	Firewood
Ceiba pentandra	Boat making, firewood
Piliostigma thonningii	Firewood, herbal medicine
Ficus thonningii	Herbal medicine, fruits, firewood, making of bee hives
Annona senegalensis	Fruit, herbal medicine, firewood
Borassus aethiopum	Herbal medicine, firewood
Diospyres mespiliformis	Herbal medicine, firewood, making of traditional serving spoon
Dombeya taylorii	Source of construction poles and withies
Thespesia dania	Fruit, firewood

Table 3.3 Income generated from biocultural enterprises in the Rabai biocultural landscape. (Source: Authors)

Bio-enterprise	Annual income (USD)
Bee-keeping	6050.00
Briquette-making	4180.00
Production of virgin coconut oil	3466.70
Broom-making	8500.00
Traditional songs and dances	670.00

gap resulting from decreased palm wine tapping, members of the community have embraced biocultural/nature-based enterprises as alternative sources of income. The enterprises are slowly picking up as demonstrated in the present findings, which established that, on average, most households earn about 200 USD per month from the selling of different biocultural products. The biocultural enterprises that generate alternative income to support local livelihoods include bee-keeping, briquette-making, production of virgin coconut oil, broom-making, and traditional songs and dances performed by the members of cultural villages located adjacent to the Kaya forests. The briquettes are made from agricultural wastes, coconut husks, and soil to ensure sustainability. The annual income generated from these enterprises is elucidated in Table 3.3. These enterprises generate income at the household level, which significantly contributes to improved living standards at both community and household levels.

Besides generating income for the community, nature-based/biocultural enterprises play a significant role in enhancing ecological connectivity and promoting biodiversity conservation. For instance, 10% of the income generated is channeled to a communal fund that supports conservation activities such as on-farm tree planting and restoration of degraded sites in Kaya forests. Planting a wide range of native trees in degraded sites within the Kaya forests and on farmlands increases the diversity of flora and enhances species interactions, hence contributing to improved ecological connectivity within the landscape. Furthermore, bee-keeping enterprises

Fig. 3.2 Beehives distributed and installed in Kaya Bomu for honey production. (Photo credit: Kenya Forestry Research Institute 2022)

enhance pollination across the Kaya forests and the agricultural landscape (Fig. 3.2). The pollinators play an important role in connecting the ecosystem and maintaining plant genetic diversity for both biodiversity and agrobiodiversity conservation while enhancing the crop yield. This ensures that the community is food-secure.

Briquette-making provides a sustainable source of fuel to meet the energy needs of households, hence reducing pressure on Kaya forests and abating the number of trees harvested from farmlands for fuelwood (Fig. 3.3). This has reduced deforestation, leading to improved conservation of Kaya forests and maintaining a high population of native trees in farmlands which, alongside agroforestry belts along the edges of Kaya forests, ensures a continuous connected and healthy landscape. The connectivity allows movement of birds and butterflies as well as pollination materials between the different Kaya forests, hence promoting ecological connectivity and biodiversity conservation. Preserving Kaya forests in a primeval condition ensures continued provision of ecosystem goods and services that support human well-being and livelihoods. The forests are sources of streams, play important roles in pollination and soil stability enhancement, and are reservoirs of plant genetic resources with their rich biodiversity, most of which is endemic to the forests, with some species being endangered. The forests also serve as a source of forest products including firewood and poles as well as non-timber products such as herbal medicine, wild fruits, vegetables, and fiber, all of which support local livelihoods.

Coconut oil production (Fig. 3.4) and broom-making ensure that coconut trees, the main raw material for these products, are conserved and sustainably utilized. Coconut trees are an integral part of the agricultural ecosystem of the Rabai landscape and play a role in enhancing ecosystem connectivity with the adjacent Kaya forests. The farmers have adopted a fast-maturing coconut variety to ensure the sustainable supply of raw materials. The hybrid variety developed by the Kenya Agricultural and Livestock Research Organization (KALRO) that has been provided to farmers since 2019 takes 2.5 to 3 years to mature and is high-yielding, producing 250–300 coconuts annually. Coconut wine derived from the trees also has a huge cultural significance as it is used in most traditional ceremonies to perform both social and natural resource-based rituals, thereby enhancing social cohesion and promoting cultural values of solidarity, collectiveness, and reciprocity,

Fig. 3.3 Briquette production in Rabai BCHT landscape. (Photo credit: Kenya Forestry Research Institute 2022)

Fig. 3.4 Virgin coconut oil produced in Rabai biocultural landscape. (Photo credit: Kenya Forestry Research Institute 2022)

which are key requirements for the successful management of the biocultural heritage territory.

On-farm planting of trees plays an important role in reducing the exploitation pressure on Kaya forests by providing an alternative source of income for local communities, as well as wood and non-wood forest products that would otherwise be derived from Kaya forests. This contributes to the protection of diverse flora and fauna species, most of which are endangered and endemic to these forests. Moreover, domestication of selected native trees in farmlands and diversification of traditional crops maintain a high diversity of pollinators (bees and butterflies), thus ensuring biodiversity conservation and landscape connectivity while safeguarding Kaya forests from illegal activities.

3.3.3 Contribution of Cultural Values to Biodiversity Conservation and Ecological Connectivity

Cultural values of solidarity (unity of purpose), collectiveness, equilibrium, and reciprocity were found to play a significant role in enhancing social cohesion, biodiversity conservation, and ecological connectivity. The cultural values promote cultivation and diversification of traditional crop varieties, tree domestication on farmlands for food and medicinal purposes, and traditional ceremonies. Commonly grown traditional crop varieties include cassava, maize, cowpeas, green grams, and sweet potatoes. Traditional crop varieties are resilient to the impacts of climate change such as increased incidences of pests and diseases and prolonged drought. Thus, traditional crop diversification provides a safety net for food security. Even in the case of failure in some crops grown in a given season, the community is assured of yield from other crop varieties that are resilient to climate change. In addition, wild plants like *Lilium orientale*, *Tamarindus indica*, *Ancylobotrys petersiana*, *Landolphia kirkii*, and *Ziziphus mauritiana* are domesticated for their fruits. Traditional crops are grown using organic fertilizers and are chemical-free and therefore provide healthy food for the community. Cultivation of traditional crops and domestication of wild fruits contribute to food security and sustain local economies. The cultivation and diversification of traditional crop varieties contribute to agrobiodiversity conservation, while the accompanying traditional farming practices such as contouring, crop rotation, and mulching maintain high soil fertility for increased crop yields. Sustainable land management and farming practices anchored on traditional knowledge also facilitate the growth of soil microorganisms and plants, hence contributing to both biodiversity conservation and ecological connectivity.

Cultural values often go hand in hand with traditional ceremonies, songs, and dances, hence providing avenues for passing over traditional knowledge (TK) from the older to the younger generation. Traditional ceremonies such as weddings, the Rabai New Year Festival, and those for circumcision promote cultural tourism, which is a source of income to the community, thus contributing to poverty reduction. These traditional ceremonies also sustain traditional cultural values and promote traditional knowledge. Some ceremonies like traditional prayers and sacrifices form part of traditional religious practices. *Sayo* and *Njuga* rituals performed during Rabai New Year celebrations are undertaken to pray for adequate rainfall, which enhances agricultural production. Closely associated with traditional ceremonies are traditional songs and dances, which are performed to promote good moral behavior and enhance social cohesion and socioeconomic development. Traditional attire is worn during the songs and dances to bring out the uniqueness of the Rabai culture, and this preserves the cultural heritage of the community, further enhancing cultural tourism.

Culture is important because it is strongly intertwined with agrobiodiversity conservation through the promotion of sustainable agroecological practices which play an important role in promoting connectivity of the landscapes. Generally, traditional ceremonies empower communities through information-sharing on health, good morals, traditional knowledge, farming systems, and environmental conservation and are therefore important in sustaining knowledge. Traditional ceremonies are also sources of income to the Rabai community. Moreover, sacred sites (*Mizimu*) in Kaya forests are also preserved through cultural values and beliefs (taboo system) that hinder access to the areas by the community; only Kaya elders are allowed to access these sacred sites. The restricted access to these sacred areas conserves biodiversity and helps keep Kaya forests in pristine condition for ecological connectivity. Rabai Cultural Village (RCV), a cultural center adjacent to Kaya Mudzi Muvya, was also found to play a significant role in the conservation of Kaya forests and associated agricultural landscapes. The RCV promotes conservation activities such as reforestation and rehabilitation of degraded areas and deters illegal activities in Kaya forests through surveillance by group members. Generation of alternative income through cultural tourism is also a deterrence to overexploitation of Kaya forest resources. Threatened tree species and important herbal plants are conserved in the cultural village. The RCV also hosts a community seed bank where collected seeds for traditional crops and native trees are stored awaiting planting season.

The level of recognition and application of cultural values and TK by different age groups was assessed. The recognition of cultural values was highest among the elderly aged 65 years and above (65%), followed by middle-aged (30%), and youth (5%). Moreover, the level of respect for elders by the youth is relatively low (40%), and this is mainly attributed to the erosion of TK, modern religious beliefs which contradict TK, focus on livelihoods/income generation, hence lack of time to focus on TK-based activities and practices, and inadequate awareness on the importance of TK. Participation of youth in traditional ceremonies was however found to be high (70%), mainly because of sensitization by RCV targeting the youth as well as the annual Rabai New Year Festival conducted by RCV in collaboration with other stakeholders that incorporates sports and fun activities to attract youth. Both the middle-aged (70%) and elderly (80%) participate in traditional ceremonies, and this was attributed to enhanced awareness on TK among this category. The low recognition of cultural values by the youth who form about 40% of the local population is likely to negate the contributions of cultural values to natural resource management, agrobiodiversity conservation, and ecological connectivity.

3.3.4 Status of Biodiversity Within the Rabai BCHT Landscape

The Kaya forests in Rabai have a rich diversity of tree species. One hundred and fourteen different tree species were recorded in the four Kaya forests (Bomu, Fimboni, Mudzi Muvya, and Mudzi Mwiru). The most common tree species was *Suregada zanzibarensis* (18.76%), followed by *Manilkara sansibarensis* (5.41%). *Zanthoxylum holtzianum* (0.11%) and *Vitex ferruginea* (0.11%) are among the tree species with low occurrence in the Kaya forests due to overharvesting from the wild for edible fruits. Moreover, 82 shrub species were found in the four Kaya forests, with *Polyspherio porrifolia* being the most dominant shrub (16.54%), followed by *Hoslundia opposita* (9.97%) and *Rytigynia celastroides* (8.31%).

Threatened and endangered endemic tree species (*Bauhinia mombassae, Combretum tenuipetiolatum, Angylocalyx braunii, Afrocanthium kilifiense, Canthium pseudoverticillatum, Coffea pseudozanguebariae, Kraussia speciosa,* and *Synsepalum subverticillatum*) were not found in most parts of Kaya forests. However, these species were found along the streams and in open areas within the forests. If the area covered by Kaya forests continues to shrink due to illegal encroachment, these species could be lost through extinction. This is because the current population of these species is very low, and the habitat range is narrow. Only three individuals of *Combretum tenuipetiolatum* were found in the four Kaya forests and six each for both *Bauhinia mombassae* and *Angylocalyx braunii*. The species also exhibited poor natural regeneration apart from *Afrocanthium kilifiense*, further demonstrating the need for aided regeneration to increase their population in the natural range and save them from extinction. The poor natural regeneration of the tree species is attributed to prolonged drought due to climate change.

The landscape has diverse species of butterflies. Fifty-one different species of butterflies were found in the four Kaya forests (Bomu, Fimboni, Mudzi Muvya, and Mudzi Mwiru). Overall, the most common species was *Charaxes varanes* (9.32%), followed by *Tirumala petiverana* (6.36%), while *Junonia trea* and *Neptis* species were the least common, accounting for only 0.42% of the butterfly species found in the Rabai Kaya forests. The most common species in Kayas Fimboni and Bomu were *Charaxes varanes* and *Graphium colonna*, which account for 8.86% of the butterfly species diversity. *Tirumala petiverana* (12.77%) and *Charaxes varanes* (10.64%) were found to be the most common species in Kaya Muzdi Muvya, while *Melanitis leda* (12.7%) and *Euphaedra neophron* (11.11%) were the most common species in Kaya Mudzi Mwiru. The population of the elephant shrew appeared to be plummeting. Only four elephant shrews were observed in Kaya Bomu/Fimboni along their trails. Furthermore, only seven nests were found in the Kaya forests: four in Kaya Bomu/Fimboni, two in Kaya Mudzi Muvya, and one in Kaya Mudzi Mwiru. Also, a total of six traps were found distributed among the four Kaya forests, with four being found in Kaya Bomu/Fimboni (Fig. 3.5). The presence of elephant shrew snares in the forests indicates that there is illegal hunting of the species by local communities for bush meat, and this poses a serious threat to the conservation of the species whose population has been on the decline.

Fig. 3.5 Snares used to trap elephant shrew in Kaya forests. (Photo credit: Kenya Forestry Research Institute 2022)

3.4 Discussion

Traditional governance system, particularly the Kaya Council of Elders, was found to play an important role in arbitration over natural resource-based conflicts, enforcement of rules and regulations for sustainable utilization of natural resources, and undertaking surveillance to deter illegal activities in Kaya forests. In addition, Kaya elders provide guidance on suitable tree species for restoration of degraded areas, thus contributing to biodiversity conservation and ecological connectivity. Moreover, community-based and nature-based/biocultural enterprises were found to play a significant role in supporting local livelihoods and therefore reducing exploitation pressure on forest resources. Furthermore, on-farm tree planting continues to play an important role in reducing exploitation pressure on Kaya forests by providing an alternative source of income and forest products. In addition, 10% of the income generated from these biocultural enterprises is channeled into a communal fund that supports conservation activities such as on-farm tree planting and restoration of degraded sites in Kaya forests. Given that 10% of the income is used for collective action for conservation while the rest is for individual households, this suggests that the benefits are equitably shared among community members.

Kaya forests are recognized for their spiritual, religious, cultural, and historical significance associated with the Mijikenda community—cultural values play a significant role in their conservation. For instance, the restricted access to sacred sites within Kaya forests helps keep Kaya forests in pristine condition, thus conserving biodiversity and enhancing ecological connectivity. Moreover, cultural values and practices promote cultivation and diversification of traditional crop varieties, tree domestication on farmlands for food and medicinal purposes, and the traditional ceremonies that provide avenues for passing over TK from the older to the younger generation. The level of recognition and application of cultural values by the members of the community seems to be declining and is relatively low due to modernity leading to the association of TK with witchcraft and backwardness, modern religious beliefs which contradict TK, focus on livelihoods/income generation, hence lack of time to focus on TK-based activities and practices, especially among the youth, and insufficient awareness on the importance of TK. There is a need for further community sensitization on the importance of cultural values for enhanced conservation and livelihood gains. The Rabai Cultural Village has had some success—a significant number of youths are likely to embrace the culture.

The landscape is rich in biodiversity; there are 196 species of plants (trees and shrubs) in the Kaya forests. The populations of threatened and endangered tree species (*Bauhinia mombassae*, *Combretum tenuipetiolatum, Angylocalyx braunii, Afrocanthium kilifiense, Canthium pseudoverticillatum*, *Coffea pseudozanguebariae*, *Kraussia speciosa*, and *Synsepalum subverticillatum*) are low and restricted to very narrow habitats within the landscape. Hence, they require urgent intervention to save them from extinction.

3.5 Conclusion and Recommendations

This chapter emphasizes the need for promoting integrated conservation and livelihood initiatives to sustainably manage the sacred Kaya forests and associated agricultural landscape to enhance ecological connectivity, conserve biodiversity, and support local economies and livelihoods. This chapter demonstrates the potential of Indigenous-led initiatives to establish a collectively governed biocultural heritage territory using decolonizing action research methodologies that can protect and revitalize traditional knowledge, cultural values and practices, Indigenous food systems, and territorialities of Indigenous Peoples and local communities to conserve biodiversity and enhance ecological connectivity in biodiversity-rich landscapes. The concept of biocultural heritage centers on the inextricable linkages between cultural and biological diversity. Moreover, this chapter has demonstrated that the Rabai community has a strong traditional governance system that is fully involved in governing and managing the landscape according to their rights, knowledge, capacities, cultural values, and institutions (Kaya Council of Elders). This chapter also highlights biocultural enterprises and products that generate income to support local livelihoods and the establishment of a communal fund, whereby 10% of the income generated from nature-based/biocultural enterprises is channeled toward supporting conservation activities such as on-farm tree planting to enhance ecological connectivity. Collective management of the Biocultural Heritage Territory in Rabai could be achieved through collaboration between chiefs, village elders, and Kaya elders, coupled with strong community participation including the youth in natural resource management, which is key for sustainability.

References

Argumedo A, Swiderska K (2014) Biocultural heritage territories. Retrieved from https://www.iied.org/g03843

Convention on Biological Diversity (CBD) (2022) Post-2020 Global Biodiversity Framework

Easdale MH, Michel CL, Perri D (2023) Biocultural heritage of transhumant territories. Agric Hum Values 40:53–64

Kaulen-Luks S, Marchant C, Olivares F, Ibarra JT (2022) Biocultural heritage construction and community-based tourism in an important indigenous agricultural heritage system of the southern Andes. Int J Herit Stud 28(10):1075–1090

Kenya National Bureau of Statistics (KNBS) (2019) Kenya population and housing census volume I: population by county and Sub-County. KNBS, Nairobi

Kefri (2022) Baseline socio-economic and ecological survey for rabai cultural landscape, nairobi, kenya. unpublished report. pp21

Kibet S, Nyamweru C (2008) Cultural and biological heritage at risk; the case of the Rabai Kaya forests in coastal Kenya. J Hum Ecol 24(4):287–295

Ndalilo L, Wekesa C, Mbuvi MTE (2020) Indigenous and local knowledge practices and innovations for enhancing food security under climate change: examples from Mijikenda communities in coastal Kenya. In: Gasparatos A, Naidoo M, Ahmed A, Karanja A, Fukushi K, Saito O, Takeuchi K (eds) Sustainability challenges in sub-Saharan Africa II. Science for Sustainable Societies, Springer, pp 63–82

Republic of Kenya (2013) Kilifi County first county integrated development plan 2013–2017. Kilifi County Government, Kilifi

Republic of Kenya (2018) Kilifi County integrated development plan 2018–2022. Kilifi County Government, Kilifi

Swiderska K, Argumedo A, Song Y, Rastogi A, Gurung N, Wekesa C (2018) Biocultural innovation: the key to global food security?. IIED briefing paper. IIED, London

Swiderska K, Argumedo A, Pimbert, M (2020) Biocultural heritage territories: key to halting biodiversity loss. IIED Briefing Paper. https://pubs.iied.org/17760IIED/

Swiderska K, Argumedo A, Wekesa C, Ndalilo L, Song Y, Rastogi A, Ryan P (2022) Indigenous peoples' food systems and biocultural heritage: addressing indigenous priorities using Decolonial and interdisciplinary research approaches. Sustainability 14(18):11311

Wekesa C, Ndalilo L (2018) Sustainable use of biodiversity in socio-ecological production landscapes and seascapes (SEPLS) and its contribution to effective area-based conservation: The case of Kaya forests on the Kenyan Coast. In: UNU-IAS, IGES (ed) Sustainable use of biodiversity in socio-ecological production landscapes and seascapes (SEPLS) and its contribution to effective area-based conservation (Satoyama Initiative Thematic Review vol. 4). University Institute for the Advanced Study of Sustainability, Tokyo, pp 85–94

Wekesa C, Ongugo P, Ndalilo L, Amur A, Mwalewa S, Swiderska K (2017) Smallholder farming systems in coastal Kenya: key trends and innovations for resilience. IIED, London

Wekesa C, Maranga EK, Kirui BK, Muturi GM, Gathara M (2018) Interactions between native tree species and environmental variables along forest edge-interior gradient in fragmented forest patches of Taita Hills, Kenya. For Ecol Manag 409:789–798

Wekesa C, Ndalilo L, Swiderska K (2021) Towards a biocultural heritage territory in Rabai cultural landscape: exploring Mijikenda cultural values and practices for sustainable development. IIED, London

Chapter 4
Bridging Worlds: Traditional Wisdom and Eco-connectivity in Nan'an

Yu-Chun Chan, Jing-Juan Hsu, Jia-Xin Yang, and Chun-En Xiao

Abstract This chapter aims to explore how an NGO, the Tse-Xin Organic Agriculture Foundation as an intermediary organization, was involved in the conservation and revitalization of socio-ecological production landscapes and seascapes (SEPLS). The case area is located on the outskirts of Yushan National Park and is surrounded by mountains. The case site is mainly inhabited by the Bunun, an Indigenous ethnic group. The Bunun's relationship with the land has changed dramatically over the last century due to colonization and forced settlement. Farming patterns have also shifted from diversified farming in mountainous areas to single rice cultivation. This has resulted in the gradual disappearance of the traditional ecological knowledge (TEK) that values a harmonious relationship between humans and nature. The land management model has a direct impact on the quality of the habitat. For example, the irrigation system of rice fields is an important habitat for the rare medaka fish *(Aphyocypris kikuchii).* Since rice farmers are the primary managers of agricultural landscapes, their recognition and participation are essential in promoting the transition to organic rice farming and wildlife conservation. Therefore, the project team has sought ways to work with farmers to protect the rice field ecosystem and the habitat of the medaka fish, including organic farming, habitat creation activities, agrifood education programs, and integrating the payments for ecosystem services policy. Our findings suggest that an NGO as an intermediary group, acting as a knowledge translator in a multi-stakeholder platform composed of farmers, ecologists, tourists, local schools, government, and others, facilitated dialogue and understanding among stakeholders of the SEPLS. Indigenous and Local Communities (IPLC) also played a crucial role in facilitating community exploration of the balance between conservation and sustainable livelihoods.

Keywords Traditional ecological knowledge · Eco-connectivity · Biocultural diversity · Diversified farming system · *Aphyocypris kikuchii*

Y.-C. Chan · J.-J. Hsu (✉) · J.-X. Yang · C.-E. Xiao
Tse-Xin Organic Agriculture Foundation, New Taipei City, Taiwan
e-mail: jj.hsu@toaf.org.tw

M. Nishi et al. (eds.), *Ensuring Ecological Connectivity in Socio-Ecological Production Landscapes and Seascapes (SEPLS)*, Satoyama Initiative Thematic Review, https://doi.org/10.1007/978-981-95-1474-8_4

4.1 Introduction

Nan'an is a community populated by an Indigenous tribe (the Bunun). It is located on the outskirts of Yushan National Park and is surrounded by mountains. Centuries ago, the Bunun people traditionally inhabited the mountains and forests at altitudes ranging from 500 to 1500 m. They adapted to the slopes by cultivating and tilling the land, practicing intercropping, and rotating a variety of crops. A small piece of cultivated land in the mountains was sufficient to feed an entire family. About one century ago, the impact of colonization and modernization forced the Bunun people to migrate from the mountains to the banks of the Rakulaku Stream at an altitude of less than 300 m.

Driven by the policies at that time and in order to meet the demands of the country and the modern market, the Bunun people reclaimed the upper terrace into rice fields and switched their agricultural production method to monoculture planting. Under the influence of Green Revolution, farmers adopted pesticides and chemical fertilizers. The conventional farmers then became heavily reliant on pesticides to control weeds, pests, and diseases, along with chemical fertilizers, to supply essential nutrients for rice growth. These chemicals and unabsorbed nutrients are washed into streams through irrigation systems, perpetuating impacts on the environment and aquatic ecosystems.

After the conversion to rice paddies and the transformation into a vast artificial wetland dominated by monoculture, the landscape of the Nan'an socio-ecological production landscapes and seascapes (SEPLS) underwent significant changes. Previously, it consisted of floodplains and upper terraces along streams, characterized by a diverse array of natural habitats such as small slow-moving streams, shallow beaches, sandbars, ponds or marshes, and riparian vegetation comprising low-stemmed grasses, tall grasses, shrubs, and trees. Now, the water level in the rice paddies is completely drained twice a year during mechanical harvesting and intermittently throughout the cultivation period. These drastic fluctuations in water level are likely to have a profound impact on the organisms inhabiting the rice wetlands. The traditional value of a harmonious relationship between humans and nature faces significant challenges (Chan 2023).

Animals that enter the paddy field environment from the surrounding streams and forests to feed or breed, and seasonal migratory birds that pass through the island, can face the risk of poisoning or death when they enter the paddy fields from the natural ecosystem. Conventional agriculture also affects the ecological connectivity of species that live in the water and land areas of farmland ecosystems. Frogs and dragonflies that need to use water and land at different life stages, as well as birds, ladybirds, and spiders that live on land but enter rice fields to feed, all have their survival affected by farmers' use of pesticides. This includes the rare and endangered medaka (*Aphyocypris kikuchii*), a fish which used to primarily inhabit streams on the island.

Since 2014, the project team from the Tse-Xin Organic Agriculture Foundation has been stationed in Nan'an at the invitation of the national park to help the Nan'an community transform their livelihoods and landscapes into more sustainable ones. The project team embarked on a series of actions to improve the quality of the landscape and habitat. The team has evolved from simple advisers to learners and peers, working with local residents and other stakeholders to protect and enhance the ecological connectivity of the area's SEPLS, and to find ways to reconcile productive and economic functions.

In 2018, the government proposed an "ecological network" policy to protect and revitalize SEPLS (Forestry and Nature Conservation Agency 2020). The policy encourages land management practices that enhance ecological connectivity, such as the conversion of farmland to eco-friendly farming. In the eastern part of the island, the policy's priorities are to improve the connectivity between streams and the surrounding canal and to enhance the connectivity of biological corridors. The Nan'an community, after several years of autonomous restoration by the community, has become one of the sites where this policy has been implemented.

It has been recognized that the involvement of Indigenous peoples and local communities (IPLC) is essential for enhancing sustainability. They live on and manage large areas of land, which are often rich in biodiversity and crucial for regional ecosystems (Tengö et al. 2017). The way local people in Nan'an manage their rice fields has a significant impact on the habitat quality and ecological connectivity of the medaka. But, how can IPLC participate in the preservation and revitalization of SEPLS? This chapter documents the three-step method of transition to a sustainable landscape. The method highlights the participation of IPLC in conserving local species and establishing ecological connectivity. It underscores how integrating Indigenous peoples' TEK and human–land relationships into ecological conservation can mitigate the serious imbalance caused by modern monocrop agricultural systems.

4.2 Method

4.2.1 Study Site

The Nan'an community is situated in Zhuoqing Village, Zhuoxi Township, Hualien County, on the eastern side of the Central Mountain Range, near the boundary of Yushan National Park. This area marks the upper reaches of Hualien's main river, the Xiuguluan, with the river terraces cultivated into farmland. The community consists of three small tribal settlements, with a total population of 1277 people (Zhuoxi Township Household Registration Office 2024) (Fig. 4.1).

This area is home to the endangered endemic fish, medaka, and Taiwanese black bears, which live in forests and are getting closer to the community. This community serves as a crucial buffer zone on the edge of the reserve (Table 4.1).

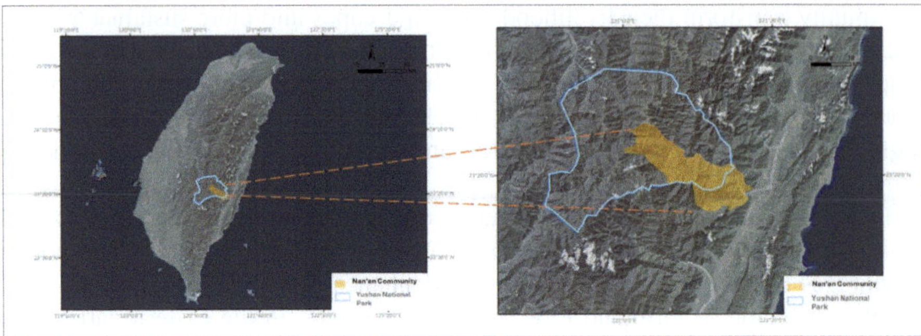

Fig. 4.1 Map of the study site. (Source: Taiwan Map Service 2024 https://maps.nlsc.gov.tw/T09/mapshow.action)

Table 4.1 Basic information of the study area

Country/Region	Taiwan, China
Province	Hualien County
District	Zhuoxi Township
Municipality	Zhuoqing Village
Size of the geographical area (hectare)	40,435
Dominant ethnicity(ies), if appropriate	Indigenous Bunun
Geographic coordinates (latitude, longitude)	23°34′46.7″N, 121°13′93.9E

Source: Authors, based on the study

4.2.2 SEPLS of the Area

The SEPLS in Nan'an consist of industrial rice paddy fields, self-sufficient home gardens, and the surrounding natural landscapes. Rice paddies are the main agricultural landscapes. They are located on upper terraces to facilitate irrigation. Nan'an and the surrounding tribes have over 100 hectares of paddy fields, of which about 17 ha have been converted to organic farming, while the rest implement conventional farming. The fields are mainly managed by large monocultures and are mostly mechanically cultivated (Fig. 4.2).

Since the irrigation water for the rice fields is taken directly from the Rakulaku Stream, water passes through the irrigation equipment, enters the rice field system, and eventually returns to the stream. The aquatic plants, animals, and microorganisms from the stream ecosystem will enter the paddies through the irrigation system. The rice paddies are like an extension of the Rakulaku Stream ecosystem.

In 2016, the project team discovered an endangered endemic species of fish, the medaka (*Aphyocypris kikuchii*), which used to primarily inhabit streams, in the rice fields. The medaka, a small freshwater fish (adult males are about 8 cm long), is endemic to the island and lives mainly in slow-flowing rivers, ditches, or ponds,

Fig. 4.2 Landscape of Nan'an community. (Source: Yushan National Park)

especially in waters with dense aquatic vegetation. It likes to move in groups, weaving between aquatic plants and underwater barriers, and feeds on insects and algae that fall to the surface. Its main habitat is the eastern part of the island, and its early population was so large that the famous railway lunch boxes used dried medaka as a side dish in the past (Chang 2022).

Medaka fish on the island face significant threats in recent years due to multiple factors (Lai et al. 2010, p. 258):

A. With the introduction of *Opsariichthys pachycephalus* from the west into the eastern part of the island, the ecological niche of *Opsariichthys pachycephalus* now overlaps with that of medaka, leading to increased competition for survival.

B. The medaka has increasingly become a common bait for anglers, resulting in overfishing and a sharp decline in its population over the years.

C. The frequency of heavy rainfall has risen, causing turbulent waters in Rakulaku Stream. This phenomenon widens the riverbed and increases the sediment content of the creek.

D. The remediation projects of rivers and lakes have solidified tributaries and channels with cement, eradicating essential aquatic plants needed for spawning. This alteration destroys the original habitat and impedes the recovery of fish populations.

4.2.3 Project Operation

Attempts to transform the landscape of the Nan'an community into a more sustainable one have been initiated since 2014, employing a four-step method to guide this transformation.

4.2.3.1 Step 1: Conversion to Organic Rice Fields and Acquisitions

As Nan'an is close to the national park, the project team began to advise farmers to switch to organic farming in 2014 on behalf of the Yushan National Park Headquarters. The project aims to create a biodiverse ecological environment through eco-friendly farming methods, providing more habitats and food for living creatures. The project team has been working with organic rice companies to purchase organic rice on a contractual basis, providing farmers with a stable income. Organic farming also reduces the environmental impact of pesticides and chemical fertilizers on the SEPLS in Nan'an.

4.2.3.2 Step 2: Actions to Enhance SEPLS by Creating Ecological Corridors to Maintain and Conserve Ecological Connectivity

We invited ecologists to conduct field inspections and discussions with farmers and developed a set of criteria to be included in the payment program. These included ridges, ditches, and ecological ponds, which were identified as more conducive to improving the ecology of rice fields. Farmers were encouraged to manually weed their fields, plant native grasses or traditional crops recommended by the Farmers' Improvement Organization, and maintain or dig ditches and ecological ponds in their fields to create diverse habitats and provide pathways for animals to move around.

4.2.3.3 Step 3: Farmland Ecology Education and Community Engagement

The management of SEPLS cannot rely solely on a few farmers or fields improving their practices but also requires the support of a market-based mechanism. Therefore, the Bunun Beans Co-op team was built up in recent years. The team organizes farmland ecology education activities, where farmers and ecological observers together lead school children to observe the ecology of the farmland. We also invite tourists to visit the farms during the Farmland Education Programs, where farmers explain to visitors how they have created their own habitats and what they have achieved. This initiative supports tribes in developing an experiential economy rooted in TEK. Activities include educational courses on Bunun culture, ecotourism that showcases local mountain wisdom, and food services using indigenous crops and unique ingredients.

4.2.3.4 Step 4: Payment for Ecosystem Services (PES)

The adoption of ecological habitat creation measures may result in a loss of production income for farmers (due to sacrificing land to create ecological ponds or ditches) or an additional labor burden (such as maintaining or planting grasses in field ridges). As such, the Forestry and Nature Conservation Agency incorporated rice paddies in the Nan'an community into the scope of benefits under the Payments for Endangered Species and Important Habitat Ecological Services Program launched in 2021. With the concept of "ecological salaries", farmers are encouraged to take actions that are beneficial to endangered species groups and important habitats.

4.3 Results

4.3.1 Conversion to Organic Rice Fields

With the goal to reduce toxic substances and excessive nutrients entering the aquatic system, together with stakeholders including rice farmers, landowners, organic rice factories, a bank, and consumers, the team guided the transformation that increased farmers' willingness to switch to organic rice. The results of the interventions are presented in Table 4.2.

Table 4.2 shows the current situation of organic rice production in the Nan'an community after the interventions aimed at transforming farms into organic rice production systems. Over the past 10 years, the area of organic farming has grown from 9 to nearly 17 hectares. In these 17 hectares of rice fields converted to organic rice fields, a total of 880,000 tons of rice was produced from 2014 to 2024. This resulted in a gained income of 870,000 USD.

4.3.2 Actions to Enhance SEPLS by Creating Ecological Corridors to Maintain and Conserve Ecological Connectivity

Implementing organic farming can reduce the introduction of toxic substances or excess nutrients into the overall system and therefore reduces the likelihood of harm to animals. This is a favorable habitat for native species, especially if ecological

Table 4.2 Features of the current situation of rice production in the Nan'an community after interventions. (Source: Prepared by the author)

Organic farming area (hectares)	16.94
Rice production (tons)	880,000
Income (USD)	870,000

corridors are created. Creating ecological corridors provides pathways for small invertebrates (e.g., insects and spiders), which helps them spread out in the field and increases the resilience of medaka in the face of natural disasters and changes in farming operations. This creates complexity in the single rice field landscape, which then also creates more microhabitats and food sources and thus increases the connectivity of the rice field landscape (animals such as birds and bees that come from the mountains or streams forage in the fields).

Initially, the project team was unaware that this area was a medaka habitat, but the initial transition to organic farming laid an important foundation for subsequent medaka conservation efforts. According to a survey, the medaka habitat in Nan'an has become the most important one in the area due to the establishment of buffer zones, ecological ponds, and other shelters that facilitate the species' propagation and reproduction (Forestry and Nature Conservation Agency 2024). It is evident from the findings that the efforts of this project have effectively enhanced and expanded the ecological connectivity for medaka and other small aquatic animals. Therefore, while converting to organic rice production, the biodiversity of rice fields was also enhanced through plant diversification on ridges and construction of small ponds as shelters to conserve native species that ecologically connect nature with rice production. The quantities of these actions are shown in Table 4.3.

Table 4.3 presents the interventions carried out to conserve native species by creating favorable living habitats for them. Interventions included growing diverse plants on ridges that currently total 600 m in length, constructing ditches equal to 160 m, and digging nine ponds as shelters for aquatic native species, including invertebrates (insects, spiders, etc.) and frogs. It helps the community spread out in the field and increases the resilience and survival of the medaka in the face of natural disasters and changes in agricultural practices. The features of the diverse ridges are shown in Figs. 4.3, 4.4, and 4.5, and shelters for aquatic species are shown in Figs. 4.6 and 4.7.

Farmers have built bamboo trellises on the ridges, planted various traditional crops or folkloric plants, and deliberately preserved various trees on the ridges, which have diversified the previous monoculture rice field environment and provided more microhabitats and food sources for organisms, thus improving the ecological connectivity of the landscape. The bamboo trellises and mixed trees have also become habitats for some forest birds, and the grains planted by the farmers (e.g., red quinoa, Fig. 4.4) or the fruits and flowers of the preserved mixed trees (e.g., *Morus alba*) also provide food sources for some forest birds.

In 2019, a total of nine farmers participated in the program and created 18 habitats. In 2020, an annual contest was begun to promote healthy competition among farmers. The good ecology and landscape of the farmlands have also provided

Table 4.3 Construction of ecological corridors in Nan'an community to increase the ecological connectivity between nature and rice fields

Length of diversified ridges (m)	600
Length of ditches (m)	160
Number of shelter ponds	9

Fig. 4.3 Farmers created an ecological ditch that preserves mixed woods and diverse grass. (Source: Authors)

farmers with materials for ecotourism and experiential activities, which bring additional income to farmers beyond primary production.

Recent ecological surveys in the Rakulaku Stream watershed, commissioned by the project team, have failed to detect any remaining trevally populations (Yang 2021). In a survey by the Forestry and Nature Conservation Agency (2024), the distribution of medaka was scrutinized. The findings revealed that medaka predominantly inhabit rice fields in the shallow mountainous regions of the eastern part of the island. Notably, they are most prevalent in irrigation and drainage ditches

Fig. 4.4 Red quinoa planted on the ridges of the rice fields. (Source: Authors)

Fig. 4.5 Ecologists surveying the medaka populations of the pond. (Source: Authors)

Fig. 4.6 Medaka fish in the rice fields. (Source: Authors)

Fig. 4.7 Farmers placing boulders in a cemented ditch to create habitats for aquatic insects. (Source: Authors)

adjacent to rice fields, indicating an overlap of their habitat with human agricultural activities.

The project team also discovered caddisfly larvae in irrigation ditches, a species typically associated with fast-flowing streams. This discovery indicates that rice paddies play a crucial role in ecological connectivity to neighboring streams. The vegetation and biodiversity of the ditches are key factors in enhancing the ecological connection of the channels.

4.3.3 *Farmland Ecology Education and Community Engagement*

With the objective of encouraging farmers to engage with the community and pass along Indigenous knowledge to youth as their successors, educational activities that connect farmers as educators with students were organized. These activities not only motivated them to become fully involved in conservation actions but also improved their confidence in maintaining farming ecosystems. Activities included observations and documentation of the ecology of their fields, enabling local students to learn about rice paddy ecology and fostering environmental awareness and hometown pride. A total of 16 farmers participated in 2023, including 12 conventional farmers. The economic benefits of the Beans Co-op associated with hosting visitors grew from USD 5300 in 2020 to USD 16,000 in 2023.

The description of the inputs and outputs of interventions for community engagement is provided in Tables 4.4, Fig. 4.8.

From 2019 to 2023, nearly 30 events on agrifood or ecology were organized, attracting nearly 1000 participants. The farmers showed energy and confidence as they demonstrated to visitors and children how to perform traditional blessing rituals and how to use traditional stone-laying techniques.

4.3.4 *Payments for Ecosystem Services*

The Payments for Endangered Species and Important Habitat Ecological Services Program has been extended to the conventional farmers of Nan'an. Their rice paddies must be watered year-round, and the use of herbicides, poisonous baits, and bird nets is prohibited. By the end of 2023, 12 conventional farmers with a total of 32 ha of rice fields in Nan'an and the surrounding area were enrolled in the payment program, with more than USD 10,000 in incentive payments issued.

Table 4.4 Inputs and outputs of interventions for community engagement in ecological landscape transformation

Inputs and outputs of interventions	Quantity and beneficiaries
Economic benefits from hosting visitors	USD 16,000 in 2023
Number of organized ecological events	30
Number of participants	1000
Engaged community members	1. Rice farmers. 2. Students. 3. Organic rice factory. 4. Forestry and nature conservation agency. 5. Eco-consultant company. 6. Observers.

Fig. 4.8 Children joining in ecological education. (Source: Authors)

4.4 Discussion and Lessons Learned

4.4.1 Organic Rice Production

Over the past 10 years, organic farming has increased to account for more than 50% of the main rice farming areas. However, the project team still faced a number of challenges when entering the area and encouraging farmers to adopt organic farming methods. The abandonment of pesticides and chemical fertilizers was at odds with the traditional rice field management practices of farmers and landowners in Nan'an, as organic fields do not allow the use of pesticides to control pests and weeds. The project team also hoped to encourage farmers to maintain the ecosystems in their fields, which would result in increased labor and management costs in the production process, including ridge maintenance and management of irrigation and drainage control.

These farmers and landowners play an important role in the SEPLS landscape. For them, effective paddy management means crop yield, weed levels, and disease control in the paddy fields. As the tribal labor force ages and disappears, the growth of organic rice fields will stagnate. Therefore, it is important to subsidize organic farmers with the necessary maintenance and management materials, large machine operation, and ridge maintenance costs to increase the farmers' willingness to switch to and maintain organic farming.

4.4.2 Embracing Local Cultural Context in Conserving Eco-Connectivity

Quantitatively, we observed an increase in biodiversity in rice fields, but more scientific research is needed to prove this.

Qualitatively, we found embracing local cultural context to be a key factor in maintaining ecological connectivity within the Nan'an landscape. Farmers tended to grow traditional crops on the ridges, particularly beans, which are often consumed by the Bunun. They were more familiar with these traditional crops than with the crops proposed by the scientists. The project team therefore tried to incorporate local knowledge into relevant habitat creation activities. Farmers were encouraged to use their own ideas and ways of creating habitats on their farmland. We observed farmers planting various traditional crops and folkloric plants on the ridges or deliberately preserving the trees on the ridges. These actions helped to diversify the previously monotonous environment of the rice fields, thus improving the ecological connectivity of the landscape.

It was important to facilitate discussions between scientists and farmers. We acted as intermediaries, translating scholarly language into accessible terms for farmers, thereby bridging the gap between expert knowledge and practical application. This approach encouraged farmers to not only comprehend but also apply this knowledge to improve their habitats. Simultaneously, external experts were re-evaluating and reinterpreting TEK within the local cultural and environmental contexts. By facilitating a mutual understanding between scientists and farmers, we tried to ensure that recommendations for habitat improvement both resonated with and were informed by local perspectives and practices, making the whole process one of knowledge co-production.

4.4.3 Fostering the Backbone of the Satoyama Economy

The Nan'an Beans Co-op serves as a hub for internal members and external resources, decentralizing responsibilities, resources, benefits, and outcomes of SEPLS operations among its members. This approach ensures that benefits are shared widely within the organization rather than being concentrated in a few hands. Moreover, the Co-op acts as a platform for connecting tribal residents and external stakeholders, fostering partnerships crucial for sustaining SEPLS ecological connectivity in Nan'an. To maintain mutual trust and a healthy partnership among the team, the income from ecotourism must be distributed very carefully. Currently, the project team functions as an intermediary organization. In the future, it is crucial to focus on establishing a system that allows community residents to take on project management responsibilities.

4.4.4 *Encouraging Farmers to Apply for PES*

The Forestry and Nature Conservation Agency's Payments for Endangered Species and Important Habitat Ecological Services Program requires applicants to attach a landowner's lease or consent form. However, some landowners have moved away from the area, or there are multiple owners of the land, making it difficult to obtain the relevant evidence. Indirectly, this has led to farmers opting to use herbicides for the sake of convenient and effective weed control with limited labor power. To increase the effectiveness of the overall ecosystem services in the Nan'an area, the project team incorporated sponsorship funds from E.Sun Bank to directly subsidize actual farmers in the Nan'an area who manage the ridges by manually cutting the weeds without using chemicals.

In addition, the application process involves many administrative tasks such as filling in paperwork and forms, which has always been a considerable burden for farmers. We will continue to provide advice to farmers on the application process and invite the younger generation of farmers to participate in the administrative process.

4.5 Conclusion

This chapter highlights the important role that Indigenous and local communities (IPLC) play in restoring and enhancing eco-connectivity with multiple stakeholders. Despite the challenges of forced settlement, monoculture, and population outflow, the Nan'an community has been able to draw increasing attention to the local SEPLS by involving stakeholders and actively collaborating with nongovernmental organizations (NGOs), research institutes, enterprises, and the government.

By integrating traditional farming knowledge and ecological expertise and economic activities rooted in traditional ecological wisdom, the Nan'an community has been able to conserve the medaka, an endangered species. This created a self-sufficient economy within the region and kept farmers motivated to continue eco-friendly farming. These multifaceted approaches helped to improve local sustainability.

Acknowledgments The project acknowledges the support from the Yushan National Park Headquarters, the E. Sun Bank, Hualien Branch, Forestry and Nature Conservation Agency, and the Satoyama Development Mechanism (SDM) for providing the funding for the implementation of the project. Thanks to our partners, Hualien District Agricultural Research and Extension Station, Ching-Feng Lin (林青峰), Chih-Chung Lin (林志忠), Jen-Jiun Perng (彭仁君), Hai-Yin Wu (吳海音), Fushin LiPaHak Eco-Farm, Taiwan Biodiversity Research Institute, National Dong Hwa University, National Taitung University, and Hualien Natural Education & Ecology Consultant Ltd., for showing their passion to Nan'an community and providing their professional assistance. We extend our thanks to Jhuole Primary School, Jhuo Cing Elementary School. We would also like to thank the Nan'an community who participated in the study and project implementation, as well as the staff and administrators of Tse-Xin Organic Agriculture Foundation who were involved in the project along the way. All their support and work are duly acknowledged.

References

Chan YC (2023) Bunun home garden in traditional farming wisdom Zhuoxi District. Green Sprouts 8:34–39

Chang MC (2022) Taiwan's endemic Medaka Fish was listed as an endangered species in 2017, PTS News, 3 May 2022, viewed 24 July 2024. Retrieved from https://news.pts.org.tw/article/579058

Forestry and Nature Conservation Agency (2020) Ecological network blueprint planning and development project report. Forestry and Nature Conservation Agency, Ministry of Agriculture, Taipei

Forestry and Nature Conservation Agency (2024) From the ecological assessment of farmland waterway to integrate the ecological and living of the East Rift Valley. Forestry and Nature Conservation Agency, Ministry of Agriculture, Taipei

Lai HT, Lin HY, Hsiung WC, Shy JY (2010) Induced spawning and larval development of Aphyocypris kikuchii (Oshima, 1919). Taiwan J Biodivers 12:251–259

Taiwan Map Service (2024) Viewed 27 February 2024. Retrieved from http://maps.nlsc.gov.tw/

Tengö M, Hill R, Malmer P, Raymond CM, Spierenburg M, Danielsen F, Elmqvist T, Folke C (2017) Weaving knowledge systems in IPBES, CBD and beyond–lessons learned for sustainability. Curr Opin Environ Sustain 26–27:17–25. https://doi.org/10.1016/j.cosust.2016.12.005

Yang ZX (2021) Investigation of Medaka fish in Nan'an. Endemic Species Research Institute, Nantou

Zhuoxi Township Household Registration Office (2024) Viewed 7 March 2024. Retrieved from https://jshr.hl.gov.tw/cl.aspx?n=580

Chapter 5
Potential of Community Forests in Maintaining Ecological Connectivity at the Landscape Level in the North of Thailand

Jeeranuch Sakkhamduang and Rattanaporn Charoenchad

Abstract Forest administration in Thailand is shared by the Department of National Parks, Wildlife and Plant Conservation (DNP) and the Royal Forest Department (RFD). The DNP manages all protected areas, while the RFD oversees reserved forests and community forests. This division creates challenges in applying a landscape approach to managing forest resources to ensure ecological connectivity. The Omkoi Wildlife Sanctuary (OKWS), a protected area under the DNP in the north of Thailand, is home to key species such as the long-tailed goral, elephant, and banteng, as well as to Karen and Lahu ethnic communities who have resided in the area since long before it was established as a wildlife sanctuary. The surrounding reserved forest harbors several community forests, while more forest lands are encroached for agricultural purposes.

The Thailand Environment Institute is working with the DNP to identify measures to reduce pressure on protected areas, particularly the OKWS. A literature review, focus group discussions with key stakeholders, and key informant interviews with DNP and RFD officials were conducted from October to December 2023 to develop suitable conservation activities and recommendations for sustainable forest landscape management.

Results indicated that promoting sustainable land use for communities inside the wildlife sanctuary, sustainable use of community forests, ecotourism or cultural tourism, and enhancing livelihoods could reduce pressures on the protected area. Strengthening collaboration between the DNP and RFD in integrated forest mangement, spatial planning, and joint patrolling are key activities to ensure the ecological connectivity and ecosystem health of protected, reserved, and community forests.

Keywords Community forests · Protected area · Ecological connectivity · Landscape approach · Mae Ping Forest Complex · Land use change

---

J. Sakkhamduang (✉) · R. Charoenchad
Thailand Environment Institute, Nonthaburi, Thailand
e-mail: jeeranuch@tei.or.th

M. Nishi et al. (eds.), *Ensuring Ecological Connectivity in Socio-Ecological Production Landscapes and Seascapes (SEPLS)*, Satoyama Initiative Thematic Review, https://doi.org/10.1007/978-981-95-1474-8_5

5.1 Introduction

After the Royal Forest Department (RFD) was divided into three departments in 2002, forest resources management in Thailand has been operated under three different authorities, namely the Department of National Parks, Wildlife and Plant Conservation (DNP), the Royal Forest Department (RFD), and the Department of Marine and Coastal Resources (DMCR). The DNP oversees protected areas, while the RFD oversees reserved forests and community forests, and the DMCR oversees marine resources, mangrove forests, and coastal areas of the country (The Amending Ministry, Sub-ministry, Department Act (B.E. 2545) 2002). The division of the RFD created disintegration in forest resources management and utilization since most of the forests under the RFD and the DNP are connected, and most of the time, wildlife and local communities are not able to recognize the boundary of areas under different jurisdictions.

Thailand has enacted various laws to protect and manage its forests, which may be categorized roughly into protected areas and other forest areas such as reserved forests and community forests. Protected areas in Thailand, on the one hand, include national parks, wildlife sanctuaries, non-hunting areas, and arboretums, which are under the responsibility of the DNP. The main legislations for managing and protecting the protected areas include the Wildlife Preservation and Protection Act, B.E. 2562 (2019), (Royal Government Gazette 2019a), and the National Park Act, B.E. 2562 (2019), (Royal Government Gazette 2019b). On the other hand, the RFD is responsible for forest areas under the Forests Act, B.E. 2562 (2019), (Royal Government Gazette 2019c), reserved forest areas under the National Reserved Forests Act, B.E. 2559 (2016) (Royal Government Gazette 2016), and community forests under the Community Forest Act, B.E. 2562 (2019) (Royal Government Gazette 2019d).

The Agricultural Land Reform Act, B.E. 2532 (1989), (Royal Government Gazette 1989), known as *Sor Por Kor* in Thai language, is one of the key national land-use policies that strongly affects forest resource management. The Act is aimed at improving the rights and holding of agricultural land, including the allocation of places for living and redefinition of the term “farmer.” Typically, *Sor Por Kor* areas are the degraded reserved forest areas encroached by people who were later granted the rights for farming and settlement under certain conditions under *Sor Por Kor*. The process of receiving a *Sor Por Kor* land title somehow encourages people to encroach on national reserved forest areas with the hope that they will be granted land titles in the future. Since December 2023, the Agricultural Land Reform Office has been working on land title deeds on *Sor Por Kor* lands for farmers throughout the country (Royal Government Gazette 2023).

Protected areas such as national parks or wildlife sanctuaries in Thailand were established based on primary and intact forests remaining in reserved forest areas in order to protect them more efficiently (ICEM 2003). The Omkoi district in Chiang Mai Province is one of the areas facing complications in forest and land resource management due to the division of authorities and disintegrated policies, plans, and practices. Fragmented policies of different authorities have resulted in conflicts in land use, deforestation, and forest degradation and have affected ecological connectivity (Phromma et al. 2019; Talerngsri 2020).

In this context, from October 2023 to April 2024, for the preparation of the FAO-GEF project "Forests for Life—Intact Tropical Forest Landscape Conservation in Thailand," a child project under the Regional IP "Indo Malaya Forest Biome Integrated Programme", the Thailand Environment Institute (TEI) collaborated with the DNP and Food and Agriculture Organization of the United Nations (FAO) to identify measures and activities to reduce pressure on protected areas, enhance forest governance and ecological connectivity in the fragmented forest surrounding protected areas, and support local livelihoods. The objective of the project is to use protected areas as a core area and reserved forests and community forests as a buffer area for protected areas, and activities will be designed accordingly. The project will be implemented from 2025 to 2029 in four forest landscapes in Thailand, including the Omkoi Wildlife Sanctuary (OKWS). Hence, this study aims to identify the current situation of forest and land use, threats to forest resources and ecological connectivity, opportunities, and proposed interventions.

5.2 Methodology

Literature review, focus group discussions (FGDs) with local communities, governmental agencies, civil society organizations, other stakeholders, and key informant interviews were the key methods employed to collect information. In total, six FGDs were conducted in the landscape. Semistructured questions were used for interviews with two DNP and two RFD officials and two representatives of civil society organizations who have worked in the landscape for a long time. The data collection was conducted between October and December 2023. The collected information was analyzed by a thematic analysis method to highlight the importance of community forests in helping decrease pressures on protected forests and their importance as a potential ecological linkage in the Forests for Life project.

5.2.1 Study Area

5.2.1.1 Omkoi District

Omkoi district, the southernmost district of Chiang Mai Province, has a total area of 209,380 ha (Table 5.1). The district is located 179 km south of the city of Chiang Mai. Within the total area, around 169,577.6 ha or around 80% is the Omkoi reserved forest, while the Omkoi Wildlife Sanctuary (OKWS) (the part that is located in the district) covers an area of 54,022.4 ha or around 25% of the district area. The majority of the population (63,224) of the district are ethnic minorities,

Table 5.1 Basic information of the study area

Country/Region	Thailand
Province	Chiang Mai
District	Omkoi
Municipality	One municipality (Omkoi) and six local administration organizations
Size of geographical area (hectare)	209,380
Dominant ethnicity(ies), if appropriate	Karen, Lahu, Hmong
Size of case study/project area (hectare)	122,400
Dominant ethnicity in the project area	Karen, Lahu
Number of direct beneficiaries (people)	1820
Number of indirect beneficiaries (people)	63,224
Geographic coordinates (latitude, longitude)	Latitude 17° 17′–17° 53′ N; longitude 98° 24′–98° 44′ E

Source: Authors, based on Department of Provincial Administration, Ministry of Interior 2023 and Wildlife Conservation Office 2018

namely Karen, Lahu, and Hmong, who mainly live and cultivate crops in the reserved forest, while a smaller number lives in the wildlife sanctuary (Department of Provincial Administration, Ministry of Interior 2023). Forest laws such as Article 121 of the Wildlife Preservation and Protection Act, B.E. 2562 (2019) prohibit community members residing in any protected area from expanding their farms beyond the area that was owned before the establishment of the protected area. Yet increasing numbers of family members in each household require more farmland to feed, although expanding the farming area is prohibited. This results in increased agrochemical inputs to sustain the yield for consumption. Some villagers opt for encroaching on adjacent reserved forest or illegal logging, which creates pressures on forest resources and ecosystems in the district and consequently livelihoods.

The main agricultural crops of the district are rice, tomato, chilli, cabbage, pumpkin, and coffee. The majority of rice produced in the district is upland rice for household consumption. Villages with enough lowland areas practice paddy rice cultivation for subsistence and selling on the market. Some other produce, such as tomato, chilli, cabbage, and pumpkin, on the one hand, are commercially grown on a large scale, using considerable amounts of agrochemicals. On the other hand, organic or agroforestry coffee from the district is in high demand by consumers due to the taste and growing practices (Omkoi Agricultural Extension Office 2021). Livestock, especially cows, act as a safety net for villagers in the district as they can be sold for cash when necessary. Many farmers who own livestock let their cows roam freely in forest areas to feed on naturally growing grasses.

5.2.1.2 Omkoi Wildlife Sanctuary

The Omkoi Wildlife Sanctuary (OKWS) is one of the nine protected areas of the Mae Ping Forest Complex, a cluster of protected areas surrounded by a reserved forest located in the northwest of the country (Fig. 5.1). The total area of OKWS is 122,400 ha, of which 54,022.4 ha is located in Omkoi district, with the rest in Sam Ngao district, Tak Province. The altitude of the wildlife sanctuary ranges from 200 to 1929 m above sea level (m.a.s.l.), with an average temperature of 25.85 °C and average annual precipitation of 1157.2 mm. Most of the area is mountainous and is part of the East Thanon Thongchai mountain range (Wildlife Conservation Office 2018; C.P. Foundation for Rural Development 2024).

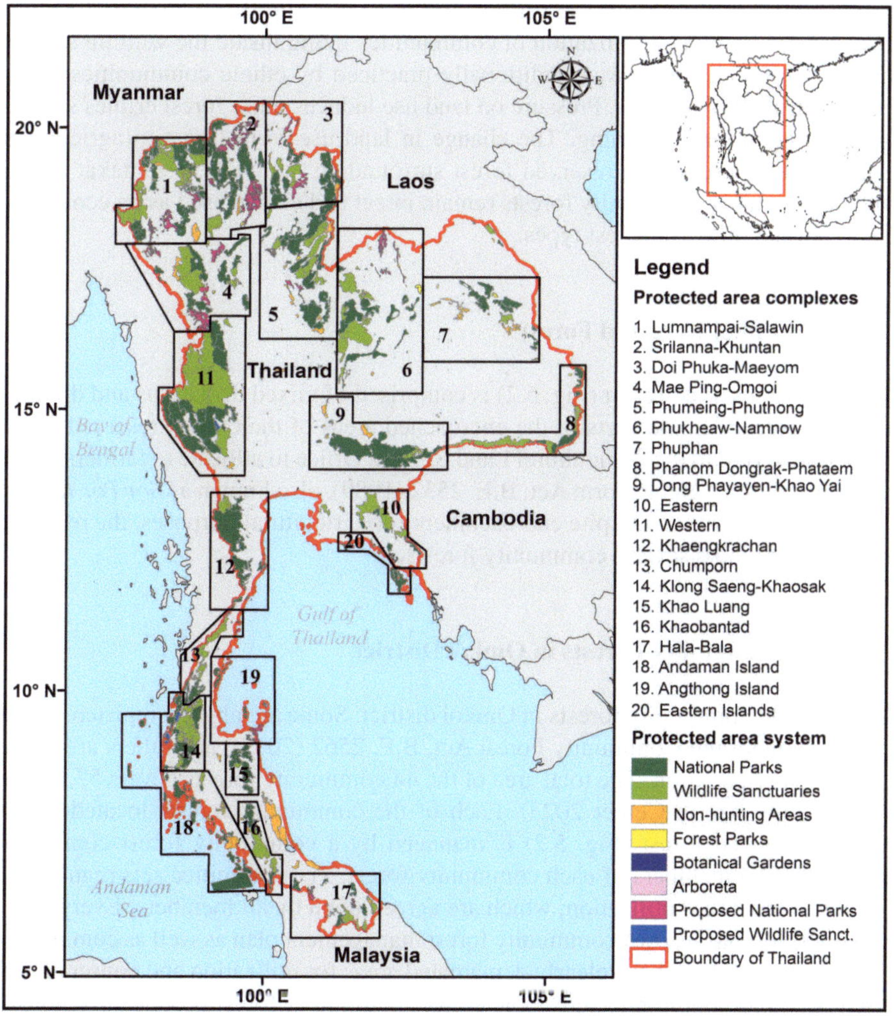

Fig. 5.1 Map showing the location of Mae Ping Forest Complex in the north of Thailand, shown in the yellow highlighted box. (Pomoim et al. 2022)

There are four forest types distributed in the OKWS, including hill evergreen forest, dry evergreen forest, mixed deciduous forest, and dry dipterocarp forest. Hill evergreen forest is situated at altitudes higher than 1200 m.a.s.l. Dry evergreen forest is found along streams between the valleys. Dry and mixed dipterocarp forest, the largest forest cover of the area, is located at altitudes lower than 1000 m.a.s.l., and in some areas patches of *Pinus merkusii* can be found associated with these forests. The main invasive species in the area include *Mimosa pigra*, *Tithonia diversifolia*, and *Ageratina adenophora* (Wildlife Conservation Office 2018).

The OKWS is home to the long-tailed goral (*Naemorhedus caudatus*), Asian elephant (*Elephas maximus*), and banteng (*Bos javanicus*), among others (Wildlife Conservation Office 2018). Besides being a wildlife habitat, the wildlife sanctuary also has six villages of Karen and Lahu ethnic communities living inside the protected area. The Wildlife Preservation and Protection Act, B.E. 2562 (2019) limits the land use and forest utilization of communities living inside the wildlife sanctuary. Shifting cultivation was traditionally practiced by ethnic communities but is now controlled by the law. Pressure on land use induces more forest crimes such as illegal logging and poaching. The change in land use from forest to agricultural lands is also observed in reserved forest surrounding the OKWS (Jantakat 1999), and only areas of community forests remain intact and can be used as an ecological link between different forest types.

5.2.1.3 Omkoi Reserved Forest

The Omkoi reserved forest (Fig. 5.2) is comprised of mixed deciduous and dry dipterocarp forests. Some parts of the encroached areas of the Omkoi reserved forest were transferred to the Agricultural Land Reform Office to allocate to farmers under the Agricultural Land Reform Act, B.E. 2532 (1989), also known as *Sor Por Kor* for agricultural purposes. Despite encroachment for agricultural purposes, the reserved forest still harbors several community forests.

5.2.1.4 Community Forests in Omkoi District

There are 44 community forests in Omkoi district. Some are already registered with the RFD under the Community Forest Act, B.E. 2562 (2019), and others are in the process of registration. The total area of the 44 community forests covers 52,700 ha (Forest Information Center 2023). Each of the community forests located in the Omkoi reserved forest (Fig. 5.2) is managed by a community forest committee elected by the members of each community forest. The committee sets regulations for forest resources utilization, which are agreed upon by all members. Every community forest has its own community forest management plan as well as community forest zoning, which have clearly demarcated areas for utilization and conservation. Members of a community forest can collect non-timber forest products for household consumption or for commercial purposes based on the regulations of each

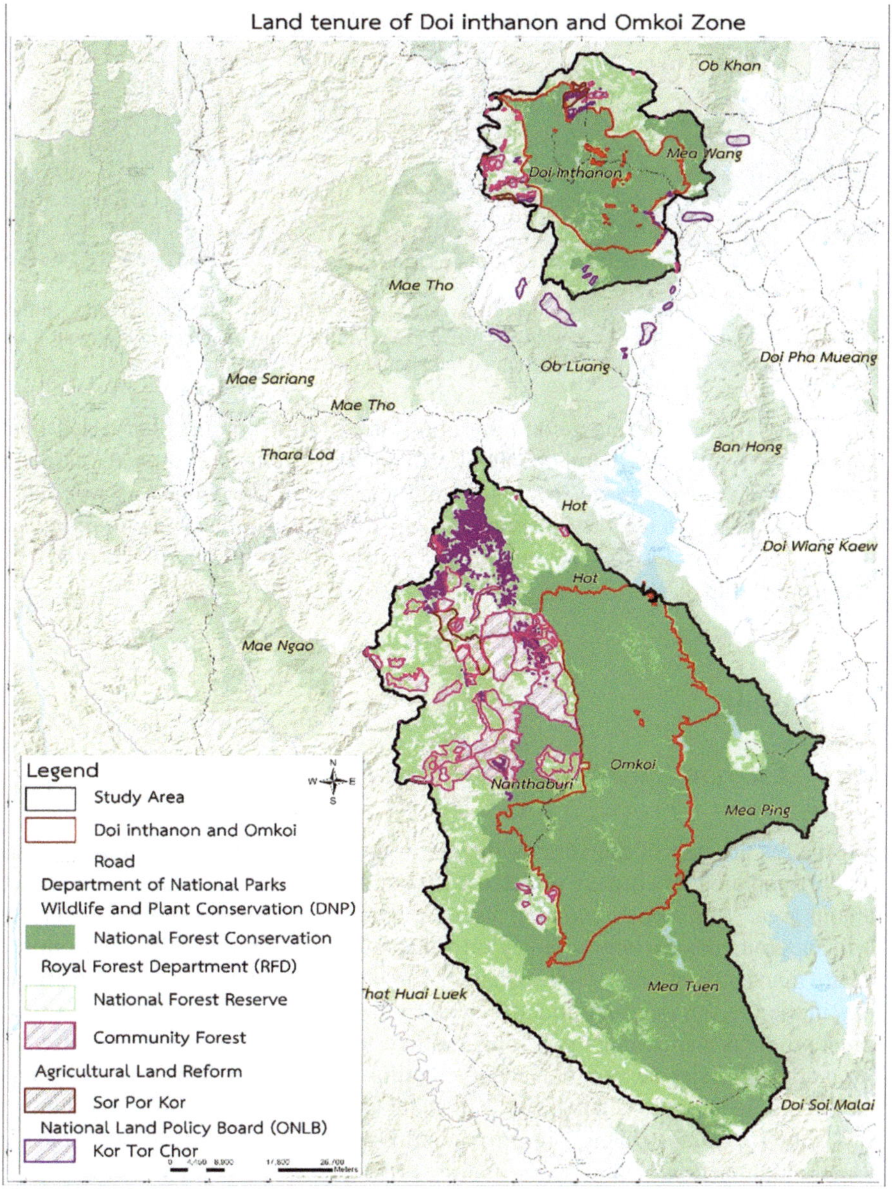

Fig. 5.2 Map showing reserved forest, community forests, and the Omkoi Wildlife Sanctuary. (Source: FAO 2024)

Fig. 5.3 Some non-timber forest products harvested from community forests, including bamboo shoots, earthstar mushroom (*Astraeus hygrometricus*), termite mushroom (*Termitomyces* sp.), and rattan shoots, respectively (Source: Jeeranuch Sakkhamduang)

community forest. The most common non-timber forest products harvested from community forests are bamboo shoot, Melientha (*Melientha suavis*), earthstar mushroom (*Astraeus hygrometricus*), termite mushroom (*Termitomyces* sp.), rattan shoot, vegetable fern (*Diplazium esculentum*), snowflake aralia (*Trevesia palmata*), Indian gooseberry (*Phyllanthus emblica*), and myrobalan (*Terminalia chebula*), for instance (see Fig. 5.3).

5.3 Results

5.3.1 Current Situation of Forest and Land Use

Forest fires are one of the serious threats to the OKWS and surrounding forests. Omkoi is an area repetitively affected by forest fires due to its location in a mountainous area, steep slopes, and long distance from fire management units. According to focus group discussions with key informants including representatives from forest fire management units, civil society organizations, and village volunteers, forest fires in the OKWS occurred in areas where villagers frequently collect non-timber forest products and conduct poaching and livestock grazing. Forest fires cause forest and biodiversity degradation while creating air pollution. There are provincial regulations such as the Chiang Mai Provincial Order for zero burning from January 1 to May 15, 2025 (Chiang Mai Provincial Department of Agriculture Extension 2024) and an action plan to suppress forest fires and curb open burning in the dry season, but enforcement is yet to be effective due to insufficient human and financial resources.

Illegal logging and poaching are still prevalent in the OKWS and Omkoi reserved forest. Burmese rosewood (*Pterocarpus macrocarpus*) is the main target tree species for loggers for the international market. There is an integrated taskforce comprised of the DNP, the RFD, and Omkoi district authority to patrol in order to prevent and stop the illegal logging using a SMART Patrol System and establishing checkpoints on the national road. However, the number of officers and rangers is not adequate to cover the entire forest area. Moreover, poverty and the demand for bush-meat are the reasons behind poaching activities.

The expansion of agricultural lands into forest areas is of great concern to relevant government agencies. In the reserved forest, in the hope of getting *Sor Por Kor* land rights, a considerable number of Lahu ethnic community members have changed their farming practices from self-sufficient multi-cropping and rain-fed agriculture to monoculture, especially cultivation of cash crops such as chilli, cabbage, ginger, and cassava. The changes have resulted in more agricultural inputs and expansion, such as an increase in seed inputs, installation of irrigation systems, and especially use of chemical fertilizers, herbicides, and pesticides, as well as forest land encroachment for farmland. Forest land encroachment by the Karen community who live and farm inside the OKWS has stemmed from an increase in family members and the low productivity of upland rice. The area of farmland inside the OKWS is protected by Article 121 of the Wildlife Preservation and Protection Act, B.E. 2562 (2019), but some community members have broken the law by expanding their farmland into the forest area to feed their families.

5.3.2 Threats to Ecological Connectivity

According to the Convention on Migratory Species (2020), ecological connectivity is defined as the "unimpeded movement of species and the flow of natural processes that sustain life on Earth." Landscape fragmentation is one of the main reasons for the creation of disconnected habitats and ecosystems. Landscape fragmentation often happens when large habitat patches are divided into smaller, more isolated fragments primarily driven by urbanization, intensive agricultural practices, and land-use changes (Arkilanian et al. 2020). Ecological connectivity is negatively impacted by these pressures that disrupt habitats, threaten biodiversity, and hamper climate change adaptation measures.

In the case of OKWS and the Omkoi reserved forest, landscape fragmentation is attributed to infrastructure construction, mainly national highways to connect the district to the provincial town of Chiang Mai, and the problem is further exacerbated by the division of forest resource administration and land-use policies. The Omkoi reserved forest, the same as other reserved forests under the RFD, receives small financial support and has inadequate personnel to oversee the area compared to that of protected areas under the DNP. Moreover, the reserved forest is a target area of the agricultural land reform policy, which aims to allocate land for agricultural purposes to poor farmers. The landscape fragmentation affects the movement of wildlife in the area, especially Asian elephants (Administration Office 2022) due to their tracks being cut by the national road, and smaller animals such as Phayre's langur (*Trachypithecus phayrei*), lar gibbon (*Hylobates lar*), smooth-coated otter (*Lutrogale perspicillata*), black-tailed giant flying squirrel (*Petaurista philippensis*), leopard (*Panthera pardus*), and Asian golden cat (*Catopuma temminckii*), among others.

5.3.3 Opportunities

There are some ecological connectivity measures being implemented in protected areas worldwide in order to reconnect fragmented landscapes. One is the stepping stone, which is defined as a small habitat patch with an area between 0.01 and 2.5 km^2; it can be natural or artificial, and it can support the dispersal of individuals (Wang et al. 2023). Another is the ecological corridor, which is defined as a strong connection between habitat patches. There are two kinds of corridors. One is a long and narrow forest area in suitable habitat patches with a width of 100 m, and the length of the corridor is the shortest distance between suitable habitats (Hilty et al. 2020). The other type of corridor is an overpass over a road in the habitat area with a width of 10 m (Wang et al. 2023). In the western region of Thailand, the Seub Nakasatien Foundation (2024) is implementing a project for ecological connectivity, applying the stepping stones concept in reserved forests to connect the fragmented reserved forest and protected areas in the western forest complex. The project establishes patches of multiple benefit trees, mainly bamboo which provides construction material, food, and shelter, in agricultural lands in reserved forests located between the Khao Sanam Preang Wildlife Sanctuary and Klong Wang Chao National Park to support wildlife movement between the two protected areas.

Community forests, both in terms of governance and management structure, are one of the effective measures to reconnect the fragmented landscapes of OKWS and the Omkoi reserved forest as there are 44 community forests in the reserved forest surrounding the OKWS. The rights of forest resource utilization and management granted by the Community Forest Act, B.E. 2562 (2019) and some subordinate legislations allow the members of community forests to collect, harvest, and utilize some restricted non-timber forest products such as urushiol from plants of the family Anacardiaceae, rattan, and sappan (*Caesalpinia sappan*), which would be illegal if collected or harvested in reserved forests. These advantages of community forests can be used to motivate forest-dependent communities to establish forest areas near their communities as community forests in order to utilize more non-timber forest products and at the same time enhance their livelihoods and ecological connectivity. The Yang Pieng community forest adjacent to the OKWS is a good example of community forest management. In 2023, it was selected as a model of successful community forest management based on activities related to headwater conservation and forest fire prevention, resulting in abundant high-price mushrooms, especially the barometer earthstar mushroom (*Astraeus* sp.). Harvesting mushrooms alone created income for community forest members of around 30,000 Thai Baht or around 884 USD per household per season (May–June) (Community Organization Development Institute 2023). However, the Yang Pieng community forest is facing the challenge of finding a suitable system to collect fees from villagers who live far away from the community forest but come to harvest mushrooms in the season.

In February 2017, the DNP announced that jungle cats *(Felis chaus)* were found for the first time in 40 years in Thailand in the OKWS (Matichon 2017). Camera traps set by DNP researchers captured the jungle cats. Although the species is listed as Least Concern in the IUCN Red List, the jungle cat is one of the protected species in Thailand (Royal Government Gazette 2003). Moreover, in November 2024, jungle cats were found in the Yang Pieng Community forest adjacent to the OKWS. The C.P. Foundation for Rural Development plans to use the species as a flagship to raise awareness of forest conservation in Omkoi District.

In Omkoi district, due to the high level of poverty and opium (*Papaver somniferum*) production (Anderson and Jongruck 2017), there are several types of support and initiatives from governmental agencies, the private sector, and nonprofit organizations to improve the situation. From the government side, there are livelihood development and poverty alleviation projects implemented to enhance the livelihoods of communities living inside forest areas. Some initiatives worth noting are under the C.P. Foundation for Rural Development and aimed at improving the livelihoods of community members in the district through the promotion of high-value agricultural products such as coffee and macadamia nut. The Highland Research and Development Institute is also promoting sustainable agriculture in the communities within the district through supporting organic vegetable production and disseminating other agriculture-related knowledge such as that concerning post-harvesting, organic fertilizer, and integrated pest management.

5.3.4 Proposed Sustainable Forest Landscape Management Activities

In order to highlight the importance of community forests to reduce pressures on protected areas and enhance forest and ecological connectivity, the research team proposed activities to be implemented inside and outside protected areas as follows.

5.3.4.1 Inside Protected Areas

The activities proposed for protected areas include effective forest fire control by involving community members in making firebreak lines, fire detection, and suppression, as forest fires are the key driver of forest degradation, which leads to ecological fragmentation. Enhancing livelihoods through promoting sustainable agriculture for ethnic communities will decrease the amount of agrochemicals applied on farmland as well as decrease further forest land encroachment. Promoting alternative livelihoods such as ecotourism or cultural tourism can enhance locals' livelihoods and awareness of the importance of the forest ecosystem. Engagement of community members in monitoring of illegal activities such as forest encroachment, illegal logging, and poaching can reduce forest degradation and biodiversity loss from those activities. These proposed activities can enhance the effectiveness of protected area management through communities' participation, sustaining ecosystem services, and ensuring ecological connectivity in the long term.

5.3.4.2 Outside Protected Area

Activities proposed for community forests include an improved value chain for non-timber forest products collected from community forests such as mushrooms and honey, as well as enhancing the capacity of existing community enterprises on marketing and product development of non-timber forest products, which generate direct benefits for community forest members. Reforestation and creation of ecological corridors in degraded forest areas in the landscapes, especially in areas connected with protected areas, would help to enhance ecological connectivity.

Sustainable agriculture and agroforestry can be promoted in agricultural areas and *Sor Por Kor* lands surrounding protected areas and reserved forests or community forests. These practices would help to increase the number of trees on farms and act as stepping stones for ecological connectivity. An integrated landscape management approach, bringing together all relevant authorities and stakeholders, would help to prevent ecological fragmentation in the future.

5.4 Conclusions

The study reveals that community forests have a high potential as linkages for ecological connectivity in the Omkoi forest landscape through creating ecological corridors in degraded forest areas and conserving natural forests inside community forests. Promoting communities' livelihoods through sustainable utilization of forest resources is the key to reducing pressure on protected areas as well as enhancing ecological connectivity. Alternative livelihoods such as ecotourism or cultural tourism could be one of the choices, and promoting high-value agricultural products for income generation would be another livelihood option for the communities in the landscape. Promoting sustainable land use for communities living inside protected areas is crucial. Furthermore, monitoring and evaluation of ecological connectivity in the Omkoi district is imperative for forest management at the landscape level.

By increasing the number and area of community forests in reserved forests and *Sor Por Kor* areas, ecological connectivity could be enhanced for both protected areas and reserved forests. This could also enhance the forest governance of community forests that promote shared rights and responsibilities of forest resource management among forest-dependent communities. Integrated landscape management, bringing together all relevant authorities and stakeholders, can reduce pressure on primary forests in protected areas. It also ensures ecological connectivity among the different types of forests and prevents further landscape fragmentation.

Acknowledgments The research team expresses their sincere thanks to the Forest and Plant Conservation Research Office, Department of National Parks, Wildlife and Plant Conservation, Omkoi Wildlife Sanctuary, Community Forest Management, Royal Forest Department, the Food and Agriculture Organization of the United Nations (FAO), and communities in Omkoi Wildlife

Sanctuary and members of community forests surrounding the wildlife sanctuary for providing information and data for this manuscript.

References

Administration Office DNP (2022) Closure of Mon Jong Peak in Om Koi Wildlife Sanctuary. Department of National Parks, Wildlife and Plant Conservation. https://news.dnp.go.th/news/11329

Anderson B, Jongruck P (2017) Forestry, illegibility and illegality in Omkoi, Northwest Thailand. For Soc 1(2):98–109

Arkilanian A, Larocque G, Lucet V, Schrock D, Denépoux C, Gonzalez A (2020) A review of ecological connectivity analysis in the region of resolution 40-3. Report presented to the Ministère de la faune, de la forêt et des parcs du Québec for the New England governors and Eastern Canadian premiers working group on ecological connectivity. 79

C.P. Foundation for Rural Development (2024) Omkoi Model, viewed 20 January 2024. Retrieved from http://www.cp-foundationforrural.org/rldf110167/(in Thai)

Chiang Mai Provincial Department of Agriculture Extension (2024) Chiang Mai provincial order for zero burning fiscal year 2024. https://chiangmai.doae.go.th/doisaket/wp-content/uploads/2024/12.pdf (in Thai)

Community Organization Development Institute (2023) Human and forest: barometer earthstar mushroom and community forest. https://web.codi.or.th/20230911-48008/ (in Thai)

Convention on Migratory Species (2020) Improving ways of addressing connectivity in the conservation of migratory species. UNEP/CMS/Resolution 12.26 (Rev.COP13)

Department of Provincial Administration, Ministry of Interior (2023) Monthly population statistic, Chiang Mai Province, December 2023, viewed 1 January 2024. Retrieved from https://stat.bora.dopa.go.th/stat/statnew/statMONTH/statmonth/#/mainpage (in Thai)

Forest Information Center (2023) List of community forest, Chiang Mai Province, Omkoi District Royal Forest Department, viewed 23 February 2024. Retrieved from https://forestinfo.forest.go.th/fCom.aspx?prov=50&zone=8&year=-1&eyear=-1&type=-1 (in Thai)

FAO (2024) Landscape Analysis for GEF 8 Thailand Child project. Presentation file, Unpublished. Potential

Hilty J, Worboys GL, Keeley A, Woodley S, Lausche B, Locke H, Carr M, Pulsford I, Pittock J, White JW, Theobald DM, Levine J, Reuling M, Watson JEM, Ament R, Tabor GM (2020) Guidelines for conserving connectivity through ecological networks and corridors, Best practice protected area guidelines series no. 30. IUCN, Gland

ICEM (2003) Thailand National Report on protected areas and development. Review of protected areas and development in the lower Mekong River Region. Indooroopilly, Queensland, p 131

Jantakat Y (1999) Land utilization trend analysis using satellite data in Omkoi Wildlife Sanctuary, Changwat Chiang Mai and Tak. Kasetsart Univeristy. Graduate School, Bangkok. (in Thai with English abstract)

Matichon (2017) DNP Announces Jungle Cats Found in Omkoi. https://www.matichon.co.th/local/news_465449 (in Thai)

Omkoi Agricultural Extension Office (2021) Data of agriculture in Omkoi District. Department of agricultural extension, viewed 24 February 2024. Retrieved from https://chiangmai.doae.go.th/omkoi/?page_id=157 (in Thai)

Phromma I, Pagdee A, Popradit A, Ishida A, Uttaranakorn S (2019) Protected area co-management and land use conflicts adjacent to Phu Kao – Phu Phan Kham National Park, Thailand. J Sustain For 38(5):486–507. https://doi.org/10.1080/10549811.2019.1573689

Pomoim N, Hughes A, Trisurat Y, Corlett RT (2022) Vulnerability to climate change of species in protected areas in Thailand. Sci Rep 12(1):5705. https://doi.org/10.1038/s41598-022-09767-9

Royal Government Gazette (1989) Agricultural Land Reform Act, B.E. 2532 (1989). 8th September 1989, Special Issue, Volume 106, Section 149 p. 13. https://alro.go.th/uploads/org/legal_aff/article_attach/article_attach_20181131542100320.pdf

Royal Government Gazette (2003) List of protected species. 1st August 2003, General Issue, Volume 120, Section 74 Gor. https://www4.fisheries.go.th/local/file_document/20200410114550_new.pdf

Royal Government Gazette (2016) National Reserved Forests Act, B.E. 2559 (2016). 24th May 2016, General Issue, Volume 133, Section 46 Gor p 12. https://forestinfo.forest.go.th/Content/file/forest_low/forest-law2559.pdf

Royal Government Gazette (2019a) Wildlife Preservation and Protection Act, B.E. 2562 (2019). 29th May 2019, General Issue, Volume 136, Section 71 Gor p 104. https://portal.dnp.go.th/DNP/FileSystem/download?uuid=9373af08-6a4f-442d-8146-5a61ffceb2cd.pdf

Royal Government Gazette (2019b) National Park Act, B.E. (2019). 29th May 2019, General Issue, Volume 136, Section 71 Gor p. 145. https://portal.dnp.go.th/DNP/FileSystem/download?uuid=f5ede081-fa7b-4e5c-8207-4232bd13fd60.pdf

Royal Government Gazette (2019c) Forests Act, B.E. 2562 (2019). 16th April 2019, General Issue, Volume 136, Section 50 Gor p. 106. https://www.ratchakitcha.soc.go.th/DATA/PDF/2562/A/050/T_0106.PDF

Royal Government Gazette (2019d) Community Forest Act, B.E. 2562 (2019). 29th May 2019, General Issue, Volume 136, Section 50 Gor p 71. https://www.forest.go.th/law/wp-content/uploads/sites/33/2019/06/%E0%B8%9E%E0%B8%A3%E0%B8%B0%E0%B8%A3%E0%B8%B2%E0%B8%8A%E0%B8%9A%E0%B8%B1%E0%B8%8D%E0%B8%8D%E0%B8%B1%E0%B8%95%E0%B8%B4-%E0%B8%9B%E0%B9%88%E0%B8%B2%E0%B8%8A%E0%B8%B8%E0%B8%A1%E0%B8%8A%E0%B8%99

Royal Government Gazette (2023) Order of agricultural land reform committee. 8th December 2023, General Issue, Volume 140, Special Section 309 Ngor. https://ratchakitcha.soc.go.th/documents/14175.pdf

Seub Nakasatien Foundation (2024) Stepping stones strategy to reconnect wildlifes, viewed 14 February 2024. Retrieved from https://www.seub.or.th/bloging/work/2024-32/ (in Thai)

Talerngsri K (2020) Forests, peoples, and governments: persistent land-use conflict in northern Thailand. Econ Peace Secur J 15(2). https://doi.org/10.15355/epjs15.2.19

The Amending Ministry, Sub-ministry, Department Act (B.E. 2545) (2002) Royal Gazette. 119:14. http://www.ratchakitcha.soc.go.th/DATA/PDF/00102282.PDF (in Thai)

Wang H, Gao Y, Li Y, Li N, Grueter CC, Xu H, Huang Z, Cui L, Xiao W (2023) Stepping stone strategy: a cost-effective way to address habitat fragmentation of endangered wildlife in montane Forest. Ecosyst Health Sustain 9:0073. https://doi.org/10.34133/ehs.0073

Wildlife Conservation Office (2018) Wildlife Sanctuary in Thailand. Wildlife Conservation Office, Department of National Parks, Wildlife and Plants Conservation, Bangkok

Chapter 6
Management in SEPLs to Ensure High-Quality Connectivity for Brown Bear in the Western Cantabrian Mountains (Northwestern Spain)

Emilio R. Díaz-Varela, Pedro Álvarez-Álvarez, José C. Pérez-Girón, José V. López-Bao, Fernando Ballesteros, María Gómez-Manzanedo, María Párraga, and Guillermo Palomero

Abstract The coexistence of wildlife species alongside human populations is inherently intricate, particularly within socio-ecological production landscapes (SEPLs), where millennia of coevolution have shaped interactions. Conservation endeavors, especially for apex predators like the brown bear (*Ursus arctos* Linn.) in the Cantabrian Mountains of Northern Spain, demand a comprehensive approach that addresses conflicts and harnesses opportunities arising from the human–wildlife interface.

This study focuses on understanding how management practices within SEPLs can contribute to maintaining high-quality ecological connectivity for brown bears

E. R. Díaz-Varela
Higher Polytechnic School of Engineering, University of Santiago de Compostela, Lugo, Spain
emilio.diaz@usc.es

P. Álvarez-Álvarez
Polytechnic School of Mieres, University of Oviedo, Mieres, Asturias, Spain

J. C. Pérez-Girón
Department of Botany, Faculty of Sciences, University of Granada, Granada, Spain
Inter-university Institute for Research on the Earth System in Andalucia (IISTA), University of Granada, Granada, Spain

J. V. López-Bao
Biodiversity Research Institute, University of Oviedo-CSIC-Principado de Asturias, Mieres, Asturias, Spain

F. Ballesteros · M. Gómez-Manzanedo · M. Párraga · G. Palomero
Fundación Oso Pardo (FOP), Santander, Cantabria, Spain

M. Nishi et al. (eds.), *Ensuring Ecological Connectivity in Socio-Ecological Production Landscapes and Seascapes (SEPLS)*, Satoyama Initiative Thematic Review, https://doi.org/10.1007/978-981-95-1474-8_6

in the western Cantabrian Mountains (NW Spain). Through a comprehensive review of existing studies, projects, and initiatives, the research aims to identify key areas where landscape connectivity can be improved while also sustaining rural activities crucial for local livelihoods.

Using Geographic Information Systems (GIS) and spatial metrics, the study seeks to characterize potential landscape-level improvements in connectivity resulting from the preservation of rural activities. Additionally, a stakeholder mapping exercise employing fuzzy cognitive mapping (FCM) analyzes the intricate social fabric within SEPLs, shedding light on inter-actor influences and potential synergies for enhancing connectivity.

The findings underscore the pivotal role of social networks and human activities within SEPLs in fostering successful conservation initiatives. Moreover, they highlight the interdependence between landscape elements and local livelihoods, emphasizing the need for a holistic understanding of socio-ecological dynamics. Such insights are instrumental in informing effective spatial planning and nature conservation management strategies tailored to the unique context of SEPLs.

Keywords Ecological connectivity · Brown bear · Agroecosystem management · Mountain areas · Ecosystem services · Socio-ecological production landscapes and seascapes

6.1 Introduction

6.1.1 Brown Bear in NW Iberian Peninsula

The brown bear (*Ursus arctos* Linn.) is a widespread species in Europe with a continuous but progressively declining presence in the Iberian Peninsula from the Pleistocene to recent times (García-Vázquez et al. 2020; Sommer and Benecke 2005; Valdiosera et al. 2008). Currently, it remains an isolated population of approximately 370 individuals as of 2020 in the Cantabrian Mountain range (NW Spain), fragmented into two subpopulations (eastern and western), separated from the beginning of the twentieth century by a wide gap (Gonzalez et al. 2016; Palomero et al. 2007). Another population of approximately 76 individuals is presently establishing itself in the Pyrenees range (NE Spain) along the border with France, after individuals from Slovenia were introduced around 1996 following the extinction of the species in the area (Palazón 2017; Palazón et al. 2002; Parres et al. 2020). Due to the population size and their isolation, the conservation status of the brown bear in Spain is considered endangered and at risk of extinction. As such, it is included in the national catalog of threatened species (*Catálogo Español de Especies Amenazadas*), as well as in the respective catalogs at the regional level in the regional governments (*Comunidades Autónomas*) of Asturias, Cantabria, and Galicia. At the European level, the species is listed in Annexes II and IV of the *Habitat Directive 92/43/CEE* and in Annex II of the *Bern Convention on the*

Conservation of European Wildlife and Natural Habitats, both reference texts for regulating the protection of wildlife species in Europe.

In response to the critical situation of the brown bear in the Iberian Peninsula, efforts toward its conservation have been developed by administrations at different levels, NGOs, and other local actors since the late 1980s. The governments of those regions with the verified presence of the species developed brown bear recovery plans including Cantabria (1989), Castilla y León (1990), Asturias (1991, revised in 2002), and Galicia (1992). At the national level, a strategy for the conservation of the Cantabrian brown bear (*Estrategia para la Conservación del Oso Pardo Cantábrico*) was issued in 1999, and a similar one for the conservation of the brown bear in the Pyrenees (*Estrategia para la Conservación del Oso Pardo en los Pirineos*) in 2006, both coordinated by specific nature conservation commissions of the former Ministry of Environment. Recently, a new strategy for the conservation of the brown bear in the Cantabrian range (*Estrategia para la conservación del oso pardo Ursus arctos en la Cordillera Cantábrica*) was issued by the current Ministry for the Ecological Transition (MITECO 2019). Nature conservation-related NGOs have also played a significant role in the preservation of the brown bear. Specifically, the Brown Bear Foundation (*Fundación Oso Pard,* FOP) has organized numerous conservation actions since its establishment in 1992. These include population monitoring, direct intervention against poaching, dissemination activities, capacity building, fruit tree planting, and land acquisition (see Sections 4.2 and 4.3 for details).

The abovementioned concerns and actions led to a significant result for the Cantabrian population of brown bear, including an increase in population to current numbers (Blanco et al. 2020; Gonzalez et al. 2016; Martínez Cano et al. 2016) as well as the restoration of species movement and gene flow between the two subpopulations (Blanco et al. 2020; Pérez et al. 2010). Also, as the brown bear is considered an umbrella species (Roberge and Angelstam 2004; Rozylowicz et al. 2011; Tosi et al. 2015; Yang et al. 2023), all conservation outcomes are expected to have a positive effect on several co-occurring species (Velasco and Romo 2010).

6.1.2 Connectivity and SEPLS

As is the case with many other wildlife species, connectivity affects different types of the brown bear's movement behavior, including foraging, reproduction, and dispersion (With 2019). The recovery of the Cantabrian population of brown bear depended to a great extent on the enhancement of high-quality connectivity for the species in the area. Two main elements should be highlighted in this sense.

First, from the initial efforts to recover the species, improvement of the interconnectivity between the aforementioned two subpopulations was a major objective in the conservation plans, and its success was crucial for regaining connectivity-related gene flow (Pérez et al. 2009, 2010). From the scientific point of view, a series of studies addressed the structural and functional connectivity and gene flow between both subpopulations (Mateo-Sánchez et al. 2015a, b; Morin et al. 2017; Eds San Miguel et al. 2012). On a more operative side, several projects were promoted by NGOs like FOP and the regional governments for the recovery of linkages between the two populations (see Sect. 4.3 for details).

Second, as previously mentioned, the Cantabrian brown bear population range is currently in expansion. The patterns for such expansion are importantly conditioned by some behavioral traits of the species (see Fig. 6.2). Specifically, dispersion patterns for the brown bear are sex-biased and density-dependent (Blanco et al. 2020; García-Sánchez et al. 2022; Odden et al. 2014; Pérez et al. 2010), with males moving across long distances and females defining more stable home ranges related to breeding areas (López-Bao et al. 2021a). In addition, landscape composition and configuration were proven factors for the different movement and dispersal habits of the species, including movements for breeding, foraging, or migration (Lamamy et al. 2019; López-Bao et al. 2021a, b; Pérez-Girón et al. 2022; Zarzo-Arias et al. 2019).

Several initiatives by different actors have been developed to facilitate the movement of the species, including the recovery of landscape elements important for the movement, diet, and reproduction of bears (Pérez-Girón et al. 2022, 2024), such as chestnut orchards and native fruit trees like buckthorns, apple trees, and cherry trees. Many of these landscape elements are also an important part of local socio-ecological production landscapes, or SEPLs (Diaz-Varela et al. 2018, 2025).

Fig. 6.2 Conceptual scheme showing how the behavioral traits of brown bears influence macroscale movements across landscapes. Male home ranges are larger than those of females, with longer dispersal distances for the former and higher philopatric behavior for the latter. Length of arrows and size of circles show a conceptual representation of home ranges and dispersal distances, respectively. See text for details. (Source: Authors; bear images by LTapsaH via Pixabay under Content License; underlying map obtained from Ministerio de Transportes y Movilidad Sostenible 2024, under license CC-BY 4.0)

Nevertheless, it is important to take into account that for social–ecological interactions with wildlife, SEPLs imply complex relationships, including synergies associated with conservation activities (e.g., enhancement of local ecosystem services supply, capacity building, strengthening of the social fabric, and support for economic development in rural areas; see, e.g., Marquínez García et al. (2021)), but also trade-offs related to conflictual interactions (predation on livestock, damage to apiaries, or fear of attacks on people; see, e.g., Ballesteros et al. 2021; Bautista et al. 2017, 2019; and Naves et al. 2018). As such, consideration of the role of SEPL management in landscape composition and configuration in both formal and informal spatial planning, and at different spatial levels, can be expected to contribute to optimizing the location of elements essential for ecological connectivity. Additionally, adopting multi-actor perspectives, including local communities, NGOs, and administrative bodies at different levels, would facilitate the implementation of conservation measures by integrating the perceptions and attitudes of the local population regarding bears and their expansion.

6.2 Objectives

This work aims to analyze how management in SEPLs may help to ensure high-quality connectivity for the brown bear in the western Cantabrian Mountains. To do so, a review of the studies, projects, and initiatives developed in the area by the NGO Fundación Oso Pardo (FOP) is conducted to analyze the activities aimed at improving connectivity for the brown bear. The resulting knowledge base is then used in a twofold approach: 1) to characterize where and how actions oriented to maintain rural activities in the area have the potential to improve connectivity at the landscape level, using GIS and spatial metrics; and 2) to develop a stakeholder map characterizing the social fabric in the SEPLs related to connectivity improvement and analyzing inter-actor influences through fuzzy cognitive mapping.

6.3 Materials and Methods

6.3.1 Study Area

This study is centered in the area formed by a total of 26 municipalities where FOP is currently implementing most of its actions related to the western subpopulation of brown bear in the Cantabrian Mountain range (see Table 6.1 and Fig. 6.3). The area, spanning 5378 km^2, is mountainous with characteristic ecosystems and differing degrees of human intervention. Coevolution of humans and nature shaped a valuable pattern of native forests (35.8%), shrublands (35.2%), agricultural crops and grasslands (12.2%), planted forests (6.7%), agricultural–forest mosaics (5.4%),

Table 6.1 Basic information of the study area

Country/Region	Spain
Province	Asturias, León, Lugo
District	N/A
Municipality	Belmonte de Miranda, Candín, Cangas del Narcea, Cervantes, Degaña, Folgoso do Courel, Grado, Ibias, Lena, Navia de Suarna, As Nogais, Páramo del Sil, Palacios del Sil, Pedrafita do Cebreiro, Peranzanes, Proaza, Quirós, Quiroga, Samos, Somiedo, Teverga, Triacastela, Vega de Espinareda, Villablino, Villafranca del Bierzo, Yernes y Tameza
Size of geographical area (hectare)	537,839.8 ha
Dominant ethnicity(ies), if appropriate	Caucasian
Size of case study/project area (hectare)	34,354.85 ha (case study 1) + 63,672.14 ha (case study 2) = 98,027 ha
Dominant ethnicity in the project area	Caucasian
Number of direct beneficiaries (people)	1716 (sub-area 1) + 9642 (sub-area 2) = 11,358
Number of indirect beneficiaries (people)	65,631
Geographic coordinates (latitude, longitude)	42° 57′ 16″ N, 6° 34′ 57″ W (centroid)

(Source: Author(s), based on geographic data from Centro Nacional de Información Geográfica 2024; demographic data from INE 2024)

and other land uses (4.7%), defining characteristic SEPLs. Sustainable agricultural and livestock practices, crucial for maintaining SEPLs and the region's identity, rely on a declining population. Additionally, these traditional activities were largely displaced by now-abandoned coal mining in the last century. Today, nature-based tourism and other alternative activities to the primary and secondary sectors are gaining relevance. All of these underscore the need for initiatives that integrate rural development and nature conservation.

At a more detailed scale, two case studies were selected for their importance in the current initiatives to recover connectivity for the brown bear. Specifically, they were selected based on the areas integrating different initiatives of FOP regarding land purchasing and stewardship. Thus, *case study 1* (Fig. 6.4) is placed in the western part of the study area and corresponds to the specific actions of the project "*LIFE Oso Courel*" (Table 6.2), developed from 2017 to 2020 and oriented to support the westbound expansion of the species (Fig. 6.5). It is constituted by a

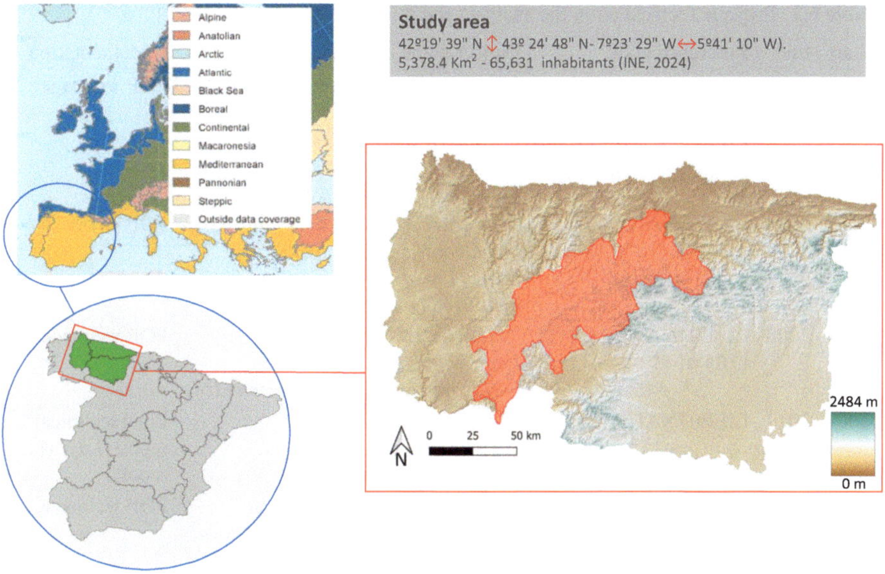

Fig. 6.3 Study area (Source: Authors, based on data from Centro Nacional de Información Geográfica 2024, under license CC-BY 4.0; biogeographical map from European Environment Agency 2017, shared under license CC-BY 4.0; demographic data from INE 2024)

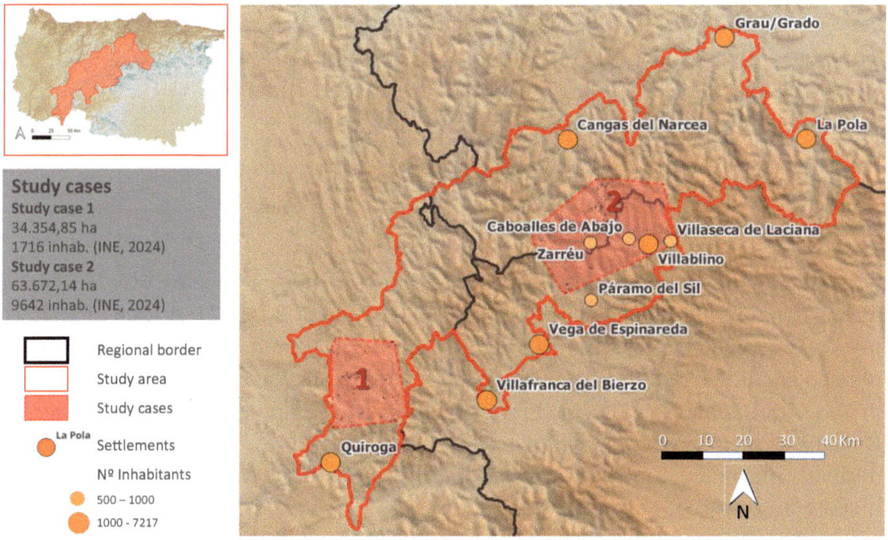

Fig. 6.4 Case studies (Source: Authors, based on aerial imagery from Centro Nacional de Información Geográfica 2024, under license CC-BY 4.0; and demographic data by INE 2024)

Table 6.2 Projects focused on in the analysis and related documents

Start	End	Reference code	Project name and main objectives	Documentation
1992	1998	LIFE92 NAT/E/014504 LIFE94 NAT/E/004830 LIFE95 NAT/E/000630	*"Acciones para la conservación del oso pardo y su hábitat en la Cordillera Cantábrica"* *Conservation of brown bear in Cantabrian Mountain range*	FOP webpage
2001	2004	LIFE00NAT/E/007352	*"Conservación del oso cantábrico y lucha contra el furtivismo"* *Brown bear conservation and fight against poaching*	FOP webpage
2004	2008	LIFE04NAT/IT/000144	*"LIFE COEX"* *Improving coexistence of big carnivores and livestock husbandry*	FOP webpage
2009	2011	LIFE07NAT/E/000735	*"LIFE + Corredores Oso"* *Connection corridors for Cantabrian brown bear*	Final technical report; Layman report;
2013	2016	LIFE12 NAT/ES/000192	*"LIFE Desfragmentación Oso"* *Habitat defragmentation for brown bear in Cantabrian Mountain range*	FOP webpage
2017	2020	LIFE16 NAT/ES/000573	*"LIFE Oso Courel"* *Actions favoring expansion of brown bear to new territories in Courel Mountains*	Final technical report; Layman report;
2020	2025	LIFE19 NAT/ES/000913	*"LIFE Osos con Futuro"* *Improvement of key trophic resources for Cantabrian brown bear in climate change scenarios*	FOP webpage
2023	2027	LIFE22-NAT-ES-LIFE	*"LIFE Coexistencia entre Humanos y Osos"* *Prevention and mitigation of conflicts in settlements to improve coexistence between humans and bears*	FOP webpage

Source: https://fundacionosopardo.org/

perimeter enclosing 142 plots where planting of different species was conducted during the autumn–winter seasons of 2019–2020 and 2020–2021. A total of 107,805 trees were planted covering 145.6 ha in plots of 1.02 ha average size. Species planted included birch *(Betula celtiberica* Rothm. & Vasc.), cherry tree (*Prunus avium* Linn.), alder buckthorn (*Frangula alnus* Mill.), apple tree (*Malus domestica* Borkh.), pear tree (*Pyrus communis* Linn.), and rowan (*Sorbus aucuparia* Linn.), normally in mixed formations (more than 58% of the area) and sometimes combined with oak tree (*Quercus robur* Linn.).

On the other hand, *case study 2* (Fig. 6.4) is placed in the approximate center of the western subpopulation area and corresponds to the area covered by projects "*LIFE + Corredores Oso*", "*LIFE Osos con Futuro*", and other projects and actions implemented from 2009 to the present time (Table 6.2) that were oriented specifically to enhance connectivity and improve resilience for the species in a future of

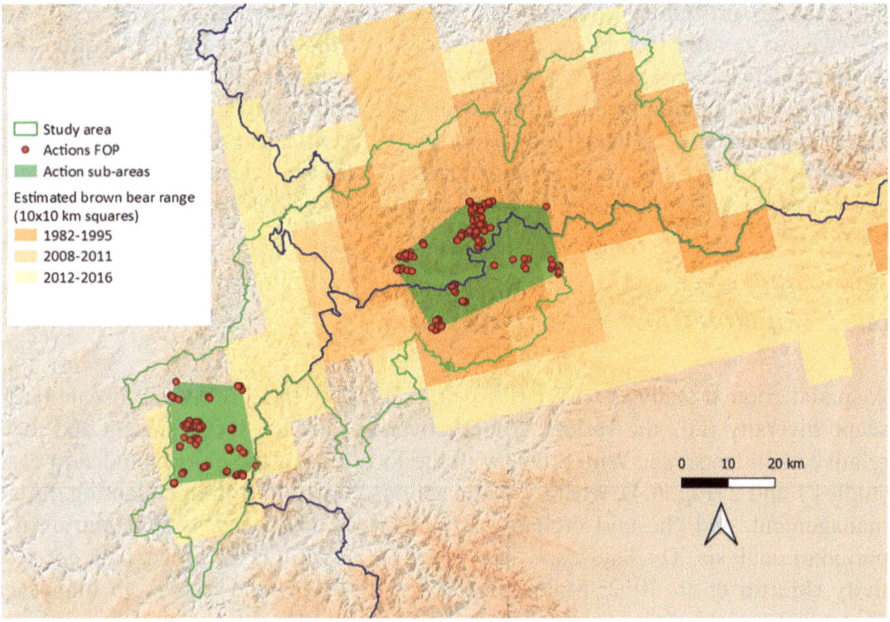

Fig. 6.5 Evolution of the estimated range and dispersion trends for the brown bear, inside and outside the study area, superimposed with specific locations for the actions undertaken by FOP and the respective sub-areas. See Sect. 3.1 for details on the brown bear ranges and a description of FOP actions. (Image source: Authors; estimated range data and location of actions obtained from FOP; underlying DEM obtained from Spain's National Geographic Institute (Instituto Geográfico Nacional 2024, under license CC-BY 4.0)

climate change. The perimeter in this case encloses 258 plots, where, in addition to the species mentioned for case study 1, whitebeam (*Sorbus aria* Crantz) and chestnut tree (*Castanea sativa* Mill.) were also included, the latter in both new plantations and recovered traditional chestnut orchards. A total of 118,296 new trees were planted mostly in mixed formations (more than 54% of the area) covering 177.3 ha in plots of 0.7 ha average size.

6.3.2 *Analysis of Reference Documents*

To address the activities of FOP in the time frame from 1992 to 2024, a series of documents generated by the organization's projects that were directly related to the conservation of the brown bear in the study area were analyzed. The selected projects were all developed within the framework of the LIFE Programme, the European Union's funding instrument for the environment and climate action. The projects and associated resources were awarded to FOP following the submission of proposals and applications corresponding to specific competitive calls, in accordance with the program procedures. The associated documents, shown in Table 6.2, were reviewed to identify factors and connections related to the stakeholders, actions for the improvement of connectivity, outcomes, and elements of the biophysical system. Also, the participation of the authors who are members of FOP allowed for direct observation of developments and outcomes of the different projects addressed in this work. Finally, official conservation plans and strategies developed at the regional and national level for the same period were also considered for contextualization.

6.3.3 Spatial Context Analysis

A spatial context analysis was performed, comparing the level of mesoscale landscape diversity (i.e., the variety of land-cover or land-use types present and their relative abundance; see With 2019) with the locations of the plots included in case studies 1 and 2 (Fig. 6.4), where specific actions related to fruit tree planting, forest management, and chestnut orchard recovery were developed by FOP during the period of analysis. The landscape diversity level may influence brown bear connectivity (Martin et al. 2012; Mateo-Sánchez et al. 2014) and thus is an important contextual variable for actions taken at the microscale.

Geographic Information System software (Q-GIS v. 3.22; see https://qgis.org/) was used for the representation and visual analysis of the evolution of the brown bear population over the years, overlying the available data in FOP for the estimated brown bear range in 10x10 km quadrats for three periods: 1982–1995, 2008–2011, and 2012–2016. The study area, conceived as the geographical setting for FOP actions, was set as a spatial reference. Spatial metrics software (Fragstats v. 4.2; see McGarigal et al. 2023), was then used to calculate the Shannon diversity index (McGarigal et al. 2023; Shannon and Weaver 1949) to a categorical land-use/land-cover map, specifically Spain's national forest map (*Mapa Forestal de España*), scale 1:25.000 (MITECO 2023), using the typologies of land-cover strata as a proxy for land use/land cover. The calculation was conducted using a moving window approach to further analyze the spatial distribution of the values of the index (Díaz-Varela et al. 2009, 2016; Gaucherel et al. 2007). To improve the interpretation of the context, the size of the moving window was adopted using a radius of 3500 m, consistent with the daily average movement of brown bear females (Penteriani et al. 2017a). The results of the Shannon diversity index were interpreted as a proxy of landscape diversity.

6.3.4 Fuzzy Cognitive Mapping

To analyze the complexity of socio-ecological relationships between the different elements of the SEPLs associated with the recovery projects performed by FOP, a fuzzy cognitive map (FCM) approach was taken. FCM is a method for the analysis of complex causal models based on a semiquantitative approach, using indicative (as opposed to predictive) values (Barbrook-Johnson and Penn 2022; Özesmi and Özesmi 2004; Vasslides and Jensen 2016), and representing variables as nodes (or "factors") in a network and causal links as arrows (or "connections"). The document analysis referred to in Sect. 3.3 and Table 6.2 was used to identify the main elements that have relevance for understanding the relationship between ecological connectivity for bears, FOP actions, and the dynamics of the SEPLs, in order to use them as factors. Then, connections between factors were identified and their degree of influence estimated. To process the map, we used an online software called Mental Modeler (see https://www.mentalmodeler.com/), due to its capacity to represent the causal relationships between stakeholders and elements of a socio-ecological system (Gray et al. 2012). The map was then synthesized to represent the main groups of factors, and the connections between them were classified as enabling, constraining, or mutually reinforcing.

6.4 Results and Discussion

6.4.1 Spatial Context of SEPL-Related Actions for Connectivity Enhancement

Figure 6.5 shows the pattern of expansion of the brown bear in the western population. The dispersion trends are highlighted taking into account the variation of the home range from the assessment made in three periods: 1982 to 1995 (a period when the population started its recovery from a continuous shrinking trend, with an important disconnection from the eastern population); 2008 to 2011 (when the increase of detected home ranges evidenced the consolidation and dispersion of the populations); and 2012 to 2016 (where there were two clear expansion trends toward the opposite sides of an east–west line). This pattern reveals important implications regarding connectivity, which in fact has guided actions for its improvement. On the one hand, there is a positive trend in the subpopulation's growth, from approximately 50–60 bears in the 1982–1995 period (Wiegand et al. 1998) to approximately 274 individuals in 2019 (López-Bao et al. 2021a). Due to the abovementioned behavioral aspects of the species (Fig. 6.2), this progressive consolidation of the brown bear is consistent with the dispersion pattern manifested on the one hand in the expansion of the brown bear to new territories toward the west and on the other hand in improved connectivity between both subpopulations toward the east, enhancing the capacity of gene flow (see Fig. 6.1 for context).

As previously mentioned, the actions implemented by FOP for the conservation of the brown bear include the plantation of fruit trees crucial to the bears' life cycle, particularly during hyperphagia—the period when bears intensify their food intake to build up fat reserves before hibernation (Pérez-Girón et al. 2022, 2024). Additionally, other tree species were planted to enhance forest cover, facilitating

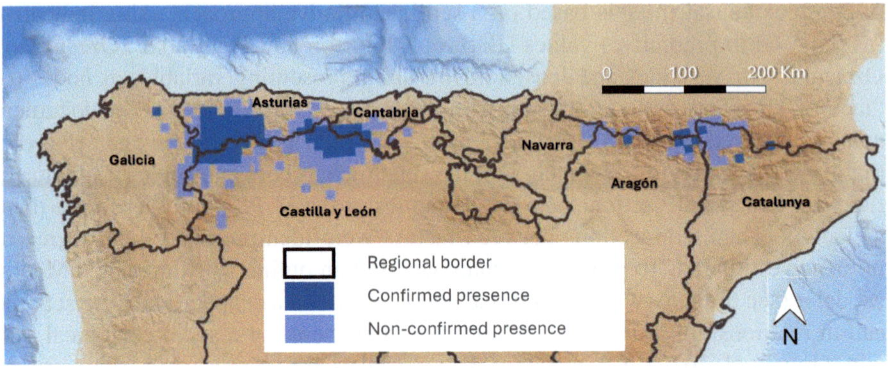

Fig. 6.1 Present distribution of the brown bear in the Northern Iberian Peninsula, showing confirmed and nonconfirmed presence. The regional administration boundaries are shown, identifying those with species presence (Source: Prepared by the authors with data from MITECO 2011)

easier movement for the species. Due to the complexity of the entire socio-ecological system and the multiplicity of associated factors, it is extremely difficult to analyze the specific contribution of these actions to the recovery of the brown bear population in recent decades. Nevertheless, the approach taken to improve bear populations by enhancing connectivity and landscape diversity through the improvement of trophic availability is consistent with the scientific evidence on the species' behavior. Brown bears perceive their environment as a gradient of habitat quality rather than categorical mosaics (Mateo-Sánchez et al. 2014), with landscape composition and human disturbance being key factors for habitat suitability on broader scales. At finer scales, bears are sensitive to ecotones that provide a mix of foraging and cover areas (Clevenger et al. 1997; Mateo-Sánchez et al. 2014). Therefore, while the landscape pattern conditions the movement of bears across the landscapes in complex ways (Lamamy et al. 2019), bears prefer landscapes of intermediate diversity and moderate human influence, combining agricultural areas with forests and shrublands (Mateo-Sánchez et al. 2014) to balance food resources and human presence (Martin et al. 2012).

Consequently, specific actions such as the plantation of species that support the brown bear's foraging needs would have an improving effect on habitat suitability for the bears at the microscale, but they also in turn may create an emergent effect on the facilitation of dispersion trends at the mesoscale. In fact, the survival of dispersing individuals may be favored by anthropogenic landscapes with good habitat quality (Díaz-Fernández et al. 2023).

Our spatial context analysis shows a relationship between the choice of action locations (i.e., those defining case studies 1 and 2) and landscape diversity, which is consistent with the scientific rationale explained above. Figures 6.6 and 6.7 show the micro- and mesoscale contexts for FOP actions. Figure 6.6 shows the landscape diversity context in the study area, together with the expansion trends of the species and the location of improvement actions at the plot level. The actions carried out in both case studies would address, at the mesoscale level, the enhancement of habitat quality with the double objective of improving dispersion and establishment. On the other hand, Fig. 6.7 shows the connection between the levels by showing the level of diversity in the location of the plots and the enhancement of microscale diversity at the species level. It can be seen how most plots (291) are placed in areas that may be perceived by bears as of intermediate diversity. A lower number of plots are placed either in areas with low (49 plots) or high (60 plots) diversity.

Box 6.1. Brown Bear Habitat Restoration, SEPLs Activities, and Multi-stakeholder Interactions

Activities developed by FOP through LIFE projects are characterized by inclusiveness in stakeholder participation. Depending on the project and its specific objectives, stakeholders may include hunters, guards, farmers, beekeepers, nature photographers, mushroom pickers, sport organizations, and different types of institutions (e.g. education centres, heritage organizations, or local authorities). The engagement of stakeholders is closely interwoven within a series of interrelations among the various social and ecological elements of the SEPLs, occurring at different levels. An example is the plantation of fruit trees and recovery of chestnut orchards in project LIFE19 NAT/ES/000913 "LIFE Osos con Futuro" (see Table 6.2), designed to improve the availability of food resources for brown bears in the face of future climate change. The activities take place on land either publicly or privately owned, but in both cases, stewardship agreements are signed. These agreements include a commitment to five years of plantation supervision by FOP and maintaining the land use unchanged for 30 years by the owner. During this period, the owner cannot cut the trees but is allowed to use the fruits produced on the plot. Prior to planting, local varieties of fruit trees or chestnuts are selected for planting or grafting with the support of universities and research institutions. This support includes genomic analysis of the varieties and capacity building for forest planting and recovery, integrating traditional knowledge with modern techniques. The location of the plots to be planted responds to multiple criteria, including the resilience to climatic variations and the deviation of the bear foraging routes from settlements by strategic placement. Finally, for plantation work in the field, local unemployed people are hired. Consequently, the involvement of different stakeholders improves the flow of information, engagement in the project, and value sharing. The co-production of knowledge serves a dual purpose: preserving local heritage and disseminating new technologies. Additionally, the diversity of local cultivated varieties is protected in locations chosen to enhance ecological connectivity and ecosystem restoration, while also reducing conflicts between local inhabitants and brown bears.

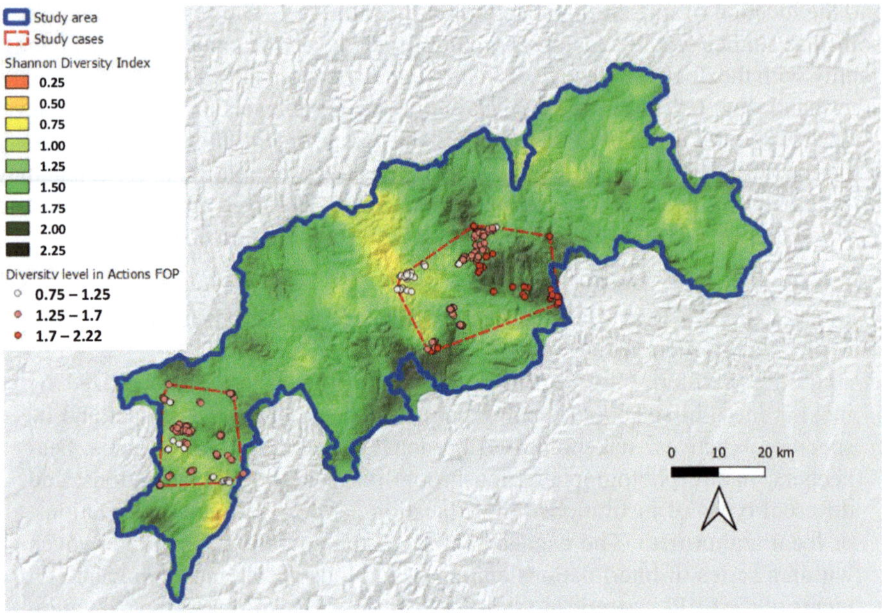

Fig. 6.6 Diversity of land-use types as a context for the connectivity improvement actions. Diversity was calculated using the Shannon diversity index. Dots representing FOP actions are classified according to their diversity values. See Fig. 6.7 for a visual representation of the value ranges. (Source: Authors; underlying DEM obtained from Spain's National Geographic Institute (Instituto Geográfico Nacional) 2024, under license CC-BY 4.0)

6.4.2 Evolution in Social–Ecological Interaction Toward Connectivity Improvement

The analysis of the LIFE projects developed by FOP allows for the detection of different thematic trends, associated with the adaptation of those projects to specific needs in the different stages of recovery of the species (Fig. 6.8; also, see Box 6.1 for a description of interrelationships associated with FOP actions). Thus, while the initial projects addressed straightforward impacts related to the continuity of brown bear individuals and populations (e.g., poaching, conflicts with farmers.) with specific measures (legal prosecution of poachers, negotiation approaches with farmers and hunters), progressively, emphasis was placed on the species' population dynamics and individual behaviors, including connectivity for movement and dispersion, as well as on the receptivity to cohabitation with bears of local inhabitants and the society in general. In these cases, emphasis on land acquisition and stewardship strategies in strategic locations for the connectivity of the species allowed the recovery of chestnut orchards and plantation of fruit trees to support the foraging habits of the bears, especially during important times of the year like hyperphagia before hibernation, as described in the previous section. In addition, capacity building for different stakeholders (including guards, farmers, beekeepers, and hunters)

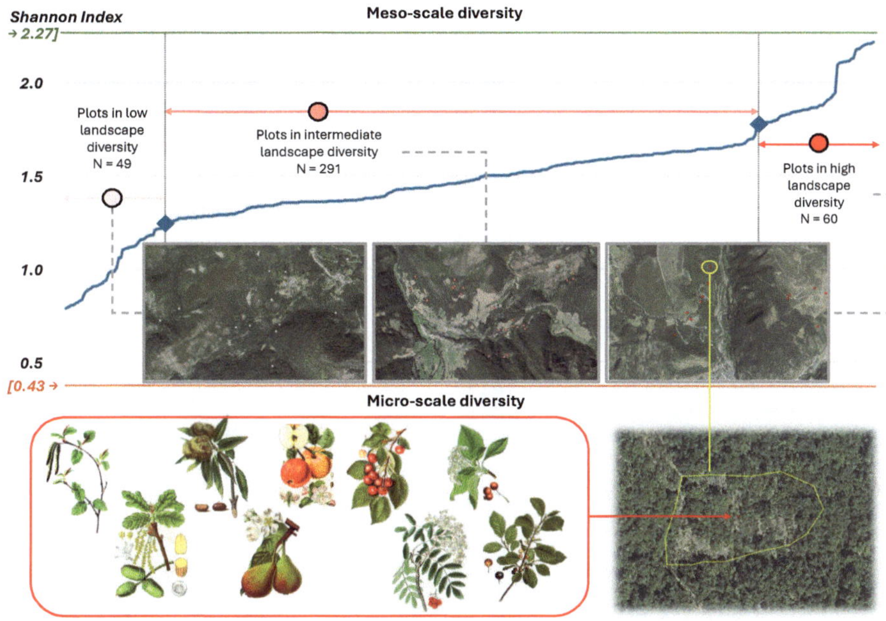

Fig. 6.7 Relationship between meso- and microscale diversity. Mesoscale diversity is shown by the ordered minimum to maximum Shannon index values. Aerial imagery is included as examples of the different land-cover diversity levels. Minimum (0.43) and maximum (2.27) values in the study area are shown for reference. Microscale diversity is exemplified by the tree species planted at the plot level. See text for details. (Source: Authors. Orthophotographs by Ministerio de Transportes y Movilidad Sostenible 2024, under license CC-BY 4.0); plant illustrations from https://creazilla.com/, license CC0 1.0 Universal)

facilitated the development of good practices for minimizing the impact of bears on socioeconomic activities, and dissemination activities in educational centers and cultural institutions were included in multiple projects to improve the social acceptance of brown bears.

From the analysis of the different projects, the FCM shows interesting interdependencies between the different stakeholders, actions, and outcomes. Figure 6.9 shows a simplified version centered on aspects directly related to the improvement of bear connectivity in the framework of the SEPLs. In the FCM, connectivity is shown to have a straightforward connection with bear population: as previously discussed, connectivity is essential for the maintenance and dispersion of the brown bear and has played an important role in the improvement of its population in the last decades. Connectivity can be influenced at multiple levels, first at the institutional level, through spatial planning and/or the definition of protected areas or other effective area-based conservation measures (see, e.g., Diaz-Varela et al. 2018), together with specific actions targeted at the conservation of the species. Also, NGOs like FOP promote actions for increasing connectivity at different scales, as described in the previous section. The plantation activities have multiple connections in the SEPLs. While their results support the improvement of ecosystem

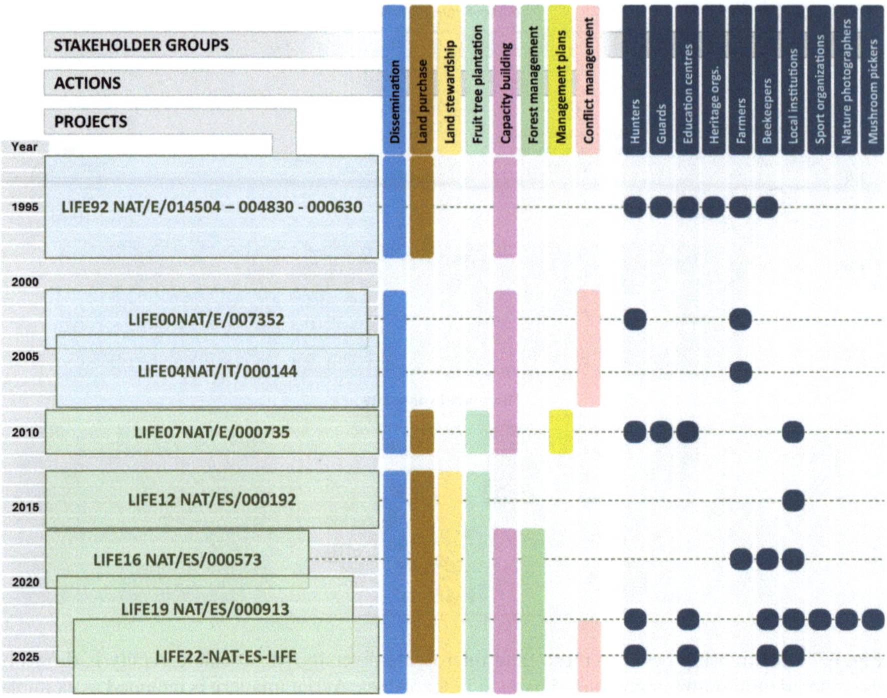

Fig. 6.8 Timeline of the development of European Commission-financed LIFE projects led by FOP, showing the main types of actions developed and the main stakeholder types targeted. Projects in light green involve specific actions addressing connectivity. See Table 6.2 for the correspondence between reference codes and project objectives (Source: Authors)

services derived from the associated benefits of the planted species in the SEPL agroecosystems, their development in plots owned or managed by FOP contributes to strengthening bear conservation as part of the social fabric, especially with the involvement of the rural population through local employment.

On the other hand, the enhancement of connectivity and the consequent improvement in brown bear populations have important consequences for bear-based economies (see, e.g., Marquínez García et al. 2021). In turn, these have mutual, strong dependencies with bear social acceptance. Through environmental educational activities, local businesses also have a mutual connection with other educational efforts, and the support of local institutions, e.g., through the maintenance of tourism infrastructures, is also important. Other connections with local activities like farm-related products can generate synergies that can, however, be affected by bear non-acceptance (see, e.g., López-Bao et al. 2021b). In this sense, capacity building for farmers and beekeepers on methods to avoid or minimize bear attacks on livestock or beehives (see, e.g., Ballesteros et al. 2021; Bombieri et al. 2019; Naves et al. 2018), and for hunters for good practices in hunting activities, could be essential to improve the perception of the brown bear. Also, guard activities and the prosecution of poaching were crucial elements for the improvement of bear populations.

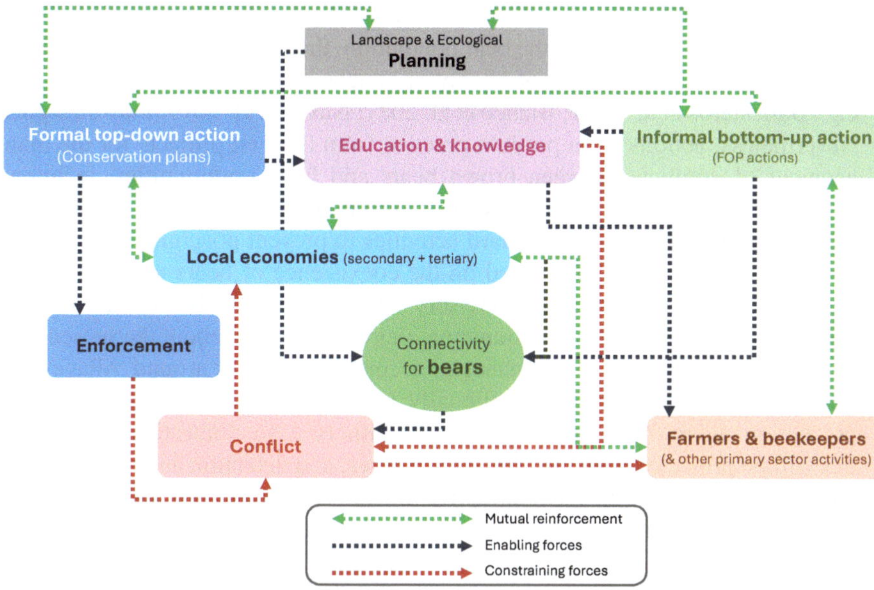

Fig. 6.9 Summary of the main group of factors, as well as relationships, identified through the fuzzy cognitive map (FCM) Mental Modeler. System interactions in the SEPLs are represented as enabling forces, constraining forces, and mutual reinforcement. Coordination between top-down and bottom-up actions, multifaceted management of conflict, and the role of education and knowledge transfer are identified among the main driving factors (Source: Authors)

6.4.3 Main Outcomes of SEPL Management for the Improvement of Connectivity

The management of socio-ecological production landscapes (SEPLs) in the western Cantabrian Mountains has yielded significant outcomes in enhancing ecological connectivity for brown bears that are described as follows:

- *Multiple benefits of SEPL management and enhancement of ecological connectivity.* Efforts to enhance connectivity within SEPLs have resulted in a more interconnected landscape, facilitating the movement and dispersal of brown bears. This improved connectivity is crucial for maintaining viable populations and promoting genetic diversity among subpopulations (Gonzalez et al. 2016; López-Bao 2020). Moreover, these efforts have included management practices implemented in SEPLs that address multifunctional aspects of social–ecological systems. Through the integration of conservation initiatives with traditional rural activities, whether through direct actions, such as planting fruit trees and improving chestnut orchards, or through indirect means such as capacity building in agriculture and livestock husbandry, SEPLs have enhanced their multifunctional capacity.

- *Trade-offs and synergies related to the management of ecological connectivity.* As a result of the increase in the bear population, the risk of negative interactions with bears (e.g., destruction of beehives, attacks on livestock or humans; see, e.g., Ballesteros et al. 2021; Blanco et al. 2021; Naves et al. 2018) may be higher. Nevertheless, actions were put into practice from the first projects for the management of conflicts between brown bears and local communities, including capacity building for farmers, beekeepers, and hunters in prevention methods, as well as the implementation of guard activities to prevent poaching. Specifically, the most recent project is focused on the coexistence of bears and society and includes establishing territorial advisory and coordination boards; creation of information channels between municipalities and residents about the presence of bears; and clearing and maintenance of security perimeters around villages and trails at risk of bear conflicts. Additionally, local institutions will be trained in the prevention and mitigation of human–bear conflicts. Also, domestic animal facilities will be protected with electrified enclosures, and detection and deterrent systems will be installed in areas attractive to bears. These actions not only reduce conflicts but also improve stakeholders' perception of bears and positively impact the local economy (Marquínez García et al. 2021). For instance, bears are an important attraction in nature-based tourism in the area (however, it is important to note that this has resulted in both positive and negative consequences for bear conservation; see Penteriani et al. 2017b), and the image of the brown bear has become a quality label for artisanal agrifood and tourism products, all contributing to community well-being. Similarly, the outcomes of SEPL management highlight the synergistic relationships between landscape elements, local livelihoods, and wildlife conservation. In fact, maintenance of economic activities in SEPLs may help to avoid the undesirable ecological consequences of human depopulation in rural areas (Martínez-Abraín et al. 2020, 2021). By fostering collaboration among diverse stakeholders and promoting sustainable practices, SEPLs have become hubs of innovation and cooperation in conservation efforts.
- *Measurement and assessment of ecological connectivity.* Connectivity analysis has been regarded as one of the main scientific topics addressed in recovery efforts for the brown bear in the Cantabrian range. Main methodologies employed in the area have been associated with the definition of species-specific movement corridors by modeling least-cost path connections, suitability maps, and similar approaches (Mateo-Sánchez et al. 2014, 2015b). Nevertheless, these methods must be used cautiously due to the influence of the spatial and thematic resolution of data, the seasonality of the species movement behavior, or the expansion pattern of the populations (Cisneros-Araujo et al. 2021; Ciudad et al. 2021). The compilation of physical evidence of presence as hair and scat samples and their use in the analysis of species population density through noninvasive techniques defining genetic profiles of individuals (López-Bao 2020; López-Bao et al. 2021a; Pérez et al. 2010) have rendered interesting results also for large-scale contextualization of connectivity for the brown bears (Fig. 6.5). On the other hand, the spatial analysis of foraging patterns allows for the definition of preferences in the use of different land covers by the species at finer scales (Pérez-Girón et al. 2022).

- *Integration of local and traditional knowledge and cultural diversity in connectivity management.* Management of SEPL elements at the micro- and mesoscale (Fig. 6.7) needs to come from local and traditional knowledge for the successful plantation of fruit trees and the recovery of chestnut orchards, especially when techniques of pruning of grafting have to be put into use (see, e.g., Diaz-Varela et al. 2018). By leveraging the expertise and practices of local communities, conservation efforts have been more effective in addressing the adaptation of local landscape elements to the needs of connectivity enhancement for the brown bears (see Box 6.1).
- *Operationalizing ecological connectivity in spatial planning and management.* To date, the operationalization of ecological connectivity in Spain has predominantly followed a top-down approach, with the inclusion of connectivity measures in spatial planning and management instruments. In 2021, a strategic document for green infrastructure, ecological restoration, and connectivity (*Estrategia Nacional de Infraestructura Verde y de la Conectividad y Restauración Ecológicas*) was established at the national level, providing a framework for similar strategies to be developed by regional governments (MITECO 2021). These regional-level strategies, currently in progress, are expected to guide more detailed, subregional initiatives. However, integrating bottom-up approaches from the SEPL level could enhance governance and implementation of measures to improve ecological connectivity (Díaz-Varela et al. 2021). Specifically regarding connectivity for brown bears, the current conservation strategy for the Cantabrian brown bear population (MITECO 2019) serves as a guiding framework for regional recovery plans in Galicia, Asturias, Cantabria, and Castilla y León. This document recognizes the collaborative efforts of various administrative levels, scientists, NGOs, and other stakeholders in the species' recovery and emphasizes conflict management, communication, social participation, and rural development—all of which are crucial for maintaining SEPLs—as key lines of action, including connectivity as one of the main efforts in the conservation and management of the brown bear's habitat. Specifically, FOP has evaluated and proposed actions to improve permeability to reduce risks for the brown bear in various road infrastructures in the Cantabrian Mountain range. In addition, FOP worked on the proper implementation of the Natura 2000 Network and the regional network of protected areas by submitting proposals directly to the relevant authorities, participating in public information periods, and, at times, through media or judicial pressure.

6.5 Conclusions and Future Directions

Stakeholder engagement and the integration of socio-ecological data are crucial for effective conservation planning, particularly for incorporating connectivity enhancement into spatial planning and management. In this regard, adopting a bottom-up approach from the management of SEPLs could be integrated with top-down strategies coordinated across different administrative levels. Initiatives such as ecosystem recovery and management at finer scales, long-term monitoring, and adaptive management strategies are essential for the sustained presence of brown bear populations in the western Cantabrian Mountains. These efforts could be significantly strengthened if embedded within an integrative framework, where the SEPLs could have an important cohesive role in the actions taken on the landscape, with the involvement of local actors in the process.

Challenges currently addressed by FOP include enhancing policies for preventing and compensating for damages caused by bear attacks, increasing public information and awareness to maintain social acceptance, mitigating the effects of climate change, ensuring safe connectivity between bear populations, updating regional recovery plans, establishing regulatory frameworks for bear tourism, and preventing and managing conflicts with habituated bears. Regarding ecological connectivity, they will focus on identifying road infrastructures that impede the movement of brown bears in the Cantabrian Mountains, a critical factor affecting the species in this range (see, e.g., Kervellec et al. 2023). Additionally, they will continue to monitor connectivity through genetic interexchange, analyzed from feces or hair samples.

Finally, future analysis of ecological connectivity may benefit from multimodel approaches, combining the mentioned genetic methods with others used to date (telemetry, modeling least-cost path connections, applying circuit-theoretic approaches, or defining suitability maps; see, e.g., Almasieh et al. 2019; García-Sánchez et al. 2022) and integrating new methodologies such as agent-based models to define animal movements at different scales (e.g., Perez et al. 2021) while integrating different behavior scenarios, including interactions with the human environment (Marley 2017).

References

Almasieh K, Rouhi H, Kaboodvandpour S (2019) Habitat suitability and connectivity for the brown bear (Ursus arctos) along the Iran-Iraq border. Eur J Wildl Res 65:57. https://doi.org/10.1007/s10344-019-1295-1

Ballesteros F, Blanco JC, López-Bao JV, Palomero G (2021) Los daños atribuídos al oso. In: Palomero G, Ballesteros F, Blanco JC, López-Bao JV (eds) Osos Cantábricos. Demografía, Coexistencia y Retos de Conservación. Ministerio para la Transición Ecológica y el Reto Demográfico. Fundación Oso Pardo, Madrid, pp 39–63

Barbrook-Johnson P, Penn AS (2022) Systems mapping: how to build and use causal models of systems. Springer

Bombieri G, Naves J, Penteriani V, Selva N, Fernández-Gil A, López-Bao JV, Ambarli H, Bautista C, Bespalova T, Bobrov V, Bolshakov V, Bondarchuk S, Camarra JJ, Chiriac S, Ciucci P, Dutsov A, Dykyy I, Fedriani JM, García-Rodríguez A, Delgado MM (2019). Brown bear attacks on humans: A worldwide perspective. Scientific Reports 9(1):8573. https://doi.org/10.1038/s41598-019-44341-w

Bautista C, Naves J, Revilla E, Fernández N, Albrecht J, Scharf AK, Rigg R, Karamanlidis AA, Jerina K, Huber D, Palazón S, Kont R, Ciucci P, Groff C, Dutsov A, Seijas J, Quenette P-I, Olszańska A, Shkvyria M, Adamec M, Ozolins J, Jonozovič M, Selva N (2017) Patterns and correlates of claims for brown bear damage on a continental scale. J Appl Ecol 54:282–292. https://doi.org/10.1111/1365-2664.12708

Bautista C, Revilla E, Naves J, Albrecht J, Fernández N, Olszańska A, Adamec M, Berezowska-Cnota T, Ciucci P, Groff C, Härkönen S, Huber D, Jerina K, Jonozovič M, Karamanlidis AA, Palazón S, Quenette P-Y, Rigg R, Seijas J, Swenson JE, Talvi T, Selva N (2019) Large carnivore damage in Europe: Analysis of compensation and prevention programs. Biol Conserv 235:308–316. https://doi.org/10.1016/j.biocon.2019.04.019

Blanco JC, Ballesteros F, Palomero G, López-Bao JV (2020) Not exodus, but population increase and gene flow restoration in Cantabrian brown bear (Ursus arctos) subpopulations. Comment on Gregório et al. 2020. PLoS One 15:e0240698. https://doi.org/10.1371/journal.pone.0240698

Blanco JC, Palomero G, Ballesteros F, López-Bao JV, Palomero G, Ballesteros F, Blanco JC, López-Bao JV (2021) Habituación, condicionamiento a las basuras y ataques a humanos. In: Palomero G, Ballesteros F, Blanco JC, López-Bao JV (eds) Osos Cantábricos. Demografía, Coexistencia y Retos de Conservación. Ministerio para la Transición Ecológica y el Reto Demográfico. Fundación Oso Pardo, Madrid, pp 65–89

Centro Nacional de Información Geográfica (2024) Modelo Digital del Terreno–MDT200. Instituto Geográfico Nacional, viewed 13 March 2024. Retrieved from https://centrodedescargas.cnig.es/CentroDescargas/index.jsp

Cisneros-Araujo P, Goicolea T, Mateo-Sánchez MC, García-Viñás JI, Marchamalo M, Mercier A, Gastón A (2021) The role of remote sensing data in habitat suitability and connectivity modeling: insights from the cantabrian brown bear. Remote Sens 13:1138. https://doi.org/10.3390/rs13061138

Ciudad C, Mateo-Sánchez MC, Gastón A, Blazquez-Cabrera S, Saura S (2021) Landscape connectivity estimates are affected by spatial resolution, habitat seasonality and population trends. Biodivers Conserv 30:1395–1413. https://doi.org/10.1007/s10531-021-02148-0

Clevenger AP, Purroy FJ, Campos MA (1997) Habitat assessment of a relict brown bear Ursus arctos population in northern Spain. Biol Conserv 80:17–22. https://doi.org/10.1016/S0006-3207(96)00081-X

Díaz-Fernández M, Naves J, Revilla E (2023) Conservation implications of range dynamics in endangered populations: an example with brown bears. Conserv Sci Pract 5:e12894. https://doi.org/10.1111/csp2.12894

Díaz-Varela E, Álvarez-López CJ, Marey-Pérez MF (2009) Multiscale delineation of landscape planning units based on spatial variation of land-use patterns in Galicia, NW Spain. Landsc Ecol Eng 5:1–10. https://doi.org/10.1007/s11355-008-0053-4

Díaz-Varela E, Roces-Díaz JV, Álvarez-Álvarez P (2016) Detection of landscape heterogeneity at multiple scales: Use of the Quadratic Entropy Index. Landsc Urban Plan 153:149–159. https://doi.org/10.1016/j.landurbplan.2016.05.004

Diaz-Varela ER, Álvarez-Álvarez P, Roces-Diaz JV, Rodríguez-Morales B (2018) The contribution of chestnut orchard recovery projects for effective area-based conservation: Two cases in Asturias (North-West Spain). In: UNU-IAS, IGES (ed) Sustainable use of biodiversity in socio-ecological production landscapes and seascapes (SEPLS) and its contribution to effective area-based conservation. Satoyama initiative thematic review, vol. 4. United Nations University Institute for the Advanced Study of Sustainability, Tokyo, pp 26–37

Díaz-Varela E, Fernández-Villar G, Diego-Fuentes A (2021) Transformative change in Peri-urban SEPLS and green infrastructure strategies: an analysis from the local to the regional scales in Galicia (NW Spain). In: Nishi M, Subramanian SM, Gupta H, Yoshino M, Takahashi Y, Miwa K, Takeda T (eds) Fostering transformative change for sustainability in the context of socio-ecological production landscapes and seascapes (SEPLS). Satoyama initiative thematic review, vol. 6. Springer, Singapore, pp 133–153. https://doi.org/10.1007/978-981-33-6761-6_8

Diaz-Varela ER, Álvarez-Álvarez P, Pérez-Girón JC, Díaz-Varela RA, Ferreira Golpe MA, García-Arias AI (2025) Chestnut production-related businesses in Courel Mountains (Galicia, NW Spain): an opportunity for biodiversity conservation, ecosystem restoration and rural development. In: Nishi M, Subramanian SM (eds) Business and biodiversity: reciprocal connections in the context of socio-ecological production landscapes and seascapes (SEPLS). Satoyama initiative thematic review, vol 9. Springer, Tokyo, pp 31–50

European Environment Agency (2017) Biogeographical Regions in Europe, viewed 13 March 2024. Retrieved from https://www.eea.europa.eu/data-and-maps/figures/biogeographical-regions-in-europe-2

García-Sánchez MP, González-Ávila S, Solana-Gutiérrez J, Popa M, Jurj R, Ionescu G, Ionescu O, Fedorca M, Fedorca A (2022) Sex-specific connectivity modelling for brown bear conservation in the Carpathian Mountains. Landsc Ecol 37:1311–1329. https://doi.org/10.1007/s10980-021-01367-8

García-Vázquez A, Pinto Llona AC, González-Fortes GM, Grandal-D'Anglade A (2020) Distribución y cronología del oso pardo (Ursus arctos L.) en la Península Ibérica durante el Pleistoceno Superior y Holoceno. Spanish J Palaeontol 30:161. https://doi.org/10.7203/sjp.30.1.17233

Gaucherel C, Burel F, Baudry J (2007) Multiscale and surface pattern analysis of the effect of landscape pattern on carabid beetles distribution. Ecol Indic 7:598–609. https://doi.org/10.1016/j.ecolind.2006.07.002

Gonzalez EG, Blanco JC, Ballesteros F, Alcaraz L, Palomero G, Doadrio I (2016) Genetic and demographic recovery of an isolated population of brown bear Ursus arctos L., 1758. Peer J 4:e1928. https://doi.org/10.7717/peerj.1928

Gray S, Chan A, Clark D, Jordan R (2012) Modeling the integration of stakeholder knowledge in social-ecological decision-making: Benefits and limitations to knowledge diversity. Ecol Model 229:88–96. https://doi.org/10.1016/j.ecolmodel.2011.09.011

INE (2024) Cifras de población y censos demográficos, viewed 13 March 2024. Retrieved from https://ine.es/

Instituto Geográfico Nacional (2024) Centro Nacional de Información Geográfica–Centro de Descargas, viewed 13 March 2024. Retrieved from https://centrodedescargas.cnig.es/

Kervellec M, Milleret C, Vanpé C, Quenette P-Y, Sentilles J, Palazón S, Jordana IA, Jato R, Elósegui Irurtia MM, Gimenez O (2023) Integrating opportunistic and structured non-invasive surveys with spatial capture-recapture models to map connectivity of the Pyrenean brown bear population. Biol Conserv 278:109875. https://doi.org/10.1016/j.biocon.2022.109875

Lamamy C, Bombieri G, Zarzo-Arias A, González-Bernardo E, Penteriani V (2019) Can landscape characteristics help explain the different trends of Cantabrian brown bear subpopulations? Mamm Res 64:559–567. https://doi.org/10.1007/s13364-019-00440-7

López-Bao JV (2020) Consistent bear population DNA-based estimates regardless molecular markers type. Biol Conserv 248:108651. https://doi.org/10.1016/j.biocon.2020.108651

López-Bao JV, Godinho R, Palomero G, Ballesteros F, Blanco JC, Jiménez J (2021a) Seguimiento de la población de oso cantábrico en un escenario de expansión. In: Palomero G, Ballesteros F, Blanco JC, López-Bao JV (eds) Osos Cantábricos. Demografía, Coexistencia y Retos de Conservación. Ministerio para la Transición Ecológica y el Reto Demográfico. Fundación Oso Pardo, pp 23–38

López-Bao JV, Planella A, Ballesteros F, Blanco JC, Palomero G, Nores C (2021b) Evolución de las osas con crías y demografía del oso en la Cordillera Cantábrica. In: Palomero G, Ballesteros F, Blanco JC, López-Bao JV (eds) Osos Cantábricos. Demografía, Coexistencia y Retos de Conservación. Ministerio para la Transición Ecológica y el Reto Demográfico. Fundación Oso Pardo, pp 1–22

Marley J (2017) Does human education reduce conflicts between humans and bears? An agent-based modelling approach. Ecol Model 343:15–24. https://doi.org/10.1016/j.ecolmodel.2016.10.013

Marquínez García J, García de la Fuente L, González Rodríguez G, Lara Gómez LM, Toro Álvarez F, López-Bao JV, Planella A, Navarro A (2021) El oso como recurso en la economía rural. In: Palomero G, Ballesteros F, Blanco JC, López-Bao JV (eds) Osos Cantábricos. Demografía, Coexistencia y Retos de Conservación. Ministerio para la Transición Ecológica y el Reto Demográfico. Fundación Oso Pardo, Madrid, pp 105–124

Martin J, Revilla E, Quenette P-Y, Naves J, Allainé D, Swenson JE (2012) Brown bear habitat suitability in the Pyrenees: transferability across sites and linking scales to make the most of scarce data. J Appl Ecol 49:621–631. https://doi.org/10.1111/j.1365-2664.2012.02139.x

Martínez Cano I, Taboada FG, Naves J, Fernández-Gil A, Wiegand T (2016) Decline and recovery of a large carnivore: environmental change and long-term trends in an endangered brown bear population. Proc R Soc B Biol Sci 283:20161832. https://doi.org/10.1098/rspb.2016.1832
Martínez-Abraín A, Jiménez J, Jiménez I, Ferrer X, Llaneza L, Ferrer M, Palomero G, Ballesteros F, Galán P, Oro D (2020) Ecological consequences of human depopulation of rural areas on wildlife: a unifying perspective. Biol Conserv 252:108860. https://doi.org/10.1016/j.biocon.2020.108860
Martínez-Abraín A, Ferrer X, Jiménez J, Fernández-Calvo IC (2021) The selection of anthropogenic habitat by wildlife as an ecological consequence of rural exodus: empirical examples from Spain. Anim Biodivers Conserv 44:195–203. https://doi.org/10.32800/abc.2021.44.0195
Mateo-Sánchez MC, Cushman SA, Saura S (2014) Connecting endangered brown bear subpopulations in the Cantabrian Range (north-western Spain). Anim Conserv 17:430–440. https://doi.org/10.1111/acv.12109
Mateo-Sánchez MC, Balkenhol N, Cushman S, Pérez T, Domínguez A, Saura S (2015a) A comparative framework to infer landscape effects on population genetic structure: are habitat suitability models effective in explaining gene flow? Landsc Ecol 30:1405–1420. https://doi.org/10.1007/s10980-015-0194-4
Mateo-Sánchez MC, Balkenhol N, Cushman S, Pérez T, Domínguez A, Saura S (2015b) Estimating effective landscape distances and movement corridors: comparison of habitat and genetic data. Ecosphere 6:art 59. https://doi.org/10.1890/ES14-00387.1
McGarigal K, Cushman SA, Ene E (2023) FRAGSTATS v4: spatial pattern analysis program for categorical maps. Computer software program produced by the authors; Available at the following web site: https://www.fragstats.org
Ministerio de Transportes y Movilidad Sostenible (2024) Infraestructura de Datos Espaciales de España (IDEE), viewed 13 March 2024. Retrieved from https://idee.es/
MITECO (2011) Bases de datos del Inventario Español de Especies Terrestres, viewed 13 March 2024. Retrieved from https://www.miteco.gob.es/es/biodiversidad/temas/inventarios-nacionales/inventario-especies-terrestres/inventario-nacional-de-biodiversidad/bdn-ieet-default.html
MITECO (2019) Estrategia para la conservación del oso pardo Ursus arctos en la Cordillera Cantábrica, viewed 13 March 2024. Retrieved from https://www.miteco.gob.es/es/biodiversidad/publicaciones/pbl-fauna-flora-estrategias-oso-cantabrico.html
MITECO (2021) Estrategia Nacional de Infraestructura Verde y de la Conectividad y Restauración Ecológicas, viewed 13 March 2024. Retrieved from https://www.miteco.gob.es/es/biodiversidad/temas/ecosistemas-y-conectividad/infraestructura-verde/infr_verde.html
MITECO (2023) Mapa Forestal de España 1:25.000, viewed 13 March 2024. Retrieved from https://www.miteco.gob.es/es/cartografia-y-sig/ide/descargas/biodiversidad/mfe.html
Morin DJ, Fuller AK, Royle JA, Sutherland C (2017) Model-based estimators of density and connectivity to inform conservation of spatially structured populations. Ecosphere 8:e01623. https://doi.org/10.1002/ecs2.1623
Naves J, Ordiz A, Fernández-Gil A, Penteriani V, Delgado MDM, López-Bao JV, Revilla E, Delibes M (2018) Patterns of brown bear damages on apiaries and management recommendations in the Cantabrian Mountains, Spain. PLoS One 13:e0206733. https://doi.org/10.1371/journal.pone.0206733
Odden M, Ims RA, Støen OG, Swenson JE, Andreassen HP (2014) Bears are simply voles writ large: social structure determines the mechanisms of intrinsic population regulation in mammals. Oecologia 175:1–10. https://doi.org/10.1007/s00442-014-2892-z
Özesmi U, Özesmi SL (2004) Ecological models based on people's knowledge: a multi-step fuzzy cognitive mapping approach. Ecol Model 176:43–64. https://doi.org/10.1016/j.ecolmodel.2003.10.027
Palazón S (2017) The importance of reintroducing large carnivores: the brown bear in the Pyrenees. In: Catalan J, Ninot JM, Aniz MM (eds) High mountain conservation in a changing world. Springer, Cham, pp 231–249. https://doi.org/10.1007/978-3-319-55982-7_10

Palazón S, Alonso M, Ruiz-Olmo J, Pomarol M, Quenette PY (2002) Resultados de una traslocación de osos pardos de Eslovenia a los Pirineos Centrales. Galemys 14:27–48

Palomero G, Ballesteros F, Nores C, Blanco JC, Herrero J, García-Serrano A (2007) Trends in number and distribution of brown bear females with cubs-of-the-year in the Cantabrian Mountains, Spain. Ursus 18:145–157. https://doi.org/10.2192/1537-6176(2007)18[145:TINADO]2.0.CO;2

Parres A, Palazón S, Afonso I, Quenette P-Y, Batet A, Camarra J-J, Garreta X, Gonçalves S, Guillén J, Mir S, Jato R, Rodríguez J, Sentilles J, Xicola L, Melero Y (2020) Activity patterns in the reintroduced Pyrenean brown bear population. Mamm Res 65:435–444. https://doi.org/10.1007/s13364-020-00507-w

Penteriani V, Delgado MDM, López-Bao JV, García PV, Monrós JS, Álvarez EV, Corominas TS, Vázquez VM (2017a) Patterns of movement of released female brown bears in the Cantabrian Mountains, northwestern Spain. Ursus 28:165–170. https://doi.org/10.2192/URSU-D-16-00012.1

Penteriani V, López-Bao JV, Bettega C, Dalerum F, Delgado MDM, Jerina K, Kojola I, Krofel M, Ordiz A (2017b) Consequences of brown bear viewing tourism: a review. Biol Conserv 206:169–180. https://doi.org/10.1016/j.biocon.2016.12.035

Pérez T, Vázquez F, Naves J, Fernández A, Corao A, Albornoz J, Domínguez A (2009) Non-invasive genetic study of the endangered Cantabrian brown bear (Ursus arctos). Conserv Genet 10:291–301. https://doi.org/10.1007/s10592-008-9578-1

Pérez T, Naves J, Vázquez JF, Seijas J, Corao A, Albornoz J, Domínguez A (2010) Evidence for improved connectivity between Cantabrian brown bear subpopulations. Ursus 21:104–108. https://doi.org/10.2192/09SC018.1

Perez AZ, Bone C, Stenhouse G (2021) Simulating multi-scale movement decision-making and learning in a large carnivore using agent-based modelling. Ecol Model 452:109568. https://doi.org/10.1016/j.ecolmodel.2021.109568

Pérez-Girón JC, Díaz-Varela ER, Álvarez-Álvarez P, Hernández Palacios O, Ballesteros F, López-Bao JV (2022) Linking landscape structure and vegetation productivity with nut consumption by the Cantabrian brown bear during hyperphagia. Sci Total Environ 813:152610. https://doi.org/10.1016/j.scitotenv.2021.152610

Pérez-Girón JC, Álvarez-Álvarez P, Ballesteros F, López-Bao JV (2024) Potential impacts of climate change on wild cherry distribution and associated consequences on brown bears. Biol Conserv 289:110390. https://doi.org/10.1016/j.biocon.2023.110390

Roberge J-M, Angelstam P (2004) Usefulness of the umbrella species concept as a conservation tool. Conserv Biol 18:76–85. https://doi.org/10.1111/j.1523-1739.2004.00450.x

Rozylowicz L, Popescu VD, Pătroescu M, Chişamera G (2011) The potential of large carnivores as conservation surrogates in the Romanian Carpathians. Biodivers Conserv 20:561–579. https://doi.org/10.1007/s10531-010-9967-x

San Miguel A, Ballesteros F, Blanco JC, Palomero G (eds) (2012) Manual de buenas prácticas para la gestión de corredores oseros en la Cordillera Cantábrica, Especies Amenazadas. Ministerio de Agricultura, Alimentación y Medio Ambiente, Madrid

Shannon CE, Weaver W (1949) The mathematical theory of communication. University of Illinois Press

Sommer RS, Benecke N (2005) The recolonization of Europe by brown bears Ursus arctos Linnaeus, 1758 after the last glacial maximum. Mammal Rev 35:156–164. https://doi.org/10.1111/j.1365-2907.2005.00063.x

Tosi G, Chirichella R, Zibordi F, Mustoni A, Giovannini R, Groff C, Zanin M, Apollonio M (2015) Brown bear reintroduction in the Southern Alps: to what extent are expectations being met? J Nat Conserv 26:9–19. https://doi.org/10.1016/j.jnc.2015.03.007

Valdiosera CE, García-Garitagoitia JL, Garcia N, Doadrio I, Thomas MG, Hänni C, Arsuaga J-L, Barnes I, Hofreiter M, Orlando L, Götherström A (2008) Surprising migration and population size dynamics in ancient Iberian brown bears (Ursus arctos). Proc Natl Acad Sci 105:5123–5128. https://doi.org/10.1073/pnas.0712223105

Vasslides JM, Jensen OP (2016) Fuzzy cognitive mapping in support of integrated ecosystem assessments: developing a shared conceptual model among stakeholders. J Environ Manag 166:348–356. https://doi.org/10.1016/j.jenvman.2015.10.038

Velasco JP, Romo H (2010) Idoneidad de la red de espacios protegidos de Asturias (España) para las mariposas diurnas y aplicación del principio de complementariedad en la selección de áreas prioritarias (Lepidoptera: Papilionoidea y Hesperioidea). SHILAP Revta lepid 38:219–234

Wiegand T, Naves J, Stephan T, Fernandez A (1998) Assessing the Risk of Extinction for the Brown Bear (*Ursus arctos*) in the Cordillera Cantabrica, Spain. Ecol Monogr 68:539–570. https://doi.org/10.2307/2657153

With KA (2019) Essentials of landscape ecology. Oxford University Press

Yang N, Price M, Xu Y, Zhu Y, Zhong X, Cheng Y, Wang B (2023) Assessing global efforts in the selection of vertebrates as umbrella species for conservation. Biology 12:509. https://doi.org/10.3390/biology12040509

Zarzo-Arias A, Penteriani V, Delgado MDM, Peón Torre P, García-González R, Mateo-Sánchez MC, Vázquez García P, Dalerum F (2019) Identifying potential areas of expansion for the endangered brown bear (*Ursus arctos*) population in the Cantabrian Mountains (NW Spain). PLoS One 14:e0209972. https://doi.org/10.1371/journal.pone.0209972

Chapter 7
Significance of SEPLs in Ecological Connectivity and Conservation of the Tropical Dry Forest: An Experience in the Dry Enclave of the Dagua River in Colombia

Andrés Quintero-Ángel, Sebastian Orjuela-Salazar, Mauricio Quintero-Ángel, Leonor Valenzuela, Diana Saavedra- Zúñiga, and Daniel Osorio-Domínguez

Abstract Globally, forest covers and many strategic ecosystems face significant threats from human activities. The middle basin of the Dagua River in Colombia, ranging from 500 to 1850 m above sea level, includes diverse ecosystems such as grasslands, cloud forests, and tropical dry forests, all of which are highly vulnerable and understudied. These forests in Colombia are critically fragmented and degraded due to historical land-use changes. In this basin, maintaining connectivity between dry and cloud forests is essential for biodiversity, species migration, climate change resilience, ecosystem preservation, and local community reliance on crucial ecosystem services. By 2018, rapid transformation in the middle basin had led to 48% of the area being converted to productive systems, reflecting a significant loss of natural cover. Fortunately, this impact was mitigated by the establishment of two Socio-Ecological Productive Landscapes (SEPLs) in the region. Since 2007, Atuncela has been promoting forest conservation and sustainable production through effective

A. Quintero-Ángel (✉) · D. S.- Zúñiga
Corporación Ambiental y Forestal del Pacifico – CORFOPAL, Valle del Cauca, Colombia
e-mail: direccionejecutiva@corfopal.org

S. Orjuela-Salazar
Fondo para la Acción Ambiental y la Niñez—Fondo Acción, Bogotá DC, Colombia

M. Quintero-Ángel
Universidad del Valle, sede Palmira, Valle del Cauca, Colombia

L. Valenzuela
WCS Colombia, Valle del Cauca, Colombia

D. Osorio-Domínguez
Corporación Ambiental y Forestal del Pacifico – CORFOPAL, Valle del Cauca, Colombia

Departamento de Ciencias Naturales y Matemáticas, Pontificia Universidad Javeriana, Cali, Colombia

M. Nishi et al. (eds.), *Ensuring Ecological Connectivity in Socio-Ecological Production Landscapes and Seascapes (SEPLS)*, Satoyama Initiative Thematic Review, https://doi.org/10.1007/978-981-95-1474-8_7

area management and the establishment of the Integrated Management Regional District (IMRD). In 2015, El Chilcal joined these efforts with its own IMRD declaration, thus safeguarding these two SEPLs. These initiatives have successfully countered the broader conversion trend. Before SEPL implementation, deforestation rates were similar inside and outside the SEPL areas. However, post-implementation, deforestation rates significantly decreased within the SEPLs, demonstrating their effectiveness in reducing deforestation. Moreover, a higher proportion of productive systems within SEPLs supports forest species connectivity compared to areas outside. The SEPLs have maintained elevational and longitudinal connectivity across the basin, aided by low deforestation rates (0% since 2018) and effective management of productive systems. These outcomes highlight how community empowerment supports sustainable production while ensuring ecosystem connectivity and biodiversity conservation.

Keywords Connectivity · Biodiversity conservation · Community empowerment · Tropical dry forest · Colombia

7.1 Introduction

Globally, wildland covers and particularly many strategic ecosystems are widely threatened by anthropogenic activities. In general, these anthropogenic pressures generate changes in land use, land covers and lead in many cases to degradation or fragmentation (De Lima et al. 2021). Particularly, fragmentation is a process that tends to disturb the ecological integrity and functionality of ecosystems and that ends up affecting the connectivity of wildland covers, for example, by the reduction of the habitat surface, the decreased size of patches, and their isolation in the landscape (Luther et al. 2020).

Particularly, in Colombia, the tropical dry forest is in a critical state of fragmentation and deterioration, attributed to landscape transformations and land-use changes that have occurred since the last century (Etter et al. 2008). Currently, the tropical dry forest ecosystem is represented by small forest fragments immersed in anthropogenic matrices, where agricultural and livestock use territories predominate, exerting strong pressure on these remnants (Pizano and Garcia 2014). On the other hand, the Andean montane forests (1000 to 3500 m.a.s.l.) have one of the highest conservation priorities due to their richness and high species endemism (Olson and Dinerstein 1997), but like the dry forest, they also show high rates of fragmentation and habitat loss due to different human activities (Etter and Wyngaarden 2000).

One way to reduce the harmful effects of fragmentation and habitat loss on biodiversity and associated ecological processes is through improving connectivity. However, for there to be dynamics between fragments, landscapes must be accessible and traversable between them (Howell et al. 2018). This connectivity is defined as the degree to which the landscape facilitates or hinders the dispersal of biodiversity, vital elements such as water, ecological interactions, and ecosystem processes between habitat remnants (Taylor et al. 1993; Crooks and Sanjayan 2006). In this field, socio-ecological production landscapes and seascapes (SEPLS), which allow interactions between ecosystems and humans that favor the creation of various ecosystem services for human well-being (Takeuchi 2010), might at the same time contribute to conservation and landscape connectivity. In this context, this chapter aims to highlight the importance of SEPLS and their articulation in territorial planning in the conservation and connectivity of the landscape, taking as a case the middle basin of the Dagua River in Colombia, which encompasses an altitudinal gradient ranging from 500 to 1850 m.a.s.l., including different types of ecosystems such as grasslands, cloud forests, and tropical dry forests.

7.2 Methodology

7.2.1 *Study Area*

The Integrated Management Regional District Subxerophytic Enclave of Atuncela (henceforth IMRDSEA) and the Integrated Management Regional District El Chilcal (henceforth IMRDEC) are two socio-ecological production landscapes (henceforth SEPLs) located in the middle basin of the Dagua River on the eastern slope of the Western Cordillera. In this area, there are rural communities involved in traditional agricultural practices, such as subsistence farming and manual cultivation techniques. Additionally, small-scale producers employ semi-technical methods, which may include the use of basic machinery, improved crop varieties, and more efficient irrigation systems. These farming practices vary in their level of technology and intensity, influencing both the local economy and the surrounding environment. Additionally, to the north, there is a subxerophytic zone characterized by low precipitation levels (Fig. 7.1).

7.2.1.1 Integrated Management Regional District Subxerophytic Enclave of Atuncela (IMRDSEA)

The Integrated Management Regional District Subxerophytic Enclave of Atuncela (IMRDSEA) is in the Corregimiento de Atuncela, Municipality of Dagua, Valle del Cauca, Colombia. It covers 2334.37 ha of protected land, declared by the Regional Autonomous Corporation of Valle del Cauca—CVC (the regional environmental authority) (Fig. 7.1 and Table 7.1). This area, previously identified as a SEPL by Quintero-Angel et al. (2025), overlaps with the Key Biodiversity Area (KBA) known as the dry enclave of Dagua (COL 36) (Critical Ecosystem Partnership Fund 2021). Situated at altitudes ranging from 500 to 1000 m.a.s.l., the IMRDSEA has an average temperature of 23.2 °C, with annual precipitation levels ranging from 700 to 800 mm. It follows a bimodal precipitation pattern, with rainy seasons typically

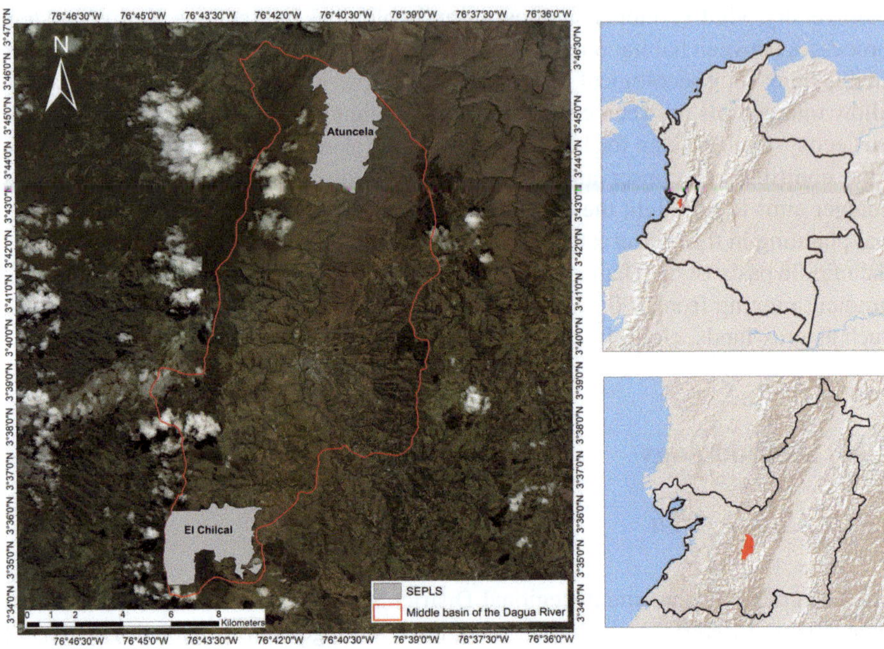

Fig. 7.1 Study area: at the extremes of the middle basin of the Dagua River are the SEPL IMRDSEA and IMRDEC. (Source: Created by the authors using ESRI basemap)

Table 7.1 Basic information of the study area (Authors' elaboration 2024)

	IMRDSEA	IMRDEC
Country/region	Colombia	
Province	Valle del Cauca	
Municipality	Dagua	
Size of the geographical area (hectare)	1011.5	912.47
Dominant ethnicity(ies), if appropriate	Atunceleños	Chilcaleños
Size of the case study/project area (ha)	15,359	
Dominant ethnicity in the project area	Mixed-race peasants	
Number of direct beneficiaries (people)	357	280
Number of indirect beneficiaries (people)	357	280
Geographic coordinates (latitude, longitude)	3° 46′ 21.2484″N, 76° 39′ 42.1272″E	3° 35′ 24.36″N, 76° 42′ 54.54″E

Fig. 7.2 Panoramic view of IMRDSEA (Source: CORFOPAL 2023)

occurring in May, October, and November and dry periods from February to March and June to August (CORFOPAL & CVC 2022a).

The IMRDSEA encompasses three ecosystems: two dry and very dry forest ecosystems in the middle and lower parts and a transition to tropical humid forest in the upper part (CVC & FUNAGUA 2010). The area contains 20 types of vegetation cover, with natural cover constituting 56.68% of the total area, including forests, streams, and successional shrubs. The remaining 43.32% consists of transformed or production-intended covers, such as pastures for livestock, crops, and residential areas (CORFOPAL & CVC 2022a) (Fig. 7.2). In the IMRDSEA, approximately 357 residents are primarily engaged in various primary economic activities, including agriculture, livestock farming, poultry farming, fish farming, panela (brown sugarcane) production, and tourism (Municipality of Dagua 2002).

The IMRDSEA showcases a diverse array of life, harboring 155 species of plants, 14 species of fish, 15 species of amphibians, 138 species of birds, 23 species of mammals, and 14 species of reptiles. Among these documented species are endangered endemics like the Ruiz's robber frog (*Strabomantis ruizi*) and Tlatepusco vanilla (*Vanilla odorata*). Additionally, species such as the Cauca poison frog (*Andinobates bombetes*), the northern tiger cat (*Leopardus tigrinus*), and the Colombian night monkey (*Aotus lemurinus*) are categorized as vulnerable. Moreover, regionally threatened species, though not globally threatened, include the Cauca lily (*Eucharis caucana*), the Loboguerrero cactus (*Melocactus curvispinus* sub. *loboguerreroi*), and the Colombian red howler monkey (*Alouatta seniculus*) (CORFOPAL & CVC 2022a).

Fig. 7.3 Panoramic view of IMRDEC. (Source: CORFOPAL 2023)

7.2.1.2 Integrated Management Regional District El Chilcal (IMRDEC)

The Integrated Management Regional District El Chilcal (IMRDEC), previously identified as a SEPL by Orjuela-Salazar et al. (2018), is in the El Chilcal village at an elevation of 1330 m.a.s.l. in the El Limonar township in the Municipality of Dagua (Fig. 7.1 and Table 7.1). IMRDEC experiences temperatures between 18 and 24 °C, with annual precipitation levels ranging from 800 to 1000 mm. This 912.47 ha area was declared a protected area in 2015 by the CVC. It features two sub-Andean dry forest ecosystems in the lower parts and a transition to tropical humid forest in the middle and upper parts (CORFOPAL & CVC 2022b). The area is characterized by very shallow soils with steep and very steep slopes, frequent landslides, high susceptibility to erosion, shallow effective depth, and low fertility (Fig. 7.3). Its agroecological conditions demand the presence of permanent forest cover (Municipality of Dagua 2002).

In the IMRDEC, approximately 280 people reside, primarily engaged in various primary economic activities such as agriculture, livestock farming, poultry farming, fish farming, mining, and tourism (Municipality of Dagua 2002). This area boasts abundant biodiversity, including 102 species of resident birds, constituting roughly 13% of those recorded in Valle del Cauca. Additionally, there are 27 mammal species, 8 reptile species, 14 amphibian species, and 261 flora species, with the majority being pioneer and early successional species (GAIA & CVC 2014; Orjuela-Salazar et al. 2019).

Among the documented species are endangered endemics, such as the Cauca poison frog (*Andinobates bombetes*) and the Ruiz's robber frog (*Strabomantis ruizi*). Other notable species include the Colombian night monkey (*Aotus lemurinus*), the glass frog (*Centrolene savagei*), the Pichinde robber frog (*Pristimantis gracilis*), and the red cedar (*Cedrela odorata*). Additionally, species like the Santa Rita rocket frog (*Colostethus fraterdanieli*) are classified as near-threatened and are endemic to the region. Furthermore, there are six species facing national or regional threats within the IMRDEC, including the bronze-winged parrot (*Pionus chalcopterus*), jaguarundi (*Herpailurus yagouaroundi*), long-tailed weasel (*Mustela frenata*), Parker's antbird (*Cercomacra parkeri*), the apical flycatcher (*Myiarchus apicalis*), and the Colombian chachalaca (*Ortalis columbiana*) (CORFOPAL & CVC 2022b).

7.2.2 *Information Gathering and Analysis*

7.2.2.1 Governance, Management, Decision-Making, and Stakeholder Engagement in the SEPLs

To address the governance, management, decision-making, and engagement of stakeholders in IMRDSEA and IMRDEC, a systematic information search was conducted following the methodology developed by Aldunce et al. (2008). This involved the detection, acquisition, and consultation of relevant literature for the research, based on predefined search criteria, information sources, and keywords. Subsequently, interviews were conducted with key informants (Geilfus 2002), and participatory observation outings (Kawulich 2005) and knowledge exchange sessions (PRATEC 2012) were implemented to gather specific information about these SEPLs.

7.2.2.2 Loss of Forest Cover

One of the main drivers of connectivity in a landscape is the loss or gain of natural areas. Therefore, the first step we took was to determine the importance of the SEPL as decelerators of forest cover loss. To do this, we compared the amount of tree cover within each SEPL before and after its declaration. These results were compared with the amount of tree cover outside the SEPL in the middle basin of the Dagua River (Fig. 7.1). The percentage loss values of tree cover were obtained from the Global Forest Watch platform (https://www.globalforestwatch.org), which monitors forest cover loss at a pixel resolution of 30 x 30 m. Statistical comparisons were made using paired Student's t tests.

7.2.2.3 Land Use and Resistance Map

Based on a CORINE Land Cover classification[1] at a scale of 1:100,000 (resolution of 30 x 30 m), we identified the different land uses present both inside and outside the SEPLs in the middle basin of the Dagua River. For land cover associated with productive systems, we calculated the percentage that still preserves natural spaces that can be used by biodiversity, increasing the permeability of the matrix. We used

---

[1] The CORINE (Coordination of Information on the Environment) Land Cover classification was initially developed by the European Environment Agency to map and categorize land cover across Europe. It classifies land into various types, such as forests, agricultural areas, and urban areas, based on satellite imagery and other data sources. Although originally created for Europe, this classification system has been adapted for use around the world, supporting environmental monitoring, land management, and spatial planning globally.

this layer to generate a resistance matrix to determine forest connectivity. We did not consider a particular species but rather focused on the general landscape characteristics that can offer costs and difficulty of movement for forest-associated species and reduce the flow of ecological processes, increasing the risk of mortality and avoidance behavior. In this sense, construction, roads, and human settlements offer the highest resistance, followed by pastures, crops, and mosaics with natural spaces.

7.2.2.4 Connectivity

To determine the longitudinal connectivity of the middle basin of the Dagua River, we conducted a least-cost path analysis between the two SEPLs as they are located at the ends of the basin (Fig. 7.1) using the Linkage Mapper Toolkit version 2.0 (McRae and Kavanagh 2011). Linkage Mapper utilizes the least-cost distance map, generated from the resistance matrix, to determine the corridors between the SEPLs. The least-cost corridor is presented as a vector layer, and other possible paths as a raster map, where the value of each cell in this raster indicates the suitability of that cell to provide connectivity.

To determine connectivity, we divided the watershed into three altitudinal bands, from 500 to 1000, 1000 to 1500, and 1500 to 1850 m above sea level. The SEPL IMRDEC fell within the band between 1000 and 1500 m.a.s.l., and the SEPL IMRDSEA between 500 and 1000 m.a.s.l. Using Sentinel-2 10 m Land Use/Land Cover Time Series images (Karra et al. 2021), we determined the forest fragments present for the years 2017 and 2021 in each of the bands, both within and outside the SEPLs, which constitute our units of analysis (Fig. 7.4). We estimated the connectance index, which is defined as the number of functional connections between the patches of the corresponding patch type, where each pair of patches is either connected or not, based on a user-defined distance criterion. Connectance is expressed as a percentage of the maximum possible connectance given the number of patches. In this analysis, we use a dispersal threshold of 30 m. We used Fragstats to calculate the connectance index (McGarigal et al. 2023).

7.3 Results

7.3.1 Governance, Management, Decision-Making, and Stakeholder Engagement in the SEPLs

According to IUCN guidelines for the type of governance, both IMRDSEA and IMRDEC have a Type A governance scheme, which is exercised by the government, as established by "*one or more government agencies*" (Borrini-Feyerabend et al. 2014). In this case, the Regional Autonomous Corporation of Valle del Cauca (CVC)

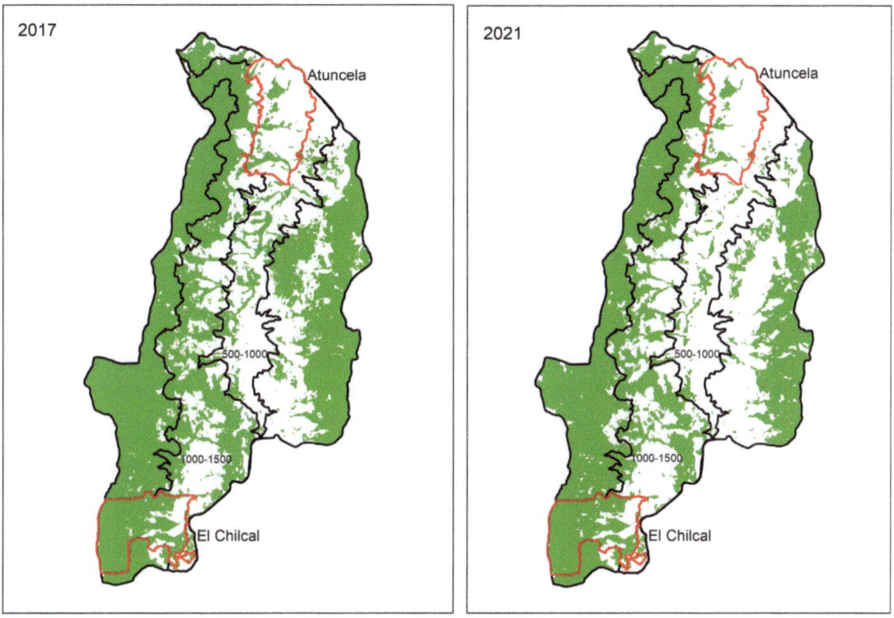

Fig. 7.4 Forest fragments for the years 2017 and 2021 in the middle basin of the Dagua River, with green color corresponding to forest fragments. (Source: Created by the authors using information from Sentinel-2 10 m Land Use/Land Cover Time Series images)

holds the authority and responsibility and is accountable for the management of the protected area, determining its conservation objectives and developing and implementing its management plan. However, the CVC has established, through ordinance and its guidelines, a participatory and binding governance scheme involving other stakeholders in the management of the protected area, known as the Co-Management Committee.

According to the provisions of Ordinance 471 of 2017, which adopts the Departmental System of Protected Areas of Valle, SIDAP Valle: "*The Committees are the social and political basis of the Governance scheme that must guarantee the participation of different actors considering elements taken from the analysis of actors of SIDAP Valle, such as competencies, local participation, benefits, and social and citizen commitment in conservation, cross-cut by ethnic-cultural and political aspects in terms of decision-making*" (CVC 2021).

In both IMRDSEA and IMRDEC, we determined that the established co-management committees have undertaken numerous governance activities, organized into five governance axes. The implementation of these axes has played a significant role in mitigating threats to the area while also fostering effective management, decision-making, and stakeholder engagement. These governance axes, along with the corresponding activities, are detailed in Table 7.2.

Another factor identified in decision-making and stakeholder engagement is the development of planning instruments for each of the SEPLs (Fig. 7.5). These instruments are elaborated through participatory processes involving workshops, social

Table 7.2 Governance axes for IMRDSEA and IMRDEC. (Authors' elaboration 2024)

Governance Axes	Description	Governance Actions	
		IMRDSEA	IMRDEC
Capacity strengthening	This axis aims to enhance the skills of the stakeholders involved in the committee to manage the area. Therefore, training has been provided on a variety of relevant topics.	Qualified participation Administration. Leadership and organization. Conflict resolution. Project formulation.	Qualified participation Leadership and organization. Conflict resolution.
Co-management and coordination	This axis aims to ensure that decisions affecting the territory are agreed upon in a timely manner within the committee and that the implementation of agreed actions is jointly coordinated.	Contextualized planning. Assignment of roles and responsibilities. Management with partners. Monitoring and evaluation. Articulation with institutional bodies.	Contextualized planning. Management with partners. Monitoring and evaluation.
Information dissemination	This axis seeks to ensure that information related to the area and decisions made are shared among committee members and the wider community.	Information dissemination. Information custody. Communication channels. Articulation with other stakeholders.	Information dissemination. Information custody. Communication channels.
Knowledge	Similar to the previous axis, the goal of this axis is to leverage local knowledge to support decision-making. In this sense, the committee collects studies and works carried out in the SEPL by other entities or its own members and keeps them for consultation if necessary.	Research and innovation. Knowledge exchange. Sectorized training.	Research and innovation. Knowledge exchange. Sectorized training. Knowledge custody.
Sustainability	This final axis seeks to ensure that decisions related to the management and use of resources within the SEPL take sustainability into account, in order to promote biodiversity conservation.	Resource management. Resource optimization. Positive reputation.	Resource management. Resource optimization. Teamwork.

mapping, and decision-making meetings. They consist of several phases: a territorial diagnosis phase where biophysical and socioeconomic information of the area is gathered, an organizational phase where conservation and management objectives are established, and a zoning exercise that identifies three main zones—preservation areas, restoration areas, and areas for sustainable use. Additionally, the type of activities allowed and prohibited in each zone is determined. Lastly, there is

Fig. 7.5 Stakeholder engagement in the development of planning instruments for each of the SEPLs: IMRDSEA (above) and IMRDEC (below). (Source: CORFOPAL 2023)

a programmatic phase where actions and projects necessary to maintain the area's integrity are identified (CORFOPAL & CVC 2022a, b). It is through the development and subsequent implementation of these planning instruments that the co-management committees manage the areas and promote biodiversity conservation and ecological connectivity.

7.3.2 *Loss of Forest Cover*

Although forest cover is not dominant in the middle basin of the Dagua River, it is one of the most important land covers in terms of biodiversity and ecosystem services. As seen in Fig. 7.6, loss of forest cover has been variable over time but has completely ceased within the SEPLs since 2018. Considering that the SEPL IMRDSEA was declared in 2007 and IMRDEC in 2015, we compared the loss of forest cover inside and outside the SEPLs before and after the declaration of the first

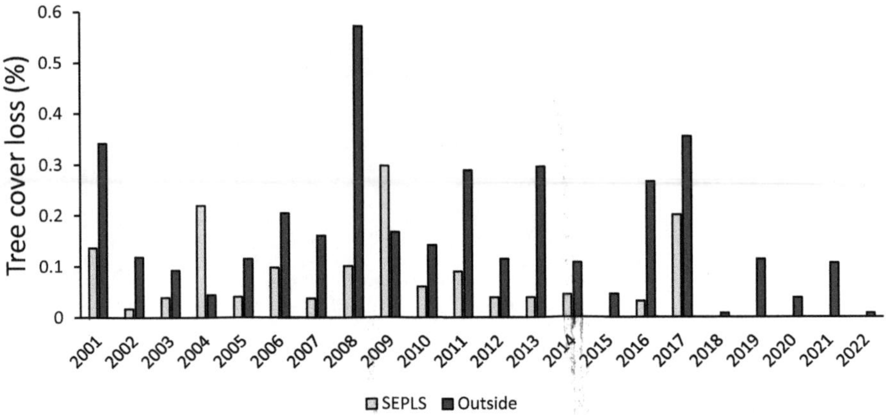

Fig. 7.6 Percentage of tree cover loss per year inside the SEPLs and outside them. (Source: Created by the authors using data from Global Forest Watch; https://www.globalforestwatch.org/map/)

SEPL (2007). In the period before the declaration, the percentage of loss on average per year within the SEPLs was 0.08% (SD = 0.07) and that outside was 0.15% (SD = 0.09), without statistically significant differences being evident ($t = -1.55$, $P = 0.19$). From 2007, a decrease in loss of cover within the SEPLs is evident with an average value of 0.06% (SD = 0.17), while outside, the percentage of forest loss remains (0.085%, SD = 0.15). This leads to significant differences in cover loss ($t = -3.2$, $P = 0.003$).

7.3.3 Land Use

According to the land cover map based on the CORINE Land Cover classification for Colombia, in the middle basin of the Dagua River in 2018, 47.9% of the area was under some type of productive system. Within the SEPLs, 58% of the area was under productive systems, while the figure for outside was 46%. However, within the SEPLs, 39.4% of the productive systems maintain areas of natural spaces compared to 33.7% outside of them.

7.3.4 Connectivity

The analysis of least-cost paths between the SEPLs shows that the path of least resistance corresponds to the upper part of the basin where sub-Andean cloud forest dominates. In general terms, the western part of the basin presents lower resistance than the eastern part. Both inside and outside the SEPLs in the western zone, the streams are the paths of least resistance (Fig. 7.7).

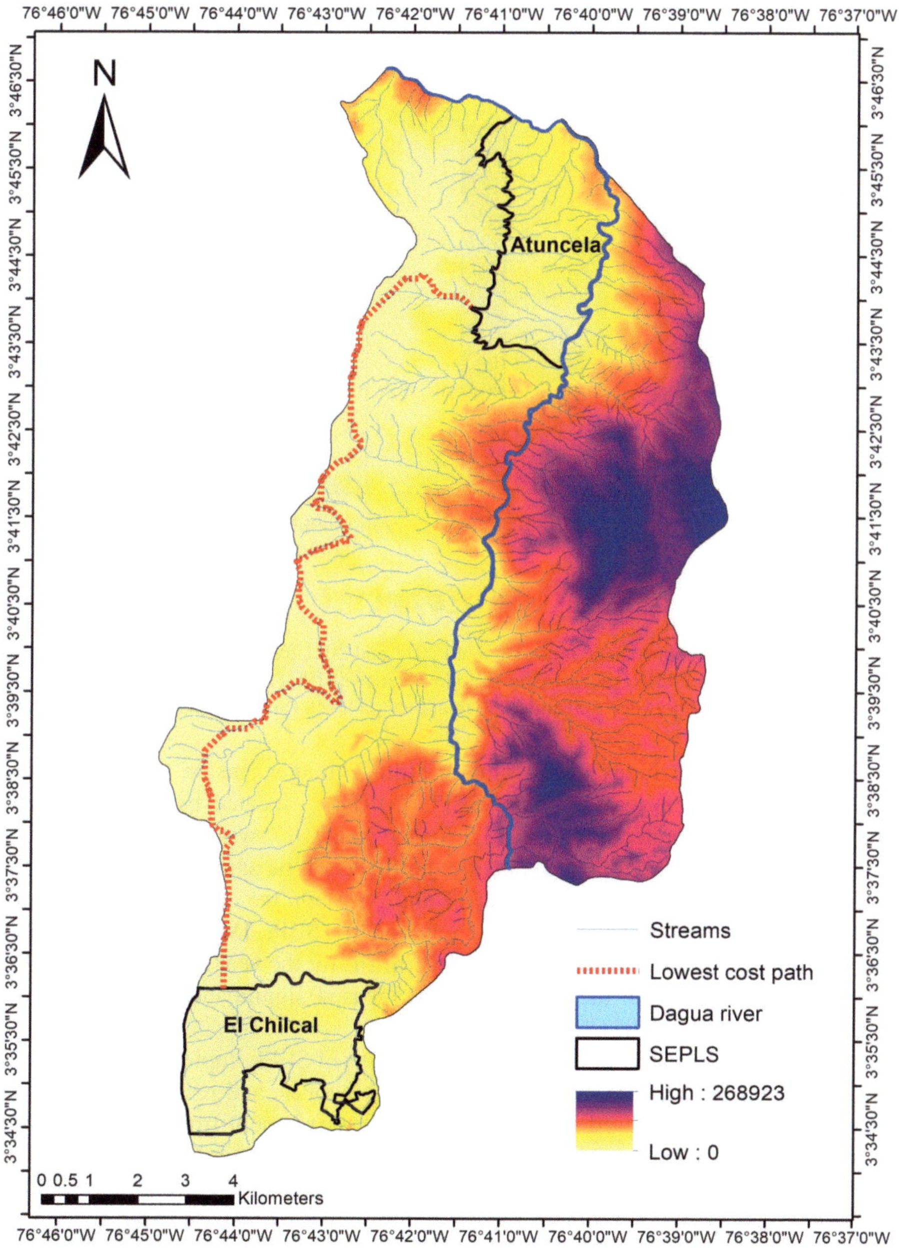

Fig. 7.7 Resistance map and least-cost routes, with light color corresponding to the lowest mobility resistance for forest species and associated flows of ecological processes. (Source: Created by the authors using spatial information provided by the CVC geovisor; https://geo.cvc.gov.co/visor_avanzado/)

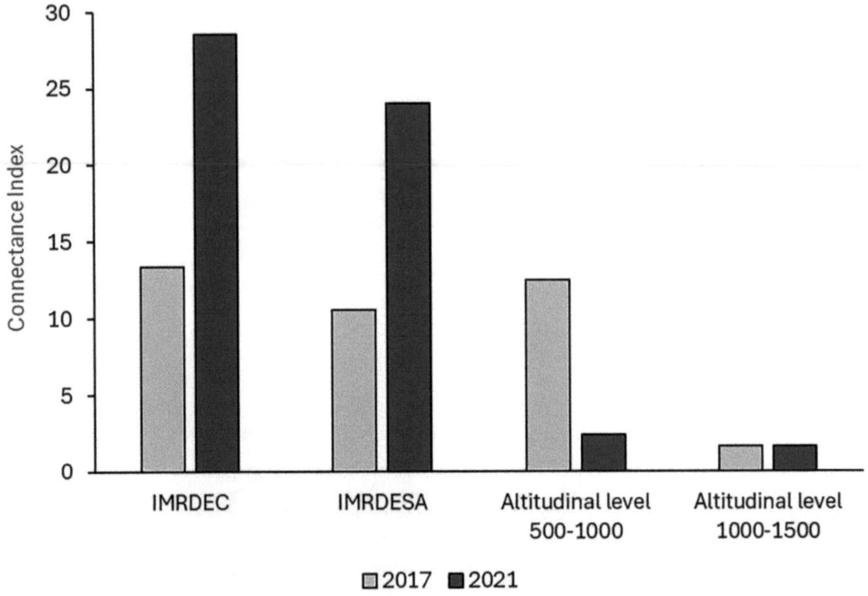

Fig. 7.8 Changes in connectance index between 2017 and 2021 for the SEPLs and the altitudinal ranges outside of them. IMRDEC is in the altitudinal range of 1000 to 1500 m above sea level, and IMRDSEA is situated between 500 and 1000 m above sea level. (Source: Created by the authors)

The connectance index varies among altitude ranges, inside and outside the SEPLs, and between years. For the SEPL IMRDEC, located between 1000 and 1500 m.a.s.l., the connectivity was 13.3 for the year 2017, while for the year 2021, it increased to 28.5. Meanwhile, outside the SEPL in the same altitude range, the connectance index was 1.6 for both 2017 and 2021. On the other hand, in the altitude range between 500 and 1000 m.a.s.l., where the SEPL IMRDSEA is located, a connectivity of 10.5 was observed within the SEPL for the year 2017, rising to 24 in 2021. Whereas, outside, the connectivity decreased significantly, dropping from 12.5 in 2017 to 2.4 in 2021 (Fig. 7.8).

This trend is primarily due to the positive effects of SEPL implementation. Inside SEPLs, better land management, conservation practices, and restoration activities have significantly increased connectivity. These areas benefit from targeted efforts to enhance ecological corridors and maintain natural habitats. In contrast, outside SEPLs, the lack of effective management or less stringent enforcement may lead to poor land management practices, resulting in minimal or even negative changes in connectivity. Human activities such as deforestation, urbanization, and agriculture further disrupt connectivity outside protected areas.

7.4 Discussion

Our analysis reveals substantial changes in forest cover, indicating significant pressure on habitat remnants in the middle basin of the Dagua River. Although forest cover is not predominant in this area, it is crucial to highlight the vital role played by these forest fragments in preserving biodiversity and providing ecosystem services in adjacent environments, considering that such impact is moderated by the isolation of the fragments in the landscape (Mitchell et al. 2014). Nevertheless, a complete halt in the loss of forest cover within the SEPLs has been observed since 2018, underscoring the effectiveness of the conservation strategies in the territory.

The comparison between the loss of cover before and after the declaration of the SEPLs reveals a statistically significant decrease in these areas, indicating a positive impact on forest preservation. These results support the effectiveness of conservation policies and local protected area management in safeguarding natural ecosystems in the Dagua River basin, aligning with previous studies that have demonstrated the importance of such policies in mitigating deforestation, as indicated by Cuenca et al. (2016).

The cessation of deforestation in the SEPLs is attributed to a combination of community and institutional conservation actions. On the one hand, the active involvement of local communities has promoted sustainable land-use practices[2] and fostered the conservation of natural resources (Orjuela-Salazar et al. 2018, 2019; Quintero-Angel et al. 2025). On the other hand, institutional measures, such as the declaration of protected areas and the implementation of a participatory governance framework, have contributed to preserving forest cover by establishing regulations and guidelines for the proper use of land. It is worth mentioning that since the rules and regulations are developed through governance frameworks within co-management committees, they are embraced and adhered to by the communities. This is because the communities feel involved in the decision-making process for the care of their territory, rather than having to comply with imposed rules established from the city without considering the local context. These combined actions have resulted in a significant reduction in deforestation (zero within the SEPLs), highlighting the effectiveness of an integrated strategy involving both local communities and government institutions in biodiversity conservation, which has also favored ecological connectivity.

The analysis of land uses in the middle basin of the Dagua River reveals that nearly half of the area is devoted to productive systems, posing challenges for

---

[2] Within the SEPLs, various sustainable land-use practices have been promoted, resulting in significant increases in connectivity. These practices include agroforestry, ecosystem restoration, organic farming, and integrated pest management. Additionally, sustainable forestry, crop rotation, watershed protection, rotational grazing, conservation agriculture, and participatory community development have been implemented. Specifically, shade-grown coffee cultivation, silvopastoral systems, and sustainable community-based panela (brown sugar cane) production have been introduced. These practices enhance biodiversity, soil health, and climate resilience, benefiting both ecosystems and local communities.

biodiversity conservation. Within the SEPLs, 58% of the area is under productive systems, compared to 46% outside of them. Despite this higher proportion within the SEPLs, it is encouraging to observe that 39.4% of the productive systems within these areas conserve natural spaces, compared to 33.7% outside of them. These findings suggest a complex interaction between productive activity and conservation, demonstrating the effectiveness of SEPLs in reconciling these objectives. Examples of productive systems within SEPLs, such as shade-grown coffee cultivation and silvopastoral systems, illustrate their positive impact on biodiversity. Shade-grown coffee integrates coffee plants under the canopy of native trees, promoting habitats for various bird species, insects, and other wildlife while maintaining forest cover, reducing soil erosion, and enhancing soil fertility. In contrast, conventional coffee plantations outside SEPLs might clear native vegetation, reducing habitat complexity and biodiversity. Silvopastoral systems combine forestry with livestock grazing, integrating trees and shrubs into pasture lands, increasing habitat diversity, supporting pollinators, and improving carbon sequestration. This practice also provides shade for animals, enhancing their welfare and productivity, whereas traditional livestock grazing without trees often leads to soil compaction, erosion, and a decline in plant species diversity. These examples highlight the effectiveness of SEPLs in balancing productive activity with biodiversity conservation by promoting sustainable practices that conserve natural spaces that favor connectivity. The importance of evaluating agricultural and productive practices within the SEPLs to ensure alignment with conservation objectives is highlighted, and future studies could explore more effective management strategies in this context.

The analysis of least-cost routes between the SEPLs in the middle basin of the Dagua River highlights the importance of conserving fragmented landscapes (Arroyo-Rodríguez and Mandujano 2006; Lion et al. 2016). It is observed that the upper part of the basin, characterized by sub-Andean cloud forest, represents the path of least resistance, underscoring the need to protect this key habitat. The western zone, with lower resistance compared to the eastern part, demonstrates the relevance of planning effective conservation strategies in specific areas of the landscape aimed at reducing pressures from productive systems on forest fragments, particularly tropical dry forest, which is one of the most threatened ecosystems and less represented in the country (Pizano and Garcia 2014).

The identification of streams as routes of least resistance, both within and outside the SEPLs, highlights the importance of preserving and restoring these natural corridors to promote altitudinal connectivity between habitat fragments, allowing for altitudinal connectivity, which is essential for preserving biological diversity, facilitating species migration, increasing resilience to climate change, conserving specific ecosystems such as subxerophytic forest, and ensuring the continuity of essential ecosystem services for local communities. Additionally, analysis of the connectance index reveals that the presence of SEPLs significantly contributes to improving landscape permeability, underscoring the need to strengthen and expand such protected areas to mitigate the impacts of fragmentation on biodiversity and ecosystem services. These results emphasize the urgency of implementing specific conservation measures in fragmented landscapes to ensure the ecological integrity and long-term sustainability of the Dagua River basin.

The establishment of governance within these two SEPLs, with the active involvement of stakeholders, was achieved through the application of a participatory action research methodology. In this approach, social actors, who are residents of the protected areas, are regarded as sentient individuals who construct knowledge based on their social and cultural contexts, as well as their personal experiences. This empirical knowledge is then engaged in a dialogue with scientific and technical expertise, encompassing both the biophysical and socioeconomic aspects of the SEPLs, to catalyze real-world transformations. These transformations lead to adjustments in social and productive practices that are in line with the conservation objectives of these landscapes. Consequently, local stakeholders play a direct role in social and environmental assessments, zoning initiatives, and decision-making processes related to environmental governance within the areas. Their involvement is on equal terms with institutional actors involved in area management. Through this collaborative process, an inclusive collective knowledge is forged, integrating local practices with the theoretical and technical underpinnings of area governance. As described by Calderon and López (2014), participatory action research represents an ongoing dialectical process whereby facts are analyzed, problems are conceptualized, and actions are planned and executed to effect transformations within the contexts and among the involved stakeholders.

7.5 Conclusions

The analysis of forest cover and land use in the middle Dagua River basin reveals significant pressure on the remaining natural habitat, emphasizing the importance of conservation strategies like SEPLs. These SEPLs have effectively halted forest cover loss and even fostered its increase, striking a balance between productive activity and conservation goals. Moreover, identifying pathways with minimal resistance, such as riparian forests, underscores the critical need to maintain landscape connectivity for preserving biodiversity and associated ecosystem services.

These findings underscore the pressing need to implement targeted conservation measures in fragmented landscapes to ensure the long-term ecological integrity and sustainability of the Dagua River basin. The establishment of SEPLs and the empowerment of the communities where they are located represent proven conservation strategies that can be effectively utilized to address these challenges.

Studying SEPLS from a spatial perspective, supported by land systems science, is highly relevant as it enables the analysis of land use and cover dynamics. It particularly contributes to advancing SEPLS monitoring and aiding decision-making and policies for their conservation. However, traditional remote sensing-based monitoring has been challenging and costly for conservation efforts. This is due to the reliance on specialized software, trained personnel, and the expense of acquiring images. These factors have restricted the use of remote sensing in conservation decision-making, particularly in developing nations. Nevertheless, in recent years, remote sensing data has become more abundant and accessible, thanks to initiatives like the Global Forest Watch platform. This increased accessibility holds promise for supporting decision-making processes.

The governance of both IMRDSEA and IMRDEC, led by the CVC as the departmental environmental authority, is crucial for achieving conservation goals and sustainability. However, the active involvement of community-based actors and organizations is essential for its effectiveness. Dialogue between environmental experts and local leaders ensures that management plans align with community needs and realities, leading to greater community ownership, establishment of effective use regulations, and, ultimately, conservation of biodiversity and ecological connectivity.

Acknowledgments We extend our sincere gratitude to the communities of Atuncela and El Chilcal for their warm hospitality and cooperation throughout this study. Special acknowledgment goes to the Co-management Committees of IMRDSEA and IMRDEC for their invaluable assistance in facilitating the logistics for fieldwork. This research was conducted as part of agreement No. 187 and No. 111 of 2021, which aimed to collaboratively update the management plans of the Subxerophytic Enclave of Atuncela and El Chilcal, respectively, located in the municipality of Dagua, Valle del Cauca. These agreements were co-funded by the CVC and CORFOPAL.

References

Aldunce P, León A, Carvajal Y, Neri C, Quintero M, Soza S (2008) Informe final « Sistematización de las políticas y estrategias de adaptación nacional e internacional al cambio climático del sector silvoagropecuario y de los recursos hídricos y edáficos » Consultoría encargada por la Oficina de Estudios y Políticas Agricolas

Arroyo-Rodríguez V, Mandujano S (2006) The importance of tropical rain Forest fragments to the conservation of plant species diversity in Los Tuxtlas, Mexico. Biodivers Conserv 15:4159–4179. https://doi.org/10.1007/s10531-005-3374-8

Borrini-Feyerabend G, Dudley N, Jaeger T, Lassen B, Pathak Broome N, Phillips A, Sandwith T (2014) Gobernanza de áreas protegidas: de la comprensión a la acción, No. 20 de la Serie Directrices para buenas prácticas en áreas protegidas. UICN. xvi + 123, Gland

Calderon J, López D (2014) Orlando Fals Borda y la investigación acción participativa: aportes en el proceso de formación para la transformación. In: I Encuentro hacia una Pedagogía Emancipatoria en Nuestra América. Centro Cultural de la Cooperación Floreal Gorini, Buenos Aires

CORFOPAL & CVC (2022a) Informe técnico final del Convenio No. 187 de 2021: Aunar esfuerzos técnicos y recursos económicos y humanos con el fin de actualizar participativamente el plan de manejo del distrito regional de manejo integrado del Enclave Subxerofitico de Atuncela, localizado en el municipio de Dagua, departamento del valle del cauca, Santiago de Cali

CORFOPAL & CVC (2022b) Informe técnico final del Convenio No. 111 de 2021: Aunar esfuerzos técnicos y recursos económicos y humanos con el fin de actualizar participativamente el plan de manejo del distrito regional de manejo integrado del El Chical, localizado en el municipio de Dagua, departamento del valle del cauca, Santiago de Cali

Critical Ecosystem Partnership Fund (2021) Perfil de Ecosistema: Hotspot de Biodiversidad de los Andes Tropicales Actualización 2021, Arlington

Crooks KR, Sanjayan M (eds) (2006) Connectivity conservation, vol 14. Cambridge University Press

Cuenca P, Arriagada R, Echeverría C (2016) How much deforestation do protected areas avoid in tropical Andean landscapes? Environ Sci Pol 56:56–66

CVC (2021) Guía detallada para la actualización y ajuste de planes de manejo de las áreas protegidas públicas del SIDAP Valle administradas por la CVC

CVC & FUNAGUA (2010) Aunar esfuerzos técnicos y económicos para realizar el análisis preliminar de la representatividad ecosistémica, a través de la recopilación, clasificación y ajuste de información primaria y secundaria con rectificaciones de campo del mapa de ecosistemas, Santiago de Cali

De Lima JA, Vieira R, Souza C, Ferreira F, De Oliveira V (2021) Effects of habitat fragmentation on biodiversity patterns of ecosystems with resource competition. Physica A 564:125497. https://doi.org/10.1016/j.physa.2020.125497

Etter A, van Wyngaarden W (2000) Patterns of landscape transformation in Colombia, with emphasis in the Andean region. Ambio: a journal of the human. Environment 29(7):432–439

Etter A, McAlpine C, Possingham H (2008) Historical patterns and drivers of landscape change in Colombia since 1500: a regionalized spatial approach. Ann Assoc Am Geogr 98:2–23. https://doi.org/10.1080/00045600701733911

GAIA & CVC (2014) Declaratoria del DRMI EL Chilcal y plan de manejo

Geilfus F (2002) 80 Herramientas Para El Desarrollo Participativo Diagnóstico, Planificación Monitoreo y Evaluación. Instituto Interamericano de Cooperación para la Agricultura (IICA), San José

Howell PE, Muths E, Hossack BR, Sigafus BH, Chandler RB (2018) Increasing connectivity between metapopulation ecology and landscape ecology. Ecology 99(5):1119–1128. https://doi.org/10.1002/ecy.2189

Karra K, Kontgis C, Statman-Weil Z, Mazzariello JC, Mathis M, Brumby SP (2021) Global land use/land cover with Sentinel 2 and deep learning. In: 2021 IEEE international geoscience and remote sensing symposium IGARSS. IEEE, pp 4704–4707

Kawulich B (2005) La observación participante como método de recolección de datos. Forum Qual Soc Res 6(2):43

Lion M, Garda A, Santana D, Fonseca C (2016) The conservation value of small fragments for Atlantic Forest reptiles. Biotropica 48:265–275. https://doi.org/10.1111/btp.12277

Luther DA, Cooper WJ, Wolfe JD, Bierregaard RO, Gonzalez A, Lovejoy TE (2020) Tropical forest fragmentation and isolation: is community decay a random process? Glob Ecol Conserv 23:e01168. https://doi.org/10.1016/j.gecco.2020.e01168

McGarigal K, Cushman SA, Ene E (2023) FRAGSTATS v4: spatial pattern analysis program for categorical maps. Computer software program produced by the authors. Available at https://www.fragstats.org

McRae BH, Kavanagh DM (2011) Linkage mapper connectivity analysis software. The Nature Conservancy, Seattle

Mitchell M, Bennett E, Gonzalez A (2014) Forest fragments modulate the provision of multiple ecosystem services. J Appl Ecol 51:909–918. https://doi.org/10.1111/1365-2664.12241

Municipality of Dagua (2002) Plan Básico de Ordenamiento Territorial 2001–2010 Acuerdo No 004_mayo 28 de 2002

Olson DM, Dinerstein E (1997) Global 2000: conserving the world's distinctive ecoregions. WWF-US

Orjuela-Salazar S, Gaitán MC, Quintero-Angel A (2018) Conservation on private lands integrating sustainable production and biodiversity in the Mid Dagua River Basin, Colombia. In: UNU-IAS and IGES (ed) Sustainable use of biodiversity in socio-ecological production landscapes and seascapes (SEPLS) and its contribution to effective area-based conservation. Satoyama initiative thematic review, vol 4. United Nations University Institute for the Advanced Study of Sustainability, Tokyo, pp 75–84

Orjuela-Salazar S, Quintero-Ángel A, Otero-Castellanos F, Larraniaga VE (2019) Estrategia de conservación comunitaria en la cuenca media alta del río Dagua. In: Mantilla C, Aroiza A, Silva A, Galán S, Solano C, Rueda AM (eds) Voces de la gestión territorial: Estrategias comple-

mentarias para la conservación de la biodiversidad en Colombia. Instituto de Investigación de Recursos Biológicos Alexander von Humboldt (IAvH), Fundación Natura, Bogotá, pp 172–179
Pizano C, Garcia H (eds) (2014) El bosque seco tropical en Colombia. Instituto de Investigación de Recursos Biológicos Alexander Von Humboldt (IAvH), Bogotá
PRATEC (2012) Diálogo de saberes: una aproximación epistemológica. AMC Editores SAC
Quintero-Ángel A, Orjuela-Salazar S, Saavedra-Zúñiga D, Castaño-Astudillo A, Borda Calvache MV, Quintero-Ángel M (2025) Participatory sustainable production of panela (brown sugar cane) and the conservation of the tropical dry and very dry (subxerophytic) forest in Colombia. In: Nishi M, Subramanian SM, Varghese P (eds) Business and biodiversity: Reciprocal connections in the context of socio-ecological production landscapes and seascapes (SEPLS). Springer, Singapore, pp 187–208
Takeuchi K (2010) Rebuilding the relationship between people and nature: the Satoyama initiative. Ecol Res 25:891–897
Taylor PD, Fahrig L, Henein K, Merriam G (1993) Connectivity is a vital element of landscape structure. Oikos 68(3):571–573. http://www.jstor.org/stable/3544927
CORFOPAL. (2023). Fotografía de campo. Archivo institucional de CORFOPAL, Cali, Colombia.

Chapter 8
Connecting Kinabalu and Crocker Range Parks for Nature and Culture

Yoji Natori, Hamelda Francisca Majit, Remmy Alfie Awang, Alfred Marinus, Fazrullah Rizally Abdul Razak, Gerald Jetony, and Alessandra Markos

Abstract There is a 10 km gap between Kinabalu Park and Crocker Range Park in Sabah, Malaysia, which is occupied by local communities and used primarily for agriculture. The way this socio-ecological production landscape is managed influences the quality of ecological connectivity between these parks. The Kinabalu Ecolinc Project, conducted during the 2014–2021 period, attempted to ensure the ecological connectivity between the parks through a four-pronged approach of protecting forests of biodiversity value by establishing community-conserved areas, restoring degraded forests, making agriculture environmentally sustainable and economically profitable, and enhancing forest-based community tourism.

To understand the current state of the project site and identify needs for future interventions, a rapid assessment was conducted in August 2023 in three communities. Community members expressed their strong wish for self-governance of land and noted that interventions that touch on land titles at various levels need to be addressed carefully. There were both active and rather reserved attitudes toward the promotion of ecotourism. Documentation of traditional knowledge, including ethnobotany, is ongoing or needs to be started. Additional stimuli seemed necessary to maintain the interventions introduced in agriculture.

For spatial planning to meaningfully operationalize ecological connectivity in the case study area, proper management of the socio-ecological production landscape is key. The local communities need to be empowered as the guardians of the landscape and participate in a multistakeholder platform that harmonizes the interests and mandates of all those concerned about the area to which they have title. Where such a platform will require a reason and motive to form, we argue for the benefit of developing and implementing a regulatory framework for other effective

Y. Natori (✉)
Akita International University, Akita, Akita, Japan
e-mail: ynatori@aiu.ac.jp

H. F. Majit · R. A. Awang · A. Marinus · F. R. A. Razak
Sabah Parks, Kota Kinabalu, Sabah, Malaysia

G. Jetony · A. Markos
Sabah Biodiversity Centre, Kota Kinabalu, Sabah, Malaysia

M. Nishi et al. (eds.), *Ensuring Ecological Connectivity in Socio-Ecological Production Landscapes and Seascapes (SEPLS)*, Satoyama Initiative Thematic Review, https://doi.org/10.1007/978-981-95-1474-8_8

area-based conservation measures (OECM). We also argue that a simple, integrative "coverage metric" can be used to facilitate interstakeholder dialogue.

Keywords OECM · Post-project sustainability · Ecotourism · Sustainable agriculture · Multistakeholder platform · Coverage metric · Borneo

8.1 Introduction

The area-based conservation target, that is, Aichi Biodiversity Target 11, was one of a few commitments the global community agreed in 2010 to achieve by 2020. Progress on the target has been assessed positively overall (Secretariat of the Convention on Biological Diversity 2020), but important quality elements have remained unfulfilled, including connectivity in protected area systems (Maxwell et al. 2020) and fairness in management (Zafra-Calvo et al. 2019). Under the new global arrangement toward 2030, the Kunming-Montreal Global Biodiversity Framework (or GBF; CBD 2022), the Parties to the Convention on Biological Diversity have committed to: ensuring all areas are covered under participatory, biodiversity-inclusive spatial planning to prevent the loss of areas of high biodiversity value (Target 1); managing areas used for primary industry sustainably (Target 10); and conserving and managing at least 30% of land and sea through effective and appropriate means inside and outside protected areas (Target 3).

Malaysia is part of the Sundaland global biodiversity hotspot (Mittermeier et al. 2004). In Borneo, highly-connected forest is concentrated in the central part of the island and well protected by existing protected areas, but Kinabalu Park is rather isolated (Proctor et al. 2011). In the state of Sabah on the island of Borneo, 26.4% of land has already been gazetted as protected areas (Sabah Forestry Department 2022). To achieve Target 3 of the GBF, spatial planning needs to identify and secure the conservation of additional areas that improve the connectivity of protected areas, most likely by other effective area-based conservation measures (OECMs; CBD 2018). This case study focuses on ongoing efforts to secure ecological connectivity in Sabah, Malaysia, and aims to discuss ways to operationalize ecological connectivity in spatial planning.

8.2 Description of Study Area

The study site is located between Kinabalu Park and Crocker Range Park in Sabah in Malaysian Borneo. The former is a UNESCO World Natural Heritage Site, and the latter is a Biosphere Reserve under the UNESCO Man and the Biosphere Programme, making this area an important bridge between two significant biodiversity sites (Fig. 8.1). The area between these parks still maintains a high forest coverage (88.9% in an area of approximately 30 km × 30 km), and there is only one major

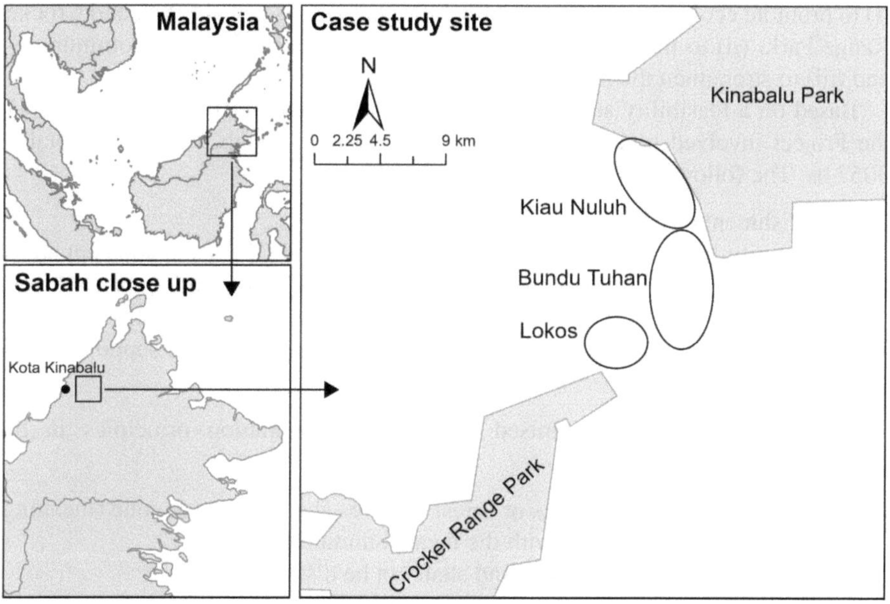

Fig. 8.1 Map of the study site. (Created by the authors in ArcGIS Pro 3.2 with ArcGIS Online resources)

road (Matsuyama et al. 2018). On the other hand, primary forest cover has been declining in the past decades, especially since 2011, within and between the two parks (Miuse and Kamlun 2019). The area is inhabited by Indigenous communities of Dusun people and used for their livelihoods. Mt. Kinabalu is worshiped as sacred. Socioeconomic options include ecotourism and highland agriculture. Shifting cultivation is the form of agriculture on the hilly topography. Community-based ecotourism is another important livelihood means that takes advantage of the area's environmental circumstances. The quality of management of such land dictates the conservation outcomes of these parks. The public visibility of Mt. Kinabalu can demonstrate livelihood opportunities that build on natural assets, rather than those that consume them. By "ecological connectivity" in this case study, we mean the continuity of habitat between the two parks established by community-conserved areas (CCAs) and ecologically sound land uses. We do not assume any specific species dictates the interventions.

8.3 Summary of Kinabalu Ecolinc Project

The Kinabalu Ecolinc Project was conducted with funding from the European Union from January 2014 to November 2021. This section summarizes the activities and achievements of this Project (Sabah Parks 2022). The Project's objectives were:

(i) to promote ecological continuity of habitats between Kinabalu Park and Crocker Range Park; (ii) to maintain ecosystem services of forests for local communities; and (iii) to strengthen the land protection system in Sabah.

Based on a feasibility study conducted between March 2010 and August 2011, the Project involved nine villages in three districts in Sabah, covering a total of 3057 ha. The following four activities were included:

1. Establishment of CCAs
2. Community-based restoration of degraded habitats in CCAs and adjacent areas
3. Development of sustainable agriculture for livelihoods and enhanced land management
4. Enhancing forest-related community tourism options to support forest management

These activities were premised on five implementation principles in the Project area:

(i) There shall be no new parks or forest reserves gazetted in this zone (meaning, the land title shall remain with the local communities).
(ii) Village land and agricultural land shall not be disturbed.
(iii) Local communities must be involved in the implementation and management.
(iv) Land managed for the Project should be secured in the long term.
(v) Any threats to biodiversity must be reduced.

The Project was able to (i) complete 95% of the tasks as outlined in the Project Work Plan (Table 8.1), (ii) gain community trust, (iii) raise awareness and understanding, and (iv) foster interest among locals on conservation aspects and issues. The proponents of the Project emphasized in the final report that self-sustained community development focusing on community empowerment and local knowledge maintenance should be the priority for future planning in the region and that the key driver of community participation in conservation efforts is the sense of belonging, security, and mutual trust.

8.4 Rapid Assessment

To understand the current state and to identify the issues that need to be addressed in future interventions, three sites were selected from the nine communities involved in the Kinabalu Ecolinc Project, namely Kiau Nuluh, Bundu Tuhan, and Lokos (Fig. 8.1 and Table 8.2). These villages were selected to represent the three districts under the Project and to cover different socioeconomic activities. All of them had their own development plans and village rules. There is a self-managed community forest reserve with potential for forest-based ecotourism activities in each site (Table 8.3). A rapid assessment, conducted from 22 to 24 August 2023, consisted of field observations and interviews with community representatives. The second and third authors were former Ecolinc Project members, and they organized site

Table 8.1 Summary of the targets and outputs of activities under the Kinabalu Ecolinc Project Work Plan

Activity	Target	Actual outputs
1. CCA establishment	≥500 ha of CCA established out of an overall 3057.2 ha	1411 ha total by *Bombon* native and customary laws (Wassai [63.5 ha], Tiong Simpodon [14 ha], Kiau Nuluh-Bersatu [485.6 ha], Bundu Tuhan [848 ha]). Additional 23.7 ha in the process of formal gazettement.
2. Community-based ecosystem restoration	≥140 ha	26.5 ha (2064 seedlings) in Bundu Tuhan, Wassai, and Toboh Lama (additional 120 ha in Tenompok Forest Reserve). Planted mainly fruit trees.
3. Sustainable farming	60 households in 3 villages benefitting	70 households participated from Bundu Tuhan, Lokos, and Kotunuan. General farming +12% revenue in the first year. Pineapple farming production doubled (Lokos and Kotunuan). Aquaculture doubled to tripled revenue (9 farmers).
4. Forest/ community tourism	≥25% increase in visitation	Kiau Nuluh cultural experience, Tiong rafflesia trail, and Wassai waterfall trail opened for visitors in April 2021. 60 local visitors as of the end of the Project (August 2021).

Created by the authors based on Sabah Parks (2022)

Table 8.2 Basic information of the study area

Country	Malaysia		
Province	Sabah		
District	Kota Belud	Ranau	Tuaran
Municipality	Kiau Nuluh	Bundu Tuhan	Lokos
Size of geographical area (ha)	40	1260	40
Size of case study/project area (ha)	40	1260	40
Dominant ethnicity in the project area	Dusun		
Number of direct beneficiaries (people)	There are no clear distinctions between direct and indirect beneficiaries in the context of this case study.		
Number of indirect beneficiaries (people) *The populations of the villages	1200 persons/145 households	3300 persons/330 households	600 persons/60 households
Geographic coordinates (latitude, longitude)	6°2′39″N, 116°29′48″E	5°59′5″N, 116°31′54″E	5°58′8″N, 116°27′43″E

Source: Authors

observations and interviews with community representatives for the first author, based on their experiences and connections under the Project, as well as provided background and supplementary information during the assessment.

The interviews (see Table 8.4 for summary) revealed, consistently across the three villages, the strong will of the communities for self-governance over their

Table 8.3 Communities covered by the rapid assessment and their community-conserved areas (CCAs)

Name of the village	Origin of the village's name	Main economic activity	Background of CCA/ Community forest	Status
Kiau Nuluh, Kota Belud	In reference to the name of the earliest settlement, Kiau-Kiau, named after the song of the *kologiau* birds that was heard there in the past.	Farming (pineapple, rice, rubber), government/ private/ religious teachers/porter work	Kiau Nuluh village and Kiau Bersatu village jointly develop and manage the Community Forest Reserve Kiau Nuluh-Bersatu under the community association GOMPITO. The Forest Reserve area was gazetted in 2021 under the Sabah land ordinance. (485.6 ha)	Conservation efforts are contributing to the water supply to the village. Ecotourism activities, such as trails in the CCA, are also being developed. The community uses and manages them.
Bundu Tuhan, Ranau	In reference to the incident and the location of the *belunu/bundu* tree that fell/collapsed.	Farming (pineapple, rice, rubber), government/ private/ religious teachers/porter work	The Community Forest Reserve/ CCA is known as the Winokok Forest Reserve. The management and care of this forest reserve is in reference to rules established in 2003. This forest was included in the village gazette in 1983, Native Residential Reserve Kg. Bundu Tuhan. (848 ha)	Conservation efforts are securing the water supply to the village. Ecotourism activities such as the Hatob Trail within the CCA can be carried out.
Lokos, Tuaran	In reference to the environment of the village area which has many small streams also known as *lokos/palu*.	Farming	Areas for CCA have been allocated by the community (area zoning) with the consent of all community members. (23.7 ha)	The community has proposed a special area for CCA but it has not yet been gazetted under the land ordinance. Requires the help of intermediaries to facilitate the process.

Source: Authors

lands. Communities were concerned that any new initiatives could become a reason for the government to disturb or take the communities' rights to their lands. These concerns were at two different levels. In one case, it was at the whole CCA level, where the community was concerned that the government might raise the issue of applying laws other than the land ordinance that they currently use to legitimize their CCA in order to assume or increase its control while limiting the community's sovereignty over the land. This is the reason why communities did not wish to consider new protected areas but were open to exploring OECM opportunities. In another case, concern was at the individual level. The absence of formal titles to individual land lots raised a sense of insecurity among individuals who may lose their right to the land at any time and for any reason.

Table 8.4 Results of interviews with community representatives in three villages

Kiau Nuluh, Kota Belud
CCA under the Land Ordinance Article 78[a]: 485.6 ha. Community interested in expanding the CCA. Remaining work from Ecolinc (bridge, visitor resthouse, signboards, species inventory). Waste management (household waste) is an issue. Ecotourism active, but limited with funding. Original trail of Mt. Kinabalu hiking. Shifting cultivation on the decline because young people leave the community; as ecotourism expands, shifting cultivation will decline further. There is a traditional knowledge documentation project by a community organization GOMPITO. In native reserves, land-use decisions go to GOMPITO (the village head is a member of GOMPITO—only village heads have the power to impose fines under written laws).
Bundu Tuhan, Ranau
CCA for water catchment (customary activities and limited ecotourism only): 848.0 ha. Three conditions for participation in future projects: (1) Make recommendations to state government on CCAs. (2) Make recommendations to the government to withdraw the claim over the community land. (3) Respect community protocol. Would welcome experimenting with new ways of agriculture. Careful about opening up to tourism; customary integrity may be eroded with outsiders. They believe in spiritual beings. Sustainable agriculture interventions not continued after Ecolinc Project.
Lokos, Tuaran
Water catchment area protected with community protocol: 23.7 ha (Community used its own protocol to designate CCA, but not yet formally gazetted by the government authority). Issue with sustainable waterflow—rivers dry up during droughts. "*Bombon*" (the native and customary resource management system) in place but not respected by outsiders. Bundu Tuhan protocol is the model for the protocol here. Want to expand ecotourism with trails, towers, and homestay program. Historical sites, sunset view, and orchids may attract tourists. "Pineapple processing center" was built but not used. Ethnobotany is present; needs to be documented. Shifting cultivation still ongoing, but on the declining trend. Youth still coming back after school. Fallow areas used for beekeeping, fruit production. Land-taking by the government is a concern, especially for those individuals who have not yet received land titles.

Source: Authors

[a]The State of Sabah Land Ordinance Article 78 provides that the Yang di-Pertua Negeri (or the head of Sabah state) can declare any area of State land as a native reserve to "protect the present and future interests and well-being of the natives of Sabah or any community thereof"

All communities knew about their natural, cultural, and historical assets. Many community members had learned the perspectives and preferences of those who come to climb Mt. Kinabalu through their experiences of working as guides or porters, so they have well-informed knowledge. However, they have differing views when it comes to how to use these assets. The original trail to the summit of Mt.

Fig. 8.2 A trail leading into the CCA from the community of Kiau Nuluh. (Photo by Yoji Natori, August 2023)

Kinabalu used to start from Kiau Nuluh, and the community members have cherished this heritage. Thus, they were positive about ecotourism activities and had created trails (Fig. 8.2) and a campsite in the CCA. The community members in Lokos, too, had aspirations for promoting ecotourism with facilities development and had a sound understanding of what may attract tourists, such as scenery and orchids, among others. The community members of Bundu Tuhan were more reserved about opening the community up to tourism because of the concern that their customary integrity might be eroded through exposure to outsiders. Belief in spiritual beings in nature existed in these communities. The forest being spiritual and sacred was explicitly mentioned during the interviews in Bundu Tuhan, and it is also written in the Community Protocol of Kiau Nuluh (Kiau Community and PACOS Trust 2018).

Regarding the livelihood activities supported by the Kinabalu Ecolinc Project, the aquaculture improvement (Fig. 8.3), which involved upgrading facilities on farmers' properties in a way they chose, seemed to be functioning. The income records during the Project showed a robust improvement, and the site observation

Fig. 8.3 An aquaculture facility in a farmer's backyard in Lokos, raising tilapia and catfish for sale. (Photo by Yoji Natori, August 2023)

this time confirmed that operations continued to be conducted in good order. The sustainable agriculture interventions, which have influence over a larger spatial extent, require further investigation on the economic and environmental aspects. The farm operation for pineapple production, as observed in Lokos, was in good condition. However, the processing facility constructed by the Project was not used, which suggested that the value-adding process was not happening in the village. The monoculture and use of the flower inducer in pineapple production need to be investigated carefully for their environmental impact. The greenhouse structure (for chemical-free kale production) in Bundu Tuhan was not in shape to fulfill its intended function (Fig. 8.4), though the practice of chemical-free farming might have continued. The different conditions of the market might have caused the difference between aquaculture and vegetable farming. Any new agricultural extension interventions of community interest need to install a sound marketing mechanism together with agricultural skills.

The communities still maintained traditional knowledge (e.g., ethnobotany) and practices (e.g., shifting cultivation; see Fig. 8.5), and some of them are documented in the community protocol (e.g., Kiau Community and PACOS Trust 2018) so that they are sure to be continued. The three communities had different levels (or perceptions) of outflow of youth. Efforts to retain traditional knowledge and practices differed among the communities accordingly.

The rapid assessment revealed that the creation of CCAs remained as a significant step forward in improving habitat connectivity between the two parks. However, elements of the Kinabalu Ecolinc Project interventions that required market integration, such as the sustainable agriculture and value-adding processing of agricultural produce, did not persist. The discontinuity in activities that comes at the end of a

Fig. 8.4 A farm in Bundu Tuhan. (Photo by Yoji Natori, August 2023). Small-scale farms of vegetables are typical of the region. White nets in the middle is the greenhouse structure established during the Kinabalu Ecolinc Project. It is partially operational but without the roof which provided shade and protection from pests

Fig. 8.5 Area under traditional shifting cultivation in Lokos. (Photo by Yoji Natori, August 2023). Patches at different stages of ecological succession under shifting cultivation cover areas near the village. Some of these patches are now planted with pineapples. Farmers practice a beautiful tradition of helping each other when opening a new lot

project presented itself as a major challenge, which reinforced the importance of post-project monitoring.

When the management of the socio-ecological production landscape (SEPL) shifts to a large scale, intensive production of cash crops, such as pineapple, and ecological connectivity in the area can be affected. Although the steep topography of this area would make such a conversion leading to a trade-off between livelihood improvements and ecological connectivity unlikely, spatial planning on what and where farming is permitted could help ensure that ecological connectivity remains. Key areas for ecological connectivity have been secured as CCAs, but similar work still needs to continue, as in the case of facilitating the gazetting of CCAs in Lokos, to further ensure long-term conservation security. What will be more challenging is putting proper production and resource-use practices in place, and anchoring them with tangible incentives, such as better access to the market, competitive advantages, and increased revenues.

From these observations from the rapid assessment, it seems worthwhile to pursue value addition that links and achieves sustainable farming (i.e., organic agriculture), ecotourism, and alternative livelihood simultaneously within the locality. What we mean by this is that local tourism operations could be used as a platform for integration of local sustainable farming, traditional knowledge and practices, and conservation activities. That could bring the market agents who are able to come (i.e., tourists) to the community, rather than the community going out to the urban market, which the remoteness could impede.

8.5 Operationalizing Ecological Connectivity in Spatial Planning

Now that most areas of high biodiversity and low human impact (including those that have always been so and those that were made so) are covered by protected areas, additional area-based conservation has to be conducted with people as an integral part of interventions. For spatial planning to be effective as a catalyst of sustainable development, it needs to harmonize the needs of diverse stakeholders appreciating synergies in the pursuit of human well-being and biodiversity conservation.

The Kinabalu Ecolinc Project area can be considered an ecological corridor connecting two globally significant protected areas (Kinabalu Park as a UNESCO World Natural Heritage Site, and Crocker Range Park as a UNESCO Biosphere Reserve) and is, therefore, a potential OECM (an area to be recognized under Target 3 of the GBF) *if* it is effectively managed for the long term. The spatial planning that is called for by Target 1 of the GBF needs to recognize such an area. For spatial planning to meaningfully operationalize ecological connectivity, however, the socio-ecological production landscape that exists there needs to be managed properly. What this means on the ground is the empowerment of local communities as

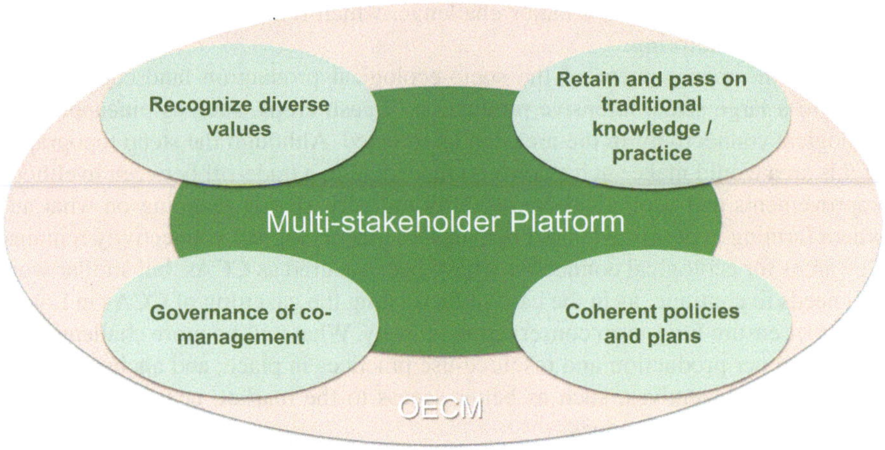

Fig. 8.6 A multistakeholder platform legitimized under OECM governance as a means to achieve elements of sustainable land use. (Source: Yoji Natori, 2023; presentation at 2nd Asia Parks Congress, Kota Kinabalu, Sabah, Malaysia)

guardians of the SEPL. As found through the rapid assessment, communities still need support in various forms. This points to the importance of multistakeholder platforms that harmonize the interests and mandates of all those concerned. Where such a platform would require a reason and motive to form, the implementation scheme of the OECM might be considered (Fig. 8.6). By the very nature of SEPLs, various stakeholders may bring diversity of interests and mandates to the landscape. Understanding and coordinating them is a critical step toward effective management of both business and conservation undertakings and projects.

8.6 Multistakeholder Strategy

Communities are tied to their landscapes, where action takers, such as producers and consumers, interact. The appropriate level of use of landscapes (the safe ecological limits or the carrying capacities)—or in the context of spatial planning, information on what activities are allowed/needed and where—needs to be determined by the research community and effectively communicated to the action takers. From another angle of sustainable landscape management, regulations by government agencies and voluntary standards by the private sector (e.g., tour operators, farmers' cooperatives, and market intermediaries) need to respect those carrying capacities, as well as ensure their application on the ground. Public opinion and consumer awareness are necessary to support the needed regulatory and voluntary standards, which may require formal and informal education. Communities should not just receive guidance and pressures from outside. They should have visions for themselves and communicate them externally to frame their development. Each of

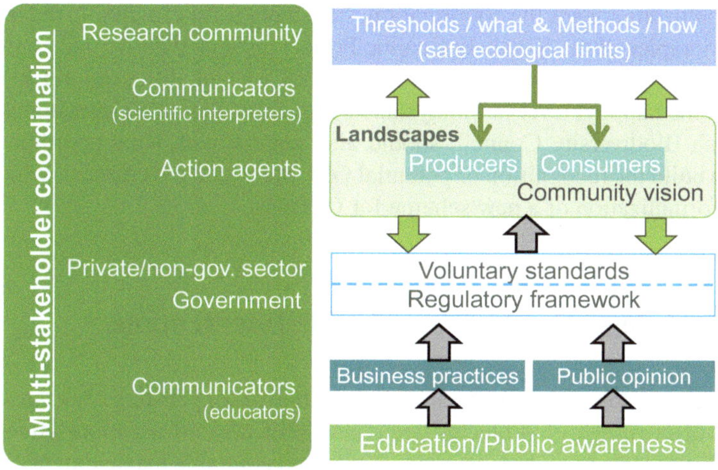

Fig. 8.7 Details of multistakeholder engagement for sustainability in socio-ecological production landscapes. (Source: Created by the authors based on a working group discussion facilitated by the first author at IPSI-3, Hyderabad, India, 2012)

these functions needs to be performed by able stakeholders. For them to operate effectively ("governance of co-management" in Fig. 8.6), they need to be coordinated ("coherent policies and plans" in Fig. 8.6) in a multistakeholder platform (Fig. 8.7). To understand the differences and similarities in perceptions and interests held by different stakeholders, assessment workshops using the Indicators of Resilience in SEPLS (UNU-IAS et al. 2014) can promote effective dialogue as reported from cases in many parts of the world (Dublin and Natori 2020; Dunbar et al. 2020).

Going forward, how might this conceptual framework be operationalized on the ground? In the case of the three villages of this study, the safe ecological limits are somewhat known as being in the corridor area of two parks. However, more clearly setting and communicating these limits with information on how to uphold them would be helpful. On the government side, the regulatory framework exists as the Sabah Biodiversity Strategy 2012–2022 (Sabah Biodiversity Centre 2012), strengthened by the GBF. Private sector activities are not yet as active as other nearby areas. The environmental awareness level is still high, thanks to the work of the Kinabalu Ecolinc Project, but additional interventions will probably be needed to maintain this level as the Project ended in 2021. Many community members work as mountain guides and porters to domestic and international visitors to Mt. Kinabalu, bringing back information that they gain from them. Such opportunities and information can serve as strategic resources for awareness and motivation for conservation. With regards to the key institution of multistakeholder coordination, community organizations like the Association for the Conservation of Traditional and Natural Heritage (GOMPITO) in Kiau Nuluh can fulfill the role at the local level. The key institution needs to convene all relevant stakeholders for multistakeholder coordination. As a local organization leads the coordination, such a platform can uphold the implementation principles of the Kinabalu Ecolinc Project presented above and communicate the community vision. At this stage where OECM modalities are yet to be determined at the national level in Malaysia, we argue that the state government, through the Sabah Biodiversity Council, should also engage with local coordination platforms to build the recognition of potential OECMs by the government, which should lead to formalization of a new scheme for OECMs.

8.7 An Integrative Metric to Guide Interventions

When it comes to conservation targets, quantitative aspects (e.g., hectares and percentage of coverage of protected areas and OECMs) attract more attention than qualitative aspects (e.g., habitat connectivity and equitable management). However, it is the delivery on qualitative aspects that matters in biodiversity conservation on the ground. Assessing a comprehensive target by breaking it into its components, as done in the Global Biodiversity Outlook 5 (Secretariat of the Convention on Biological Diversity 2020), may misinform and misguide actions as it allows for choosing "convenient" metrics. This is especially of concern where diverse stakeholders need to be involved. Simpler metrics that capture both quantitative and qualitative aspects need to be adopted. For spatial planning and area-based conservation, we propose using the coverage of areas of biodiversity importance that are effectively managed (in their totality, without fragmenting into components). First, spatial planning identifies the areas to conserve (including key areas for ecological connectivity). This will establish spatially explicit, quantitative targets, as well as who has jurisdiction and interest in the management of the areas (i.e., stakeholders), what needs to be done (i.e., activity planning), and what resources are available and needed (i.e., financial and capacity planning). Second, responsible site managers (either from the government or from the community, or both) assess the status of progress against the criteria that need to be met for effective management, which are determined site-specifically as part of their regular work. Only those areas identified in the first step that meet the minimum standard for effective management, as determined in the second step, will be counted in the metric. This way, quality assurance and necessary monitoring of area-based conservation can be instituted. This simple, single metric can inform decision-makers on the quantitative and qualitative status together, and day-to-day management activities on site can also be kept in check. The "coverage metric" can contribute to operationalizing ecological connectivity in spatial planning, as well as improving fair and effective area-based conservation.

8.8 Conclusion

The Kinabalu Ecolink Project left a legacy of creating CCAs and starting improved livelihood means, which contributes to enhancing ecological connectivity in the area between Kinabalu Park and Crocker Range Park. Ecological interventions rooted in local communities' culture, such as community protocol, are necessary for the effective management of areas outside the parks. The positive outcome of the Project needs to be monitored for continuity of the activities initiated, and shortcomings need to be addressed to ensure synergies and minimize trade-offs between environmental and livelihood interests. The implementation of OECM in the project area may bring relevant stakeholders together to operationalize proper management of the socio-ecological production landscape leading to creating and enhancing the ecological connectivity of the area. A metric (i.e., the "coverage metric") that combines quantitative and qualitative aspects of area-based conservation can be used to monitor progress toward this end.

References

Convention on Biological Diversity (CBD) (2018) Decision 14/8: protected areas and other effective area-based conservation measures. Secretariat of the Convention on Biological Diversity. Retrieved from https://www.cbd.int/conferences/2018/cop-14/documents

Convention on Biological Diversity (CBD) (2022) Decision 15/4. Kunming-Montreal Global Biodiversity Framework. Secretariat of the Convention on Biological Diversity. Retrieved from https://www.cbd.int/doc/decisions/cop-15/cop-15-dec-04-en.pdf

Dublin DR, Natori Y (2020) Community-based project assessment using the indicators of resilience in SEPLS: lessons from the GEF-Satoyama Project. Curr Res Environ Sustain 2. https://doi.org/10.1016/j.crsust.2020.100016

Dunbar W, Subramanian SM, Matsumoto I, Natori Y, Dublin D, Bergamini N, Mijatovic D, Alvarez AG, Yiu E, Ichikawa K, Morimoto Y, Halewood M, Maundu P, Salvemini D, Tschenscher T, Mock G (2020) Lessons learned from application of the "Indicators of Resilience in Socio-ecological Production Landscapes and Seascapes (SEPLS)" under the Satoyama Initiative. In: Saito O, Subramanian SM, Hashimoto S, Takeuchi K (eds) Managing socio-ecological production landscapes and seascapes for sustainable communities in Asia. Science for sustainable societies. Springer, Singapore, pp 93–116. https://doi.org/10.1007/978-981-15-1133-2_6

Kiau Community & PACOS Trust (2018) Protokol Komuniti Dusun Kiau, Kota Belud: Amalan Adat dan Protokol Hutan Simpan (Community Protocol, Kiau Community, Kota Belud: customary practices and forest reserve protocols). PACOS Trust, Sabah, Malaysia

Matsuyama S, Hashimoto K, Nishimura K, Mita G, Kaneko M, Vairappan CS (2018) Reports on forest distribution and current status outside Kinabalu Park forest reserves. J Rakuno Gakuen Univ 42(2):141–146

Maxwell SL, Cazalis V, Dudley N, Hoffmann M, Rodrigues ASL, Stolton S, Visconti P, Woodley S, Kingston N, Lewis E, Maron M, Strassburg BBN, Wenger A, Jonas HD, Venter O, Watson JEM (2020) Area-based conservation in the twenty-first century. Nature 586(7828):217–227. https://doi.org/10.1038/s41586-020-2773-z

Mittermeier RA, Gil PR, Hoffmann M, Pilgrim J, Brooks T, Mittermeier CG, Lamoreux J, da Fonseca GAB (2004) Hotspots revisited: earth's biologically richest and most endangered ecoregions. CEMEX, Mexico City

Miuse CF, Kamlun KU (2019) Multi-temporal land-cover classification of Kinabalu Eco-Linc Site and the Protected Park areas. J Phys Conf Ser 1358:012034. https://doi.org/10.1088/1742-6596/1358/1/012034
Proctor S, McClean CJ, Hill JK (2011) Protected areas of Borneo fail to protect forest landscapes with high habitat connectivity. Biodivers Conserv 20(12):2693–2704. https://doi.org/10.1007/s10531-011-0099-8
Sabah Biodiversity Centre (2012) The Sabah Biodiversity Strategy 2012–2022 - Shared Heritage, Shared Responsibility, Bornean Biodiversity and Ecosystems Conservation Programme (BBEC) Phase II
Sabah Forestry Department (2022) Sabah Forestry Department annual report 2022. Sabah Forestry Department, Sandakan
Sabah Parks (2022) Kinabalu Ecolinc Project 2014–2021: achievement and record of activities. Sabah Parks, Kota Kinabalu
Secretariat of the Convention on Biological Diversity (2020) Global biodiversity outlook 5. Secretariat of the Convention on Biological Diversity, Montreal. Retrieved from https://www.cbd.int/gbo/gbo5/publication/gbo-5-en.pdf
UNU-IAS, Bioversity International, IGES, UNDP (2014) Toolkit for the indicators of resilience in socio-ecological production landscapes and seascapes (SEPLS). https://satoyama-initiative.org/featured_activities/indicators-of-resilience/
Zafra-Calvo N, Garmendia E, Pascual U, Palomo I, Gross-Camp N, Brockington D, Cortes-Vazquez JA, Coolsaet B, Burgess ND (2019) Progress toward equitably managed protected areas in Aichi Target 11: a global survey. Bioscience 69(3):191–197. https://doi.org/10.1093/biosci/biy143

Chapter 9
Enhancement of Ecological Connectivity Through UNESCO's Man and Biosphere Reserve: Case of Kinabatangan Landscape, Sabah, Malaysia

Gerald Jetony, Isabelle Lackman, Alessandra Markos, and Vivi Sandra Victor

Abstract The Kinabatangan Landscape is located in the eastern part of Sabah, Malaysia, within the greater Kinabatangan basin. It is situated on both sides of the Kinabatangan River, comprising fragmented protected areas (PAs), vast plantation areas, and villages. The area is renowned for its rich biodiversity and unique ecosystem, making it a hotspot for ecotourism. Fragmentation of PAs has occurred due to land clearance for agricultural activities and other land uses starting from the 1980s and 90s. The existing PAs are too small to accommodate sustainable wildlife populations. Hence, the Kinabatangan Landscape needs a platform for stakeholders to work together to achieve common goals, that is, sustainable development. The state government of Sabah approved the proposed nomination of the Kinabatangan Landscape as a UNESCO Man and Biosphere (MAB) Reserve in March 2022. The main purpose of this nomination was to provide a platform for a shared vision for the stakeholders to coexist by encouraging them to synergize their operations toward the UNESCO MAB concept and objectives of sustainable development. The aim was to develop a shared vision as a win–win situation. In preparing the dossier for the nomination, a series of consultations, workshops, meetings, and seminars were held to make the process a win–win both for the people and the nature. Under the UNESCO MAB Programme, involvement in the initiative must be on a voluntary basis. Stakeholders have various objectives for participating in this initiative, which, as much as possible, need to be aligned with government agencies' objectives for integrated management and sustainable development in the Kinabatangan landscape. Therefore, engagement strategies must be tailored to the needs and motivations of the respective stakeholders while also being opportunistic enough to offer benefits from the trade-offs of these conservation opportunities. After more than 2 years of consultation, the perimeter of the biosphere reserve, together with zona-

G. Jetony (✉) · A. Markos · V. S. Victor
Sabah Biodiversity Centre, Chief Minister's Department, Kota Kinabalu, Sabah, Malaysia

I. Lackman
Hutan, Unit 510-512, Kota Kinabalu, Sabah, Malaysia

M. Nishi et al. (eds.), *Ensuring Ecological Connectivity in Socio-Ecological Production Landscapes and Seascapes (SEPLS)*, Satoyama Initiative Thematic Review, https://doi.org/10.1007/978-981-95-1474-8_9

tion, was agreed upon by stakeholders. In addition to providing a corridor for wildlife, the nomination of Kinabatangan landscape as a MAB reserve will also have the potential to promote sustainable development by synergizing socioeconomic needs with environmental considerations and promotion of sustainable development.

Keywords Kinabatangan landscape · Sustainable development · MAB reserve · Biodiversity · Tourism

9.1 Introduction

Socio-ecological production landscapes and seascapes (SEPLS) can be located within a larger landscape, and some SEPLS are critically important for the overall sustainability of fauna and flora in the broader landscape. One such area is the Kinabatangan floodplain in Sabah, Malaysia. The Kinabatangan floodplain is home to a number of iconic species in Sabah, such as the orangutan (*Pongo pygmaeus*), North Bornean gibbon (*Hylobates funereus*), proboscis monkey, Bornean banteng (*Bos javanicus lowi*), hornbill, pygmy elephant, numerous birds, and rare flora species of renowned conservation value. This biodiversity has become an attraction for tourists, making the Kinabatangan landscape one of the hotspots for ecotourism in Sabah. The landscape is also fertile because the Kinabatangan River periodically overflows, bringing nutrients to the surrounding area. As a result, most of the areas along the Kinabatangan River have now been developed for agriculture, particularly oil palm. The combination of biodiversity richness and socioeconomic activities has made the Kinabatangan landscape a very good research area in the field of conservation.

Before the 1960s, the Kinabatangan area was covered by thick forest inhabited by wildlife, with a few Orang Sungai settlements scattered along the Kinabatangan River. Orang Sungai, or "people of the river", are one of Sabah's native or Indigenous peoples who predominantly live along rivers on the east coast of Sabah. They depend on the river and the surrounding areas' resources, including wildlife, for their livelihoods. They have a rich cultural heritage, particularly in traditional ecological knowledge, which has developed over a long period in their quest for livelihoods in harmony with river ecosystems. They are involved in various socioeconomic activities, such as agriculture, fishing, swiftlet nest farming, aquaculture, and ecotourism (Ancrenaz et al. 2007).

The timber boom in the 1970s, reaching its peak in the 1990s, altered the Kinabatangan landscape forever (Sabah Biodiversity Centre 2023). Furthermore, the oil palm revolution in Sabah, beginning in the early 1990s, transformed the logged areas into agricultural land. Large swathes of natural forest are now limited to a few fragmented protected areas (PAs) scattered along the Kinabatangan River, which have become refuge areas for the wildlife that originally lived outside them. However, these PAs are too small and fragmented to support all the wildlife within their areas. The Lower Kinabatangan Wildlife Sanctuary (26,103 ha), the largest among the PAs, is also being encroached upon from time to time by the local communities. Seven more small PAs along the river include the Gomantong, Materis, Bod Tai, Keruak, and Pangi Forest Reserve (1816 ha), the Sg Lokan Forest Reserve (1852 ha), the Pin Supu Forest Reserve (4523 ha), the Gomantong Forest Reserve (3297 ha), the Gologob Forest Reserve (7900 ha), the Balat Damit Wildlife Reserve (4140 ha) and Balat Damit (Extension) Wildlife Reserve (670 ha), and the Trusan Kinabatangan (Extension) (649 ha).

In terms of forest-type distribution within these PAs, there are freshwater swamp forest (19.70%), peat swamp forest (9.24%), beach forest (0.1%), mangrove forest (18.88%), ultramafic forest (0.38%), mixed dipterocarp forest (28.86%), limestone forest (2.32%), seasonal swamp forest (19.89%), and heath forest (0.63%) (Sabah Biodiversity Centre 2023). The diversity of geology and geomorphology directly influences the diversity of fauna and flora in the area (Sabah Biodiversity Centre 2024).

In the last few decades, human–wildlife conflict, especially with elephants and orangutans, has become more rampant. As the PAs are fragmented, the wildlife, particularly the elephants that require large roaming areas, frequently encroaches into oil palm estates and villages. In addition to human–wildlife conflicts, there are also pollution problems stemming from agricultural activities that flow into the river system. Environmental degradation from the discharge of effluents from palm oil mills and nonpoint sources of pollution has a negative impact on aquatic life in the river system, directly affecting the livelihoods of the local communities. There are also complaints from the local communities that they are left behind in socioeconomic development, such as tourism and agriculture-related activities, even though they are local and have been living in the area for generations. Currently, there are about 166,980 people living in Kinabatangan, mostly foreign nationals working in the oil palm sector (Sabah Biodiversity Centre 2024).

At this critical juncture, finding a win–win solution for the coexistence of all stakeholders operating in the Kinabatangan landscape is crucial. It is not merely an option but a necessity to find an appropriate approach to mediate the increasing conflicts and resource clashes within the landscape. With the support of state agencies, several nongovernmental organizations (NGOs) are currently working to promote sustainable development in the landscape. To sustain the NGOs' activities, support is required not only from land users but also from government agencies, local communities, civil society, and international bodies.

Efforts by NGOs to establish ecological connectivity in the early 2000s failed to gain traction due to a lack of support from the government and growers. However, in 2022, the Sabah government agreed to a proposal by the Sabah Biodiversity Centre (SaBC) to nominate the Kinabatangan landscape as a UNESCO Man and Biosphere (MAB) Reserve. UNESCO's MAB is very relevant in this matter because its main objective is promoting sustainable development within the perimeters of PAs. Adopting a MAB approach in this landscape will provide a framework for a common vision of all the actors in the areas. SaBC subsequently appointed HUTAN, a Kinabatangan-based NGO, as a consultant to prepare the dossier for the nomination.

The government's decision to nominate Kinabatangan was formally announced during the State-Level International Biodiversity Conference 2022 held in Kinabatangan. The preparation of the nomination dossier took approximately two-and-a-half years to complete, with the dossier being formally submitted in early July 2024. The dossier, endorsed by both the state and federal governments, is now with UNESCO MAB for consideration.

The MAB concept was chosen as an integrating platform for the Kinabatangan landscape because the approach to sustainable development under it aligns well with the issue of conflict of land uses which we aim to address in this area. Namely, the MAB Programme promotes a holistic approach to addressing the interrelated issues of sustainable conservation, development, and logistical support. By employing the MAB concept, which includes a participatory approach, promotion of research, and zonation, SaBC aspires to integrate and synergize activities in the Kinabatangan landscape toward a shared vision. This vision includes the establishment of ecological corridors, sustainable agriculture, improved livelihoods for local communities, and more inclusive tourism.

As part of the dossier preparation, core, buffer, and transition zones had to be identified and established. Additionally, UNESCO requires that consultations comply with the principle of Free, Prior, and Informed Consent (FPIC).

9.2 Methodology

The proposed vision statement for the nomination outlined the aim to establish a "resilient, interconnected Lower Kinabatangan landscape where people coexist peacefully with nature, integrating flourishing regenerative economies rooted in a shared commitment to sustainable development principles with the preservation of a globally renowned natural and cultural heritage". A dossier was prepared for submission to UNESCO's Man and Biosphere Programme. The dossier included the zonation and draft management plan of the proposed biosphere reserve. Details on the preparation of the dossier are as follows.

9.2.1 Compilation and Documentation Data

Data included recent satellite imagery, locations of the communities, demographic data, land ownership, locations and boundaries of PAs, land uses, administrative data, fauna and flora, and other relevant information. Data on demographic profiles, local issues, and the willingness of stakeholders to participate was also collected. This data was collected through a consultancy process with the stakeholders including local communities, oil palm growers and palm oil mill owners, tour operators, government agencies, NGOs, and civil society. Oil palm growers were divided into three categories—big plantations, medium-sized growers, and smallholders. Consultation with the communities was guided by FPIC principles.

Fig. 9.1 Location maps and the proposed Kinabatangan Biosphere Reserve (KBR). (Source: Based on Final Dossier, Sabah Biodiversity Centre 2024)

Table 9.1 Basic information of the study area

Country/Region	Malaysia
Province	Sabah
District	Kinabatangan
Municipality	Kota Kinabatangan
Size of geographical area (hectare)	413,253 ha
Dominant ethnicity(ies), if appropriate	Orang Sungai
Size of case study/project area (ha)	413,253 ha
Dominant ethnicity in the project area	Orang Sungai
Number of direct beneficiaries (people)	12,700
Number of indirect beneficiaries (people)	154,280
Geographic coordinates (latitude, longitude)	5°23′14.008″ N, 118°4′57.672″ E

9.2.2 *Mapping of the Proposed Biosphere Reserve*

All data obtained through this data collection, mapping, and engagement process was used to design the management plan and prepare the nomination dossier. This data helped to outline the proposed zonation of the biosphere reserve, including defining the objectives for the zonation of each core area, buffer zone, and transition zone. The proposed Kinabatangan Biosphere Reserve (KBR) is indicated by a red dot (see Fig. 9.1 and Table 9.1).

9.2.3 *Stakeholder Engagement*

Certain principles were followed in the consultation process with various stakeholders. Consultation procedures with local communities followed these guiding principles: (a) respect communities' rights by getting their consent before the consultation process; (b) be transparent and share the necessary information; (c) conduct the negotiation in good faith by respecting their rights and interests; (d) be sensitive to cultural practices, beliefs, and traditions; (e) promote participatory engagement by allowing the local communities to express their views freely; and (f) the engagement must be closely monitored. For the oil palm sector, the first step involved identifying the mills and growers within the transition zone. Subsequently, the concept was socialized, and stakeholders were involved in drafting the management plan. For the tourism sector, individual interviews were conducted with all operators, and industry-wide workshops and sharing sessions were also held.

9.3 Findings of Stakeholder Consultations on Proposed Kinabatangan Landscape Biosphere Reserve

Engagement with the stakeholders was quite complex due to the diverse actors involved, each with different expectations and opinions on the proposed nomination. It required perseverance from both SaBC, as the project proponent, and the consultant.

9.3.1 Engagement with Local Communities

No fewer than 25 individual meetings with village heads and district offices have been held. A joint meeting with the communities and district officer was also conducted in May 2023. To ensure that local community leaders could fully participate and understand the objectives of the nomination, meetings were scheduled at their convenience. HUTAN developed a community engagement plan in accordance with the FAO's FPIC Manual and the United Nations Declaration on the Rights of Indigenous Peoples. During the engagement process, the HUTAN team made every effort to ensure that as much as possible all stakeholders were consulted.

Initially, there was a negative connotation associated with the term "reserve" among the local communities, as it could imply government reserve land or the potential gazetting of their land into protected areas, thereby limiting their access to natural resources. However, after detailed explanations, the local communities became supportive of the nomination. They were optimistic about the potential benefits of the nomination for their livelihoods, the preservation of their cultural heritage, and the protection of traditional lands, rights, and practices. They also looked forward to reviving Orang Sungai music and craft-making traditions, which had been declining. Therefore, integrating the cultural heritage of the Orang Sungai is essential for the sustainable management of the Kinabatangan landscape.

9.3.2 Engagement with the Oil Palm Sector

Involvement and commitment from the oil palm sector are critical for successfully achieving the objectives of nominating the Kinabatangan landscape as an MAB Reserve. Oil palm plantations are the major land users in the area, with many operations extending to the fringes of the protected areas (PAs). It is anticipated that approximately 60% of the proposed buffer zone will consist of oil palm plantations. Therefore, it was essential to identify and engage all oil palm growers and mills within the perimeter of the proposed buffer and transition zones. Oil palm operators in the landscape can be categorized as large, medium-sized, and smallholders, each with distinct challenges, perspectives, and approaches. Consequently, the engagement plan must be tailored to these factors. Large operators typically run integrated operations, including mills and possibly refineries, and are required by palm oil certification bodies to establish sustainability units. Engagement with these operators tends to be more straightforward, as they are already familiar with sustainability practices, such as identifying and establishing High Conservation Values (HCVs) within their plantation areas. Additionally, many large operators have adopted Environmental, Social, and Governance (ESG) principles in their business practices, making this nomination a good fit for their sustainability objectives and commitments.

Medium-sized growers own agricultural lands ranging from 40 to 2000 ha. They do not face the stringent requirements imposed directly by palm oil buyers, as their operations are primarily downstream. They sell their oil palm fruits to large operators who own palm oil mills and sometimes refineries. However, the Malaysian Palm Oil Certification Council (MPOCC) mandated that all oil palm operators in Malaysia must be certified by 2019. As a result, all oil palm growers are required to obtain Malaysian Palm Oil Certification to sell their fruits to Malaysian Sustainable Palm Oil (MSPO)- or Roundtable on Sustainable Palm Oil (RSPO)-certified mills.

Engaging with medium-sized growers involves collaborating with regulatory authorities, such as the Malaysia Palm Oil Board (MPOB), the Sabah Jurisdictional Approach for Sustainable Palm Oil (JASPO), and Malaysia's Department of Environment, the Jabatan Alam Sekitar, to demonstrate government commitment to the KBR initiative. Both MPOB and JASPO have endorsed the KBR dossier. By participating in this initiative, medium-sized growers can also receive support from government agencies, such as assistance in flood mitigation efforts, which requires cooperation with government bodies and other estate owners to facilitate debris removal from river systems.

Connecting medium-sized operators to the MAB initiative required significant effort in explaining the long-term sustainability benefits of their operations. For smallholders, motivation to participate in the MAB initiative often comes from the prospect of obtaining alternative or additional livelihood opportunities.

9.3.3 Engagement with the Tourism Sector

The consultant developed an engagement plan for the tourism sector with the objective of increasing the understanding of the nomination process. For tourism operators, interest in participating is driven by opportunities for business growth through branding, promotion, and a regulatory framework for sustainable practices. Interviews were conducted with tour operators, ranging from community-based homestays and bed and breakfasts (B&Bs) to high-end lodges. Additionally, a roundtable discussion was held with relevant government agencies. Accordingly, the Tourism Management Plan was developed and integrated into the KBR management plan. This plan encompasses, among other objectives, more systematic promotion of tourism in the area, products, carrying capacity, local community participation in tourism, and a proposed one-stop information center for MAB. Kinabatangan is a popular area for ecotourism in the region. By integrating Orang Sungai culture and research, tourism has the potential to become an added value for the communities.

9.3.4 Engagement with Government Agencies

Government agencies, such as the Sabah Wildlife Department, Sabah Forestry Department, Department of Drainage and Irrigation, Department of Fisheries, Department of Environment, and many others, will benefit from this initiative. The management objectives of the MAB will become the guiding principles for any proposed development in the area. It will indirectly reduce potential conflicts for any future development initiatives across the landscape.

9.3.5 Nongovernmental Organizations (NGOs)

NGOs play a crucial role as boundary brokers in synergizing activities for sustainable development in the proposed KBR. They are also expected to provide capacity-building for the community, as well as to bridge and communicate the local community's concerns to government agencies and corporate owners. NGOs and the Sabah Biodiversity Centre played an important role as boundary brokers in the negotiation of KBR. A boundary broker must be trusted not only by the government but also by the private sector and local community. Most of the stakeholders have had issues with enforcement agencies in the past such as the Sabah Wildlife and Forestry Departments. NGOs and the Sabah Biodiversity Centre can bridge the trust of all the stakeholders which include state agencies.

9.3.6 Researchers and Academia

Research and education are pillars of the MAB Programme. In order for the proposed MAB reserve to become a laboratory for sustainable development, a research component is a necessity in achieving the goals. As the Kinabatangan landscape is an area rich in biodiversity with human-altered landscapes, it can be an excellent study area for researchers. Research is also needed to identify future requirements of the KBR and to establish facts for the purpose of negotiation with the landowners.

9.3.7 International Partners and Donors

We are aiming to establish a robust model of sustainable development in areas where land use conflicts are occurring. Securing international support is crucial for the successful implementation of the action plan. Therefore, international partners and donors must provide financial contributions to ensure its success.

The proposed core area of the Kinabatangan Biosphere Reserve (KBR) aims to ensure the protected area's integrity by providing optimal protection for ecosystems and biodiversity. This includes creating a safe haven for threatened species and connecting all the protected areas along the Kinabatangan River, linking the Heart of Borneo in the west with the Lower Kinabatangan-Segama Wetland (LKSW) Ramsar site in the east. The PAs are Gomantong, Materis, Bod Tai, Keruak and Pangi Forest Reserve, Sg Lokan Forest Reserve, Pin Supu Forest Reserve, Gomantong, Gologob Forest Reserve, Balat Damit Wildlife Reserve, Balat Damit (Extension) Wildlife Reserve, and Trusan Kinabatangan (Extension). Through this connectivity, we hope to establish a wildlife corridor from the interior of Sabah to the east coast. Due to the complexity of the landscapes, the first 3 months of the initiative were dedicated to socializing the concept with all stakeholders.

The buffer zone area is designed to establish a connection or linkage between the protected areas (PAs) mentioned above. The model will be developed to ensure that only sustainable economic activities with minimal or no impact on the environment are permitted in the core areas. It also aims to protect important species, scenic beauty, and the wilderness of the area, positioning it as a premier global ecotourism hotspot.

The initiative aims to provide a buffer zone with a minimum width of 200 m bordering the PAs. This 200 m width requirement is also extended along the Kinabatangan River. This can be achieved by leveraging the existing mandatory environmental and sustainability standards for agriculture. The buffer zones will include legally constituted riparian reserves along the Kinabatangan River and its tributaries. With a width of 50 m, these reserves will act as a sponge or filter, stabilizing riverbanks and aiding in flood mitigation.

The proposed buffer zone will also encompass riverine villages such as Kg Bilit, Abai, Batu Puteh, Sungai Lokan, Sukau, and Balat. Pressured by human–wildlife conflicts, some plantation estates have opted to establish conservation areas and wildlife corridors within their plantation areas. This approach will serve as the model for establishing and managing the buffer areas.

The total area of the proposed buffer zone is 12,473 ha, with 54% covered by oil palm plantations, 30% consisting of unprotected riverine forest, 2% as riparian reserve, and 13% designated for other land uses. Through negotiation with local communities, efforts will be made to bypass these villages to mitigate human–wildlife conflicts, although decisions will be made on a case-by-case basis through a consultation process. In the long term, the buffer zone will be promoted for research and monitoring on ecosystem regeneration, with the hope of expanding the buffer zone area if anticipated benefits are realized.

The transition zone of the proposed KBR is a mosaic of land use, predominantly dominated by oil palm plantations, which cover 77% of the zone. This area is home to over 154,200 workers, most of whom are migrants. It also contains 19 villages, predominantly inhabited by the Orang Sungai, with an estimated population of about 9300. In these areas, sustainable management will be promoted by ensuring that both the ecological integrity of the KBR and the socioeconomic well-being of the local people are well maintained. Activities with minimal impact, such as ecotourism and swiftlet farming, will be encouraged.

9.4 Discussion

Sustainable development in the Kinabatangan landscape can only be realized if all actors are involved in the process and share a common vision for the future of the area. Therefore, nominating sites such as the Kinabatangan landscape as an MAB Reserve can serve as a platform to achieve this common vision. However, dealing with complex landscapes where there are clashing land uses, such as conservation, tourism, fishing, and agriculture, poses significant challenges.

The first step is to convince political leaders of the value of the KBR initiative. Before the project commenced, SaBC, as the project proponent, obtained state government endorsement for the initiative. Government commitment is critical in convincing stakeholders, especially players in the palm oil industry, to participate in this initiative.

Second, it is essential to communicate ideas to all parties involved. SaBC, with the support of consultants, designed communication strategies for each sector. It was also necessary to involve the right people for specific tasks. Specialists in local community engagement, palm oil sustainability, and local tourism were brought into the nomination process. These experts were crucial for communicating the initiative's objectives and aligning them with the needs and management objectives of stakeholders to create a win–win situation between conservation and socioeconomic development.

Several sectoral workshops and forums were organized to address current issues and concerns and to explore the benefits of this initiative. A multisectoral workshop was held to refine the strategies and action plans for a common understanding and vision of the landscape.

For integrated management, we need to consider all aspects—economic, social, political, and ecological—and identify opportunities that may arise from trade-offs. In the Kinabatangan landscape, despite clashing land uses, many opportunities can be leveraged, such as the commitment to ecological passages, the rich cultural heritage of the Orang Sungai, and the potential for tourism. Therefore, we need to identify and exploit these opportunities for the sustainable management of the area (Moon et al. 2014). For example, the community's pride and traditions can be aligned with the conservation objectives of the KBR. Promoting a sense of belonging will enhance their commitment to protecting the fauna and flora in the surrounding environment.

In establishing a shared vision to be translated into the management plan, interventions with local communities, especially the Orang Sungai, must align with their perspectives, identity, and values. It is crucial to evaluate the benefits and economic costs and integrate them from the beginning of the initiative (Adams et al. 2010). For local communities, the main motivation for participating is the opportunity to improve their livelihoods. A significant issue for these communities is poverty due to marginalization in socioeconomic activities in the area. However, any intervention on livelihoods must align with their sociocultural values, which are deeply intertwined with riverine ecosystems (Waylen et al. 2010). The communities' personal ties to the surrounding ecosystems are rooted not only psychologically but also in their social and cultural values (Vining and Tyler 1999).

Recognizing and integrating the cultural values of the Orang Sungai in this landscape are crucial for gaining their support for the new management approach. We must also address other social factors, such as enhancing their sense of belonging to the initiative, building confidence in the government, encouraging membership in local and broader networks, and fostering a willingness to collaborate (Knight et al. 2010). By highlighting the uniqueness of the ecosystems and the richness of species in their areas, we can increase the sense of pride and belonging among local communities and motivate them to cooperate in preserving the area (Drew and Henne 2006). They feel victimized by past experiences when the government alienated land to outside companies or designated it as protected areas without proper consultation. Therefore, in consultations with local communities, we need to address and overcome the mistrust that arose from these past actions.

The nomination process of the MAB Programme, which required consultation based on FPIC (Free, Prior, and Informed Consent) principles, convinced the local communities of the government's sincerity. The guiding principles for consultation were instrumental in securing their commitment. Their participation in the process motivates them to view themselves as part of the solution rather than the problem for the Kinabatangan landscape. To ensure the sustainability of their commitment to the MAB initiative, it is essential for them to collectively agree on activities that directly affect them in the landscape. During the consultation process, they expressed a desire to participate in socioeconomic tourism and be involved in the management of the areas. This aligns well with the MAB objectives of sustainable development and empowering the local community. Empowering local communities in monitoring and, to some extent, in enforcement will enhance their sense of ownership of the protected areas and surrounding regions. This approach will also lead to better protection of the areas, as they become partners rather than problems or objects of conservation.

Furthermore, for state agencies responsible for managing protected areas (PAs), empowering the local community in monitoring activities can reduce the management costs and shift the balance of monitoring costs to benefits (Walker 2009). Collective choice rules, such as the traditional Tagal system, should be promoted because communities are generally more committed to enforcing rules when they can prevent outsiders from accessing resources they are empowered to manage (Ostrom 2009).

For local communities to participate effectively, their capacity must be enhanced. They will only be able to engage in the initiative if they have capable leaders who can organize them and advocate their interests with government agencies and other stakeholders, such as tour operators, plantation owners, and international funding agencies. In addition to bridging gaps and building trust in the MAB initiative, NGOs can also play a role in enhancing the capacity of local communities.

In dealing with oil palm growers, it is important to clearly communicate the issues, challenges, and available options for the government in addressing land use conflicts in these areas. The oil palm sector is certainly looking for benefits from participating in this initiative. They should understand the current situation faced by the government in managing conflicting land uses in the area and the rationale for designating the Kinabatangan landscape as a biosphere reserve. At the same time, they must realize that other options are still available to the state government, such as imposing more stringent regulations on operations if the MAB initiative fails. Two regulating bodies overseeing oil palm activities in Sabah, namely MPOB and JASPO, have endorsed the initiative by signing the dossier of the nomination. This endorsement aims to ensure that all parties involved view the initiative objectively. However, fairness and openness from the government are essential in gaining the commitment of oil palm growers, as trust is critical for the long-term success of the proposed integrated management of the landscape and must be fostered from the initial stage.

The support and commitment of oil palm growers are vital because more than 60% of the buffer zone and 77% of the transition zone consist of oil palm areas. Therefore, identifying conservation opportunities within these palm oil areas is crucial, as the landscape lacks a natural buffer zone due to most plantation areas being on the fringes of protected areas. During the consultation process, negotiations with oil palm growers have focused on providing wildlife connectivity within the oil palm areas. For example, the MAB reserve initiative can offer plantations the opportunity to leverage their existing environmental and sustainability commitments under MSPO certification. This benefits both the plantation sector and conservation efforts in the landscape. By establishing wildlife corridors within their oil palm plantations, the occurrence of human–wildlife conflicts and damage to palm oil can be reduced. Additionally, wildlife visibility in these corridors may present new business opportunities for plantation owners. They are also encouraged to consider the potential branding benefits of the KBR if they participate in wildlife conservation efforts.

The tourism sector is a major beneficiary of the MAB Programme initiatives. There is significant potential for branding and promotion through the initiative. The cultural aspects of the Orang Sungai can be a key attraction for tourism if properly designed and aligned with expected visitor arrivals. Therefore, engagement with the Sabah Tourism Board is crucial for promoting the enhanced tourism products of Kinabatangan. To ensure the sustainability of the MAB initiative, it is essential to empower local communities to actively participate in tourism activities. Providing them with a stake in this sector will motivate them to cooperate in protecting the PAs and natural resources as a whole (Ancrenaz et al. 2007). Without their involvement, they may be demotivated to conserve biodiversity because it would not benefit them directly. Thus, in addition to promoting Orang Sungai culture, it is necessary to employ them as tour guides for Kinabatangan. The Sabah Biodiversity Centre, through the proposed MAB Information Centre, should play a critical role in introducing the culture and empowering local communities (Shukla 2004). A Tourism Management Plan for Kinabatangan has been developed and annexed to the Proposed Management Plan for KBR.

Research and education will play a crucial role in ensuring the success of the MAB Programme in Kinabatangan. Data is essential to support the decision-making process, and scientific findings are needed to assess the effectiveness of the action plans developed under the initiative. The Danau Girang Research Centre, located in Kinabatangan, will be pivotal not only in conducting research but also in determining the type of research needed, collecting data, and disseminating results to the management authorities of the MAB reserve and other stakeholders for their capacity-building and effective action (Roux et al. 2006). Local communities, especially the younger generation, need to be given a strong scientific foundation to understand the consequences of not implementing the MAB Programme in the landscape. Integrating modern science with traditional knowledge will help find solutions that are politically, socially, and culturally acceptable to all parties involved (UNU-IAS & IGES 2016). Additionally, the media must actively participate in writing and publishing the success stories from this initiative. This will provide recognition to the actors involved and motivate them to achieve new heights of sustainability.

After consultation with the stakeholders, the total area of the KBR covers approximately 413,866 ha. The zones within the KBR are as follows: (a) core zone—50,950 ha; (b) buffer zone—12,473 ha; and (c) transition zone—350,444 ha. The core zone will provide a safe refuge for wildlife and protect threatened ecosystems. The buffer zone will insulate the protected areas (PAs) from outside threats and serve as a connecting corridor between the PAs. The transition zone is intended to support the conservation goals of the MAB Reserve, promote sustainable development, and create synergy with conservation efforts in the core and buffer zones. Activities in the transition zone impact both core and buffer zones, as well as the livelihoods of the local communities, since discharges from the transition area flow into the riverine system. There is potential for more plantation estates to join the initiative in the future, especially if the expected benefits, such as achieving sustainability in their operations, are realized.

9.5 Conclusion

We can conclude that UNESCO's MAB Programme is a valuable tool for synergizing activities in areas with overlapping or conflicting land uses. The MAB Programme can serve as a unifying platform for stakeholders to achieve sustainability in their activities within the landscape by fostering a shared vision with interlinked goals and multiple objectives, such as in Kinabatangan (UNU-IAS & IGES 2016). The success of connecting the Ramsar site in the east with the Heart of Borneo in the west under this initiative is due to a good understanding of the problems and interests of the stakeholders. However, it will depend on the commitment of all involved parties. This initiative also will provide an opportunity for co-management, where power in the decision-making process is shared among the government, local communities, private sector, NGOs, researchers, and civil society.

The stakeholders in the areas have overlapping and competing interests, and these naturally will lead to misunderstandings on societal and ecological environments and cause conflicts between parties. Therefore the government, especially SaBC, must have the final say in enforcement and monitoring in co-managed areas (Berkes 2007; Ostrom et al. 1999). Benefits generated by the initiative must be equitably distributed among all stakeholders, particularly the local communities. The management plan must be designed to ensure that all layers of society, especially the poor and underprivileged, are involved in both the initial consultation process and the implementation stage. There should also be no monopoly of benefits by one village, privileged group, or elite.

Connecting the experiences of the stakeholders in pursuing sustainability in the landscape with international conservation discourse will help promote the area to the outside world and recognize their commitments (Stokes et al. 2010). Transparency in management is vital for the success of this initiative. All major decisions in these areas must be discussed by a committee overseeing the management plan, which should include representatives from industry, local communities, and civil society. To ensure sustainability and promote active participation from all stakeholders, a MAB Information Center must be built at a location accessible to everyone. This center will serve as the starting point where visitors can obtain first-hand information on the biodiversity, culture, programs, and activities in the Kinabatangan landscape.

The Sabah Biodiversity Centre plays a critical role in integrating the activities and programs designed under the management plan. In doing so, it must earn the trust of all the stakeholders operating in the Kinabatangan landscape.

References

Adams VM, Pressey RL, Naidoo R (2010) Opportunity costs: who really pays for conservation? Biol Conserv 143(2):439–448

Ancrenaz M, Dabek L, O'Neil S (2007) The costs of exclusion: recognizing a role for local communities in biodiversity conservation. PLoS Biol 5(11):e289. https://doi.org/10.1371/journal.pbio.0050289

Berkes F (2007) Community-based conservation in a globalized world. Proc Natl Acad Sci 104(39):15188–15193

Drew JA, Henne AP (2006) Conservation biology and traditional ecological knowledge: integrating academic disciplines for better conservation practice. Ecol Soc 11(2):34

Knight AT, Cowling RM, Difford M, Campbell BM (2010) Mapping human and social dimensions of conservation opportunity for the scheduling of conservation action on private land. Conserv Biol 24(5):1348–1358

Moon K, Adams VM, Januchowski-Hartley SR, Polyakov M, Mills M, Biggs D, Knight AT, Game ET, Raymond CM (2014) A multidisciplinary conceptualization of conservation opportunity. Conserv Biol 28(6):1484–1496. https://doi.org/10.1111/cobi.12408

Ostrom E (2009) A general framework for analyzing sustainability of social-ecological systems. Science 325(5939):419–422. https://doi.org/10.1126/science.1172133

Ostrom E, Burger J, Field CB, Norgaard RB, Policansky D (1999) Revisiting the commons: local lessons, global challenges. Science 284(5412):278–282

Roux D, Rogers K, Biggs H, Ashton P, Sergeant A (2006) Bridging the science–management divide: moving from unidirectional knowledge transfer to knowledge interfacing and sharing. Ecol Soc 11(1):4

Sabah Biodiversity Centre (2023) Interim Report on Nomination of Kinabatangan Landscape as a Biosphere Reserve, MAB, UNESCO. Unpublished report

Sabah Biodiversity Centre (2024) Draft nomination Dossier for Kinabatangan Biosphere Reserve. Unpublished report

Shukla SR (2004) Strengthening community-based conservation through traditional ecological knowledge. In: Indigenous knowledge: implications for sustainable development. Proceedings of the international conference on indigenous knowledges: transforming the academy, 27–28 May 2004, Penn State University, pp 44–58

Stokes DL, Hanson MF, Oaks DD, Straub JE, Ponio AV (2010) Local land-use planning to conserve biodiversity: planners' perspectives on what works. Conserv Biol 24(2):450–460

UNU-IAS & IGES (ed) (2016) Mainstreaming concepts and approaches of socio-ecological production landscapes and seascapes into policy and decision-making, Satoyama Initiative Thematic review, vol 2. United Nations University Institute for the Advance Study of Sustainability, Tokyo

Vining J, Tyler E (1999) Values, emotions and desired outcomes reflected in public responses to forest management plans. Hum Ecol Rev 6(1):21

Walker KL (2009) Protected-area monitoring dilemmas: a new tool to assess success. Conserv Biol 23(5):1294–1303

Waylen KA, Fischer A, Mcgowan PJK, Thirgood SJ, Milner-Gulland EJ (2010) Effect of local cultural context on the success of community-based conservation interventions. Conserv Biol 24(4):1119–1129

Chapter 10
Connecting the Dots: Enhancing Connectivity Within and Across SEPLS Through Implementation of Taiwan Ecological Network in Hualien County

Paulina G. Karim, Hwa-Ching Lin, Chium-Tse Huang, Mei-Yin Chen, Yuan-Jyun Wang, Yu-Chuan Lo, Chia-Tzu Chen, Kuang-Chung Lee, and Chih-Chin Shih

Abstract Taiwan Ecological Network (TEN) was introduced in 2018 as an island-wide cross-sectoral biodiversity-focused spatial planning strategy aimed at (I) restoration and support of resilient, biodiverse, connected, and well-functioning ecosystems within its 45 conservation corridors and (II) conservation and sustainable use of biodiversity in SEPLS and promotion of community resilience. Promotion of the Satoyama Initiative and landscape conservation makes up one of TEN's strategic action tasks. Our study is a retrospective overview of how application of community-based resilience assessment workshops (RAWs) *as a tool*, operationalization of the integrated landscape–seascape approach (ILSA) in SEPLS *as an approach*, and their alignment with TEN *as a strategy* has helped to ensure and enhance connectivity *within* and *across* SEPLS over time. For our analysis, we employ a mixed-method, multilayered, and cross-temporal approach and look at the RAWs–ILSA–TEN processes (2018–2023) in four SEPLS (Fengnan, Xinshe, Gangkou, and Luoshan) located within two of TEN's conservation corridors (the Xiuguluan River and the Northern Section of the Coastal Mountain Range) in Hualien County of eastern Chinese Taipei. We then discuss these processes from the perspective of the "tool–approach–strategy" bundle, multi-stakeholder interactions,

P. G. Karim · K.-C. Lee
National Dong Hwa University (NDHU), College of Environmental Studies and Oceanography, @ScapesLab, Hualien County, Taiwan

H.-C. Lin · Y.-C. Lo · C.-T. Chen · C.-C. Shih (✉)
Forestry and Nature Conservation Agency (FANCA), Taipei City, Taiwan
e-mail: ccshih@forest.gov.tw

C.-T. Huang · M.-Y. Chen · Y.-J. Wang
Forestry and Nature Conservation Agency, Hualien Branch (FANCA-Hualien), Hualien City, Hualien County, Taiwan

M. Nishi et al. (eds.), *Ensuring Ecological Connectivity in Socio-Ecological Production Landscapes and Seascapes (SEPLS)*, Satoyama Initiative Thematic Review, https://doi.org/10.1007/978-981-95-1474-8_10

and multiple meanings of connectivity in SEPLS, keeping in mind associated synergies and trade-offs. We conclude by outlining lessons learned and ways forward.

Keywords Taiwan Ecological Network · Connectivity · Spatial planning · Multi-stakeholder partnerships · Integrated landscape · seascape approach · Resilience assessment workshop · Facilitation

10.1 Introduction

10.1.1 Understanding Ecological and Sociocultural Context

Taiwan Island is located on the western edge of the Pacific Rim of Fire at the junction of the Philippine Sea Plate and the Eurasian Plate within subtropical and tropical climatic zones (Teng et al. 2000). From its highest peak (Yushan Mountain, 3952 m), it reaches the shores of Taiwan Strait in the west and the Pacific Ocean in the east within a short distance of 80 and 50 km, respectively. These unique geological, geomorphological, and climatic conditions have fostered its rich biodiversity (over 60,000 species, or 3.4% of the world's total) and a high level of endemism (Chung and Shao 2025). Geographic location, bountiful natural forests, fertile agricultural land, and abundant coastlines have further shaped the island's unique sociocultural pattern (Karim et al. 2024).

Research shows that there is a high overlap between wildlife biodiversity hotspots and satoyama landscapes in Chinese Taipei, with as many as 55% of the island's protected animal species and 64% of its Red List plant species distributed across socio-ecological production landscapes and seascapes (SEPLS) (FANCA 2025). This speaks to at least two important points. One is the inextricable role that the island's Indigenous Peoples and local communities (IPLCs) have played in the sustainable use of biodiversity in SEPLS and passing on this knowledge across generations. The second is the existing and potential threat to biodiversity as SEPLS shrink in size, become depopulated and mismanaged, or even disappear. Since the early 1970s, high population density, urban sprawl, conventional agriculture, human–wildlife conflicts, and road kills have been the main causes of biodiversity loss and habitat fragmentation on the island (Shiu et al. 2022).

The Satoyama Initiative entered the island in 2011 for the purpose of socio-ecological revitalization of SEPLS (Lee 2016). Integrated landscape and seascape approaches (ILSA) have been promoted to operationalize the initiative in local contexts, while the indicators of resilience in SEPLS and resilience assessment workshops (RAWs) have been used as a tool for the identification of priority issues in SEPLS (Karimova and Lee 2023; Lin et al. 2023). The Taiwan Partnership for the Satoyama Initiative (TPSI) was established in 2015 to foster island-wide networking among SEPLS practitioners (Lee et al. 2020b). A nationwide strategy for biodiversity-focused spatial planning, however, was needed to align SEPLS efforts with the conservation agenda.

10.1.2 Taiwan Ecological Network

In 2018, the Taiwan Ecological Network (TEN) was introduced by the Forestry and Nature Conservation Agency (FANCA, former Forestry Bureau) as the first of its kind island-wide program that combines biodiversity data-based spatial planning and cross-sectoral cooperation for an effective area-based conservation in the shallow mountains, plains, wetlands, and coastal areas (Lin et al. 2018). Two main objectives of TEN focus on (I) restoration and support of resilient, biodiverse, connected, and well-functioning ecosystems and (II) conservation and sustainable use of biodiversity in SEPLS and promotion of community resilience (Shih et al. 2021). TEN completed its first phase in 2018–2021 and is currently in its second phase (2022–2025).

Ecological connectivity in the context of TEN aims at ensuring the size and quality of habitats that can support healthy population sizes as well as averting and opening ecological bottlenecks for species migration (Chen 2022). TEN implies that ecological connectivity may occur within conservation corridors depending on the species and their needs. A conservation corridor serves to maintain vital ecological or environmental connections by providing physical—linear, landscape, or stepping stones—linkages between the conserved sites (Bennett and Mulongloy 2006; Shiu 2020). As key areas of sustainable biodiversity use which also supply a wide array of ecosystem services, SEPLS play a crucial role in enhancing connectivity. This also explains why promoting the Satoyama Initiative and landscape conservation make up one of TEN's seven strategic action tasks (Chen 2023).[1] Community-based conservation, eco-friendly production activities in SEPLS, and multi-stakeholder and cross-sectoral partnerships are imperative for achieving TEN's objectives and fostering connectivity across its 44 priority biodiversity areas and 45 conservation corridors (Fig. 10.1).

The objective of our study is to demonstrate how application of community-based RAWs (as an issue identification tool), operationalization of ILSA in SEPLS (as a management approach), and their alignment with biodiversity-focused spatial planning—TEN (as a spatial strategy)—has helped to ensure and enhance connectivity *within* and *across* SEPLS over time. For our analysis, we look at the RAWs–ILSA–TEN processes in four SEPLS located within two conservation corridors of TEN in Hualien County. We then discuss these processes from the perspective of the "tool–approach–strategy" bundle, multi-stakeholder interactions, and multiple meanings of connectivity in SEPLS, and share the main lessons learned and ways forward.

[1] Seven strategic action tasks of TEN (2022–2025): improving TEN database and planning; restoring native plants and removing invasive alien species; connecting ecological corridors and constructing animal passages; identifying high-risk areas and conserving endangered species; fostering eco-friendly production environment; promoting the Satoyama Initiative and landscape conservation; enhancing communication, education, and public awareness (CEPA) (FANCA 2025).

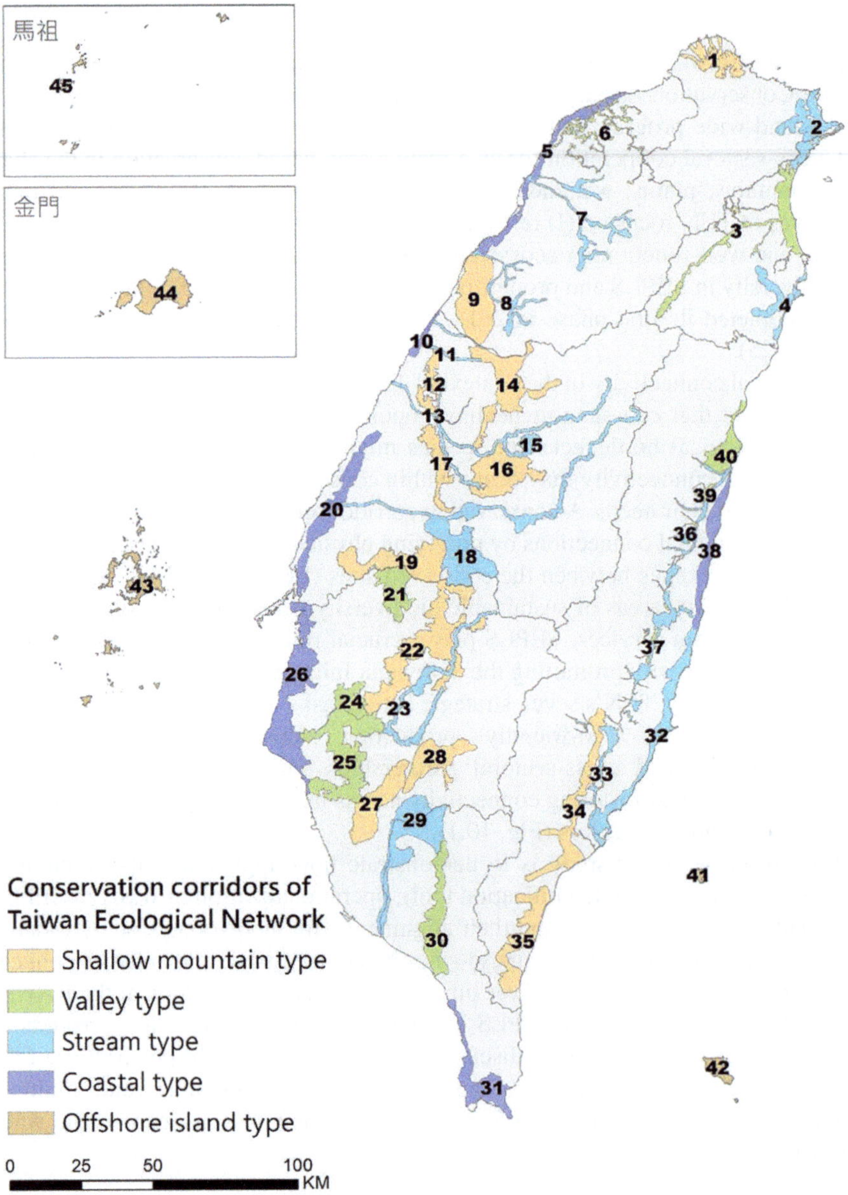

Fig. 10.1 Map of 45 conservation corridors of Taiwan Ecological Network (TEN). (Source: FANCA 2025)

10.2 Methods

10.2.1 Description of the Conservation Corridors

The Xiuguluan River conservation corridor (#37, stream type) and the Northern Section of the Coastal Mountain Range conservation corridor (#38, coastal type) were selected for this study (Fig. 10.2). Both corridors face a set of ecological and socioeconomic challenges common for the island's eastern part. Water quantity and quality, aquatic biodiversity, and freshwater habitats are threatened due to unsupervised human activities, such as nutrient pollution, garbage disposal, river dredging, poaching, and concrete river slope engineering. Depopulation and aging are particularly serious factors causing abandonment and unsustainable management practices in production farmlands and coastal paddy terraces. This further leads to biodiversity decline, ecosystem fragmentation, and encroachment of invasive alien species.

The goal of TEN within the Xiuguluan River and the Northern Section of the Coastal Mountain Range conservation corridors is to ensure connectivity of the landscape–seascape elements *at a watershed scale*, by linking terrestrial and aquatic as well as freshwater and marine habitats and enhancing biodiversity and ecosystem services. A detailed list of relevant ecosystem types, conservation objectives, strategic action tasks, and priority conservation species within the two corridors is provided in Appendix B.

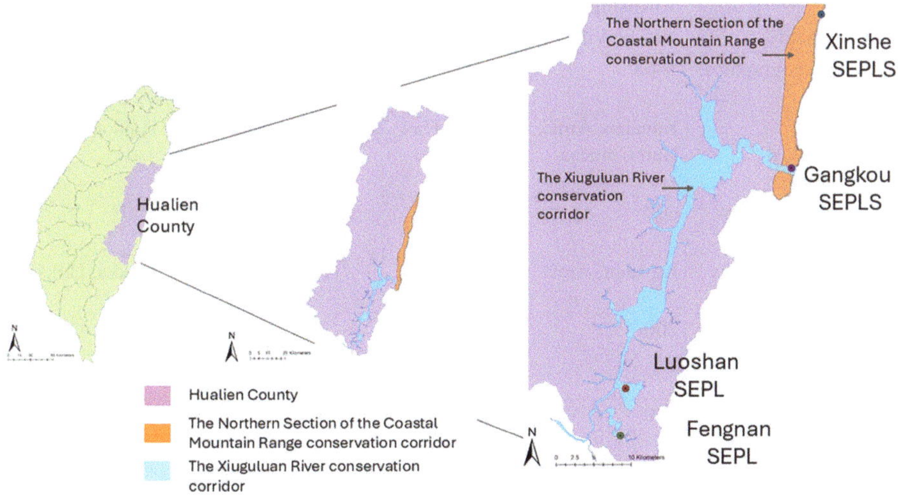

Fig. 10.2 Map of the Fengnan SEPL, Xinshe SEPLS, Gangkou SEPLS, and Luoshan SEPL in their relation to the Xiuguluan River conservation corridor (light blue) and the Northern Section of the Coastal Mountain Range (orange) conservation corridor in Hualien County, Taiwan. (Source: Prepared by the authors) (For abbreviations used in the figures, please refer to Appendix A. List of abbreviations)

10.2.2 Description of Case Study Sites

The Fengnan SEPL (Fuli Township), Xinshe SEPLS (Fengbin Township), Gangkou SEPLS (Fengbin Township), and Luoshan SEPL (Fuli Township) are the four sites selected for our study. Figure 10.2 shows their location in relation to the two conservation corridors. Table 10.1 provides the basic information about each SEPLS.

Xinshe and Gangkou SEPLS are located within the Northern Section of the Coastal Mountain Range conservation corridor facing the Pacific Ocean and demonstrate a classic version of forest–river–village–ocean landscape–seascape connectivity. They are both situated within the watersheds of independent coastal streams (of the Dabuan/Jialang Stream[2] and the Shitigang Stream, respectively)—a typical feature of the island's eastern coastal terrain (Fig. 10.2). Xinshe SEPLS is home to the Indigenous Amis Dipit and Kavalan pateRungan tribes, while Gangkou SEPLS belongs to the Indigenous Amis Makotaay tribe. Terraced rice paddy

Table 10.1 Basic information about the Fengnan SEPL, Xinshe SEPLS, Gangkou SEPLS, and Luoshan SEPLS

	Xinshe SEPLS	Gangkou SEPLS	Fengnan SEPL	Luoshan SEPL
Country/Region	Chinese Taipei	Chinese Taipei	Chinese Taipei	Chinese Taipei
Province	Hualien County	Hualien County	Hualien County	Hualien County
District	Fengbin Township	Fengbin Township	Fuli Township	Fuli Township
Municipality	Xinshe Village	Gangkou Village	Fengnan Village	Luoshan Village
Size of geographical area (ha)	1463	1682	3289	2003
Dominant ethnicities, if appropriate	Kavalan, Amis, Han Chinese	Amis, Han Chinese	Amis, Hakka, Hokkien	Hakka, Hokkien
Size of case study/ project area (ha)	600	1682	3289	2003
Dominant ethnicity in the project area	Kavalan, Amis	Amis	Amis	Hakka, Hokkien
Number of direct beneficiaries (people)	200	200	200	150
Number of indirect beneficiaries (people)	600	800	670	460
Geographic coordinates (latitude, longitude)	23°65′67″N 121°53′95″E	23°46′88″N 121°49′84″E	23°14′30″N 121°27′26″E	23°20′08″N 121°27′98″E

[2] The dual name of the Dabuan/Jialang Stream is explained by the different ways of naming its upstream and midstream sections (Dabuan) by the Amis locals and the downstream section (Jialang) by the Kavalan locals.

farming, home gardens, growing of edible and non-edible native crops, traditional weaving practices, wild plant foraging, fishing, and intertidal seaweed foraging are common human–nature interactions in the area. Located in close vicinity and within similar geo-ecological settings, Xinshe and Gangkou SEPLS share many cultural, economic, and relational ties. Notably, Gangkou SEPLS is also situated within the river mouth of the Xiuguluan River and therefore forms an important river–ocean estuary connectivity point at the easternmost end of the Xiuguluan River conservation corridor (Fig. 10.2).

The Xiuguluan River is one of the major waterways of the eastern part of the island. The Fengnan and Luoshan SEPL are its two inland connectivity points. Fengnan SEPL is located within the watershed of the Turtle River—the southernmost upstream tributary of the Xiuguluan River, and Luoshan SEPL within the watershed of the Luozai Stream and the Jiu'An Stream watersheds that are midstream tributaries of the Xiuguluan. Over 60% of Fengnan SEPL residents are Indigenous Amis, well-known for their traditional hunting, farming, and foraging skills as well as their knowledge of wild plants. Luoshan SEPL is predominantly ethnically Hakka and Hokkien, culturally rooted in rice paddy farming, bamboo harvesting, tofu making, and fruit growing. In 2002, it was given the title of the first organic village on the island.

10.2.3 Mixed-Methods Approach and Conceptualization

Our study employs a mixed-methods, multilayered, and cross-temporal approach for its conceptualization within the theme of this volume. We schematically demonstrate and briefly explain the contents and relationship of the methods in Fig. 10.3, which is then used as a conceptual foundation for each case study in the Results section. We apply community-based RAWs as a tool[3] for the identification of ecological connectivity needs and priorities in each SEPLS, ILSA[4] as an approach to their operationalization by the means of adaptive co-management (ACM) and multi-stakeholder collaboration, and TEN as a spatial strategy for identification and cross-sectoral promotion of strategic objectives in each site. Participatory and collaborative approaches are central to each method. We acknowledge several supporting methods that exist to facilitate the "tool–approach–strategy" bundle, such as TPSI regional exchange workshops, cross-SEPLS RAWs–to–ACM experience exchanges, on-site visits, and others. They are, however, not within the scope of this study.

---

[3] We acknowledge that from a broader ILSA perspective, RAWs as a tool may be viewed as an intrinsic part of ILSA as an approach. We do, however, keep them separate in our narrative to make the role of each more pronounced.

[4] The presence or absence of 'S' in SEPLS and ILSA (e.g. SEPL and ILA) depends on the presence or absence of the seascape component at a given site.

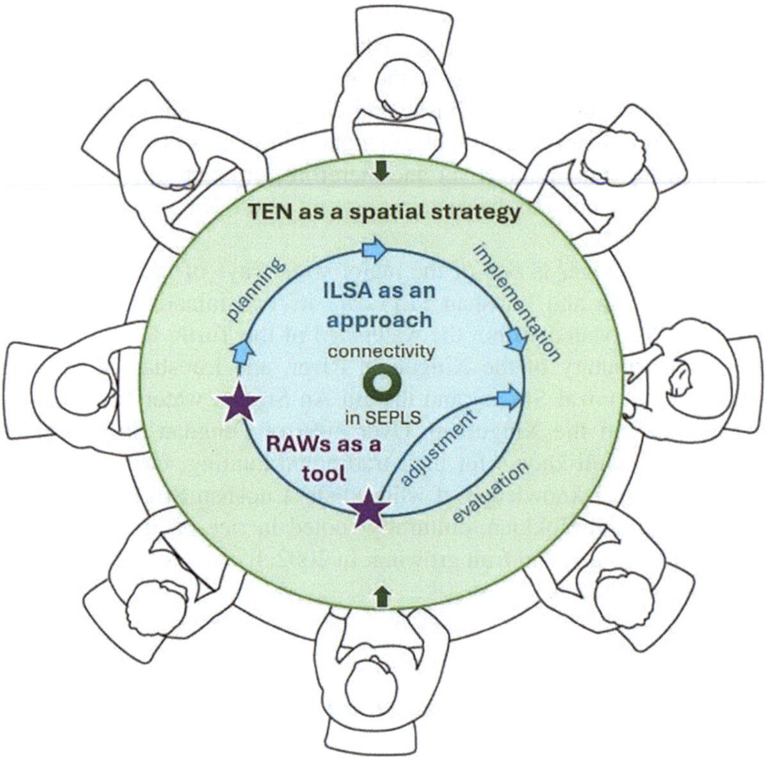

Fig. 10.3 "Tool–approach–strategy" methodological conceptualization for our study. (Source: Prepared by the authors. Image source of sitting persons: iStock by Getty Images, 2024)

Resilience assessment workshops, RAWs (as a tool, purple stars):

Five socio-ecological perspectives (ecosystem health and connectivity (A), eco-friendly production activities (B), weaving of Indigenous and local knowledge and modern science (C), community-based co-management systems (D), nature-based income-generating activities, and sustainable livelihoods (E)), 20 indicators; international (Bergamini et al. 2014) and localized (Sun et al. 2020) versions of indicators (in our analysis, we focus on Perspective A)

ACM integration: an evaluation tool for planning and adjustment of the action tasks in SEPLS

Community-based and participatory approach

Integrated landscape-seascape approach, ILSA (as an approach, blue arrows):

Adaptive collaborative management (ACM) cycle: issue identification–planning–implementation–evaluation–adjustment

Multi-stakeholder platform: Indigenous peoples and local communities (IPLCs), government agencies, academia, private companies, etc.

Adaptive and iterative learning process

Taiwan Ecological Network, TEN (as a spatial strategy, green space):

2018–2021: planning and data mapping; identification of priority biodiversity areas (44) and conservation corridors (45)

2022–2025: implementation across FANCA administrative network based on seven strategic action tasks (SEPLS focus)

Multi-stakeholder collaboration and cross-sectoral partnership (as partners at the negotiation table)

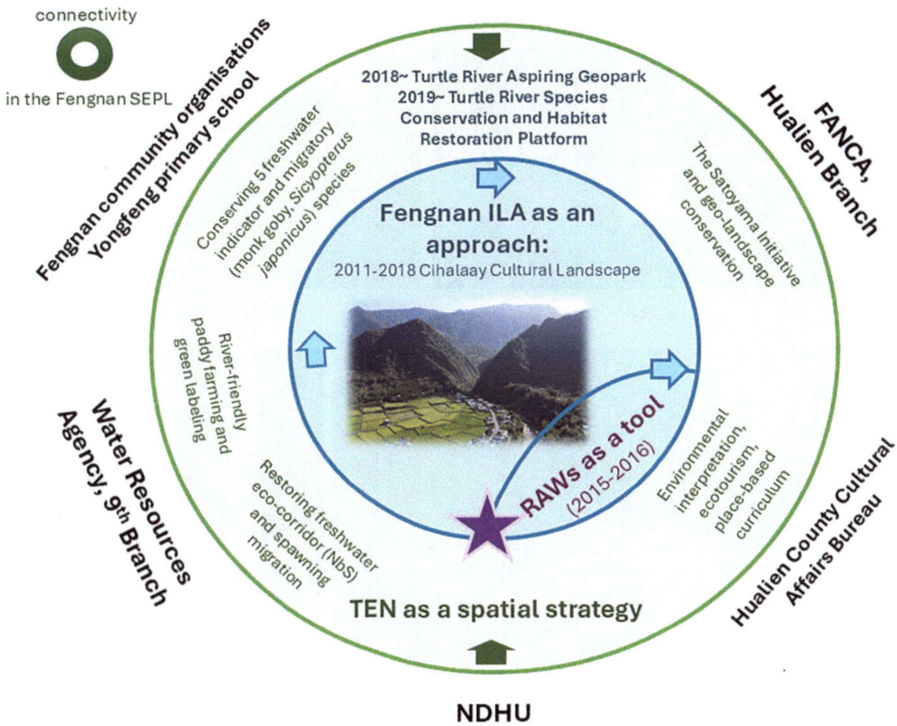

Fig. 10.4 RAWs–ILA–TEN processes and multi-stakeholder engagement in the Fengnan SEPL. (Source: Prepared by the authors. Photo credit: Hualien Natural Education & Ecology Consultant Ltd.)

10.3 Results

In this section, we describe the RAWs–ILSA–TEN processes in the four SEPLS focusing on the 2018–2024 period. We visually support the description of results with the "tool–approach–strategy" as depicted in Figs. 10.4, 10.5, 10.6, 10.7, 10.8, 10.9, 10.10, and 10.11.

10.3.1 Fengnan SEPL

Due to its southernmost upstream location within the Xiuguluan River watershed, unique geological and geomorphological features, and a century-old tradition of nature–culture interactions, the Fengnan SEPL has been a pioneer site in all three "tool–approach–strategy" dimensions (Lee and Karimova 2021). Its ILA in the form of the Cihalaay Cultural Landscape was established in 2011 and became one of the earliest examples of implementation of the Satoyama Initiative on the island

Fig. 10.5 Photos of the Fengnan SEPL: (**a**) Community-based monitoring and conservation of Chinese softshell turtle (*Pelodiscus sinensis*). (Photo credit: Hualien Natural Education & Ecology Consultant Ltd.); (**b**) *Hala* (monk goby, *Sicyopterus japonicus*). (Photo credit: Hualien Natural Education & Ecology Consultant Ltd.; Bai-Kai Chiu); (**c**) Stone Gate irrigation channel—a scenic spot on the Turtle River aspiring Geopark. (Photo credit: FANCA-Hualien); (**d**) Turtle River Rice. (Photo credit: FANCA-Hualien)

(Lee et al. 2016; Figs. 10.4 and 10.5). The international Toolkit for the Indicators of Resilience in SEPLS (Bergamini et al. 2014) was also first tested here in 2015–2016. At the time, RAWs identified the following community-based ecological connectivity priorities: (1) promotion of ecological and environmentally-friendly farming methods; (2) improvement of landscape diversity in rice paddies and surrounding hedges; (3) enhancement of connectivity between natural streams and irrigation channels; and (4) management of human–wildlife conflict (wild boars, Formosan macaques) and removal of alien species. These perspectives were integrated into the Cihalaay Cultural Landscape action plan (Lee and Yan 2019).

Notably, the Fengnan ILA period preceding the introduction of TEN (2011–2018) elicited the sociocultural and ecological importance of *hala* (monk goby, *Sicyopterus japonicus*)—an endemic migratory fish species of the Turtle River. It also explains the name Ci**hala**ay given to the area by its early Indigenous Amis inhabitants (Lee

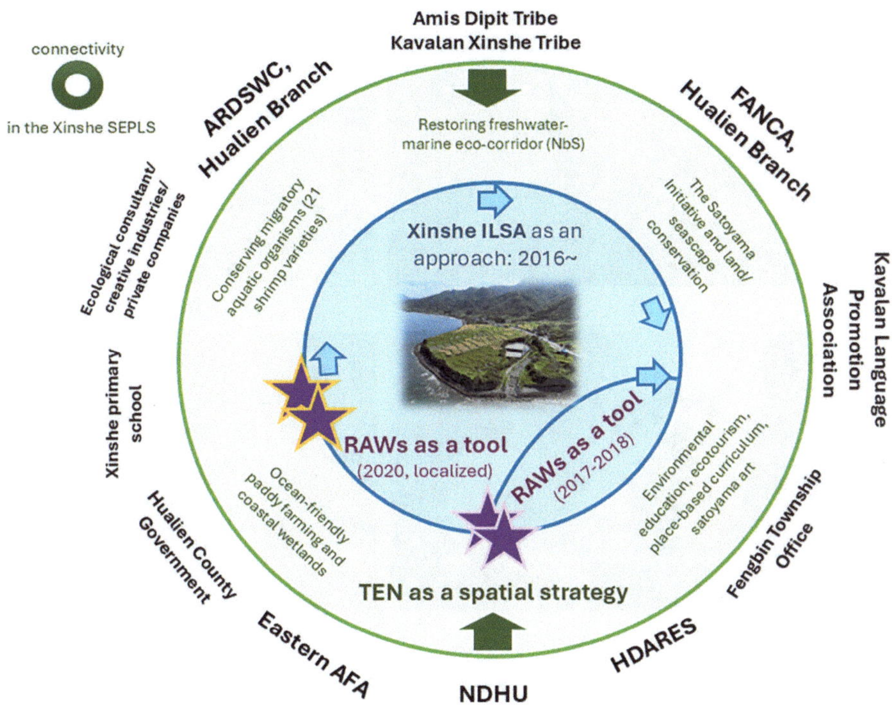

Fig. 10.6 RAWs–ILSA–TEN processes and multi-stakeholder engagement in the Xinshe SEPLS. (Source: Prepared by the authors. Photo credit: Vision Way Communication Co. Ltd.)

et al. 2016). The locals see a close correlation between the promotion of eco-friendly paddy rice farming and nurturing of healthy freshwater habitats for the *hala* fish. As a result of their efforts, the Hala Rice brand was established in 2016 and received the Green Conservation Label award from FANCA and the Tse-Xin Organic Agriculture Foundation (TOAF 2017).

Starting from 2018, the Fengnan ILA was repurposed to better consider geodiversity and biodiversity conservation objectives as well as sustainable local livelihoods within the framework of TEN. Thus, following a series of community-based resource identification activities (2018–2019), the Turtle River Aspiring Geopark Collaborative Planning Platform was established in 2020 (2020–2023). Supported by FANCA and National Dong Hwa University (NDHU), it has been working on zoning, geo-tourism and geo-product development, and the environmental interpretation capacity of Fengnan locals for the future gazettement of the Turtle River aspiring Geopark (planned for 2024).

In parallel with this process, in 2019, the Hualien Branch of FANCA (FANCA-Hualien) and the Ninth River Management Branch of the Water Resources Agency established the Turtle River Species Conservation and Habitat Restoration Platform.

Fig. 10.7 Photos of the Xinshe SEPLS: (**a**) River weir at the Dabuan/Jialang Stream mouth before dismantling, November 2020. (Photo credit: Observer Ecological Consultant Co.); (**b**) River weir at the Dabuan/Jialang Stream mouth after dismantling, December 2020. (Photo credit: Observer Ecological Consultant Co.); (**c**) Cross-sectoral and multi-stakeholder negotiations on dismantling of the river weirs in the Xinshe SEPLS, March 2020. (Photo credit: Observer Ecological Consultant Co.); (**d**) "Look! Our shrimp is back!" (Photo credit: Hanna Deng); (**e**) Indigo dye—promotion of eco-friendly farming and satoyama art. (Photo credit: Amis Dipit Fuxing Tribe)

It identified five indicator species—monk goby (*Sicyopterus japonicus*),[5] medaka (*Aphyocypris kikuchii*), tiny-spotted Rhinogoby (*Rhinogobius delicatus*), Taitung

[5] Monk goby (*Sicyopterus japonicus*) is the indicator migratory species for the entire Xiuguluan River conservation corridor—connecting Fengnan SEPL, Luoshan SEPL, and Gangkou SEPLS all the way from the source to the estuary.

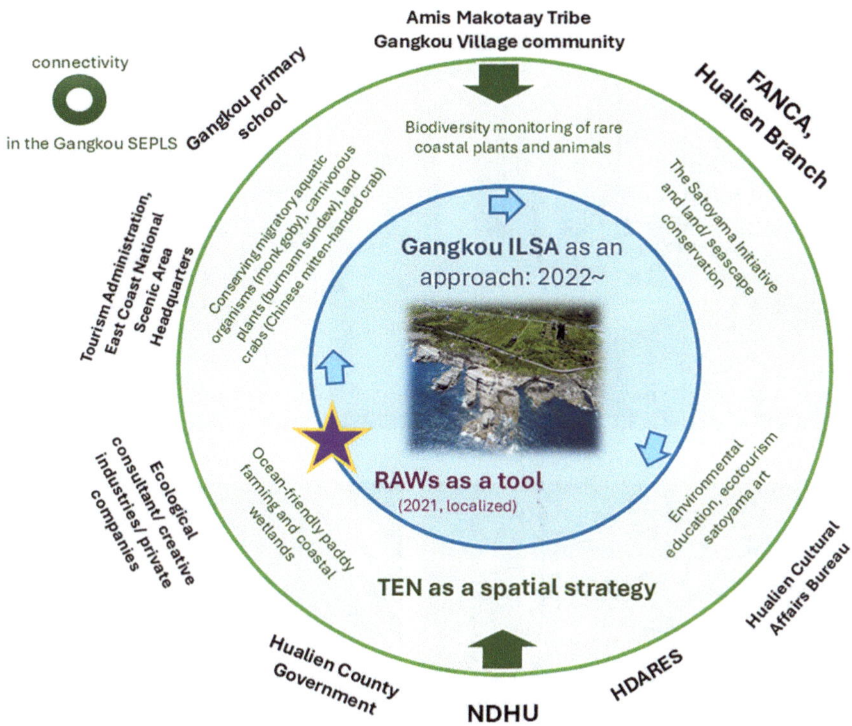

Fig. 10.8 RAWs–ILSA–TEN processes and multi-stakeholder engagement in the Gangkou SEPLS. (Source: Prepared by the authors. Photo credit: FANCA)

river loach (*Hemimyzon taitungensis*), and Chinese softshell turtle (*Pelodiscus sinensis*)—and their habitats as the main conservation objectives of the Platform (Fig. 10.4). Fengnan SEPL community engagement in both platforms—through eco-friendly river engineering projects, paddy rice farming, and co-management of common resources—resulted in the establishment of the Turtle River Rice brand in 2022 (FlyingV 2023).

Over the years, the Fengnan SEPL has demonstrated a highly dynamic relationship between the RAWs–ILA–TEN components and has had a substantial ripple effect on the other three SEPLS.

10.3.2 Xinshe SEPLS

The Xinshe Forest–River–Village–Ocean Eco-Agriculture Initiative (the Xinshe Initiative) is the most written about ILSA on the island. It was established in 2016 for a 10-year period, including a short-term period (2016–2020) and mid-term period (2021–2026). To date, its multi-stakeholder platform (MSP) consists of two

Fig. 10.9 Photos of the Gangkou SEPLS: (**a**) Multi-stakeholder platform meeting in the Gangkou SEPLS showing Director of FANCA-Hualien Mr. Chium-Tse Huang, Tribal Chief of Makotaay Gangkou Tribe Mr. Qing-Jin Lin, and Professor Kuang-Chung Lee (NDHU). (Photo credit: FANCA-Hualien); (**b**) Community-based coastal patrol and beach clean-up event. (Photo credit: Jie-Min Chen); (**c**) Chinese mitten-handed crab (*Platyeriocheir formosa*). (Photo credit: Hualien Natural Education & Ecology Consultant Ltd.); (**d**) Burmann's sundew (*Drosera burmannii*). (Photo credit: Jie-Min Chen)

local tribes, four government agencies, a local primary school, an Indigenous language promotion association, and a bridging stakeholder (NDHU). Over the years, other organizations have also contributed to the Xinshe ILSA (Karimova and Lee 2022; Figs. 10.6 and 10.7). For the initial design of its short-term action plan (2017–2020), it learned from the Fengnan ILA (Cihalaay Cultural Landscape) with its action tasks based on the five perspectives of the Satoyama Initiative (Lee et al. 2019). To date, the Xinshe ILSA demonstrates one of the most mature ACM models of all the island's SEPLS.

RAWs for the Xinshe ILSA have been employed twice: in 2017–2018 for adjustment of the short-term action plan, and in 2020 for the development of the mid-term action plan (2021–2026). The international Toolkit for the Indicators of Resilience in SEPLS was used during the first round of RAWs in both tribes (Lee et al. 2020a),

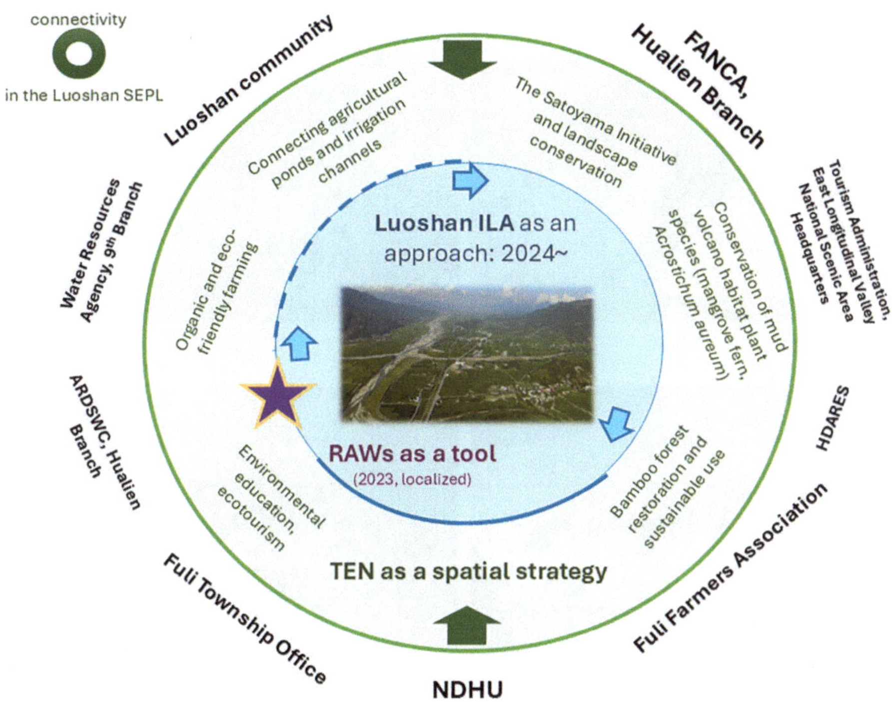

Fig. 10.10 RAWs–ILA–TEN processes and multi-stakeholder engagement in the Luoshan SEPL. (Source: Prepared by the authors. Photo credit: Ying-Ming Huang)

and a localized set of indicators was used for the second round (Karimova et al. 2022).[6] This was the first time for the 20 indicators to have been localized on the island—an example that has been followed ever since.

The main connectivity priorities of 2017–2018 RAWs in the Xinshe SEPLS focused on the need to better understand (1) the ecological interactions between forest–farmlands–river–ocean elements of the Xinshe SEPLS, (2) the role of the Dabuan/Jialang Stream in SEPLS connectivity, and (3) a biodiversity inventory in SEPLS. These community-based concerns were addressed through adjustments to the short-term action plan such as (among others) conducting inventory and monitoring of marine, terrestrial, and agricultural biodiversity; revitalization of fallow farmlands and promotion of eco-friendly farming in coastal paddy terraces; and application of nature-based solutions (NbS) to enhance spawning migration of shrimp.

In 2020, following a successful implementation of the first 3 years of the Xinshe ILSA, RAWs identified more in-depth connectivity topics, which included (1)

[6] For more information on the development and application of the localized resilience indicators in the Xinshe SEPLS, please see Karimova et al. (2022) in the Satoyama Initiative Thematic Review Volume 7.

Fig. 10.11 Photos of the Luoshan SEPL: (**a**) Resilience assessment workshops (RAWs) in the Luoshan SEPL. (Photo credit: FANCA-Hualien); (**b**) Multi-stakeholder platform meeting in the Luoshan SEPL. (Photo credit: FANCA-Hualien); (**c**) Mud volcano tofu, a local specialty of the Luoshan SEPL. (Photo credit: FANCA-Hualien); (**d**) Bamboo forest restoration and eco-tourism at the Luoshan SEPL. (Photo credit: FANCA-Hualien)

improving household water supply and ensuring irrigation water quality, (2) preventing excessive and illegal fishing in the Dabuan/Jialang Stream, and (3) fostering landscape–seascape connectivity with attention to sustainable use of ocean and coastal resources. These newly identified freshwater-marine connectivity priorities took center stage in the mid-term action plan. They, for example, included such action tasks as establishment of a community-based cross-tribal river protection mechanism; protection of the Dabuan/Jialang Stream's water source; eco-friendly and consultative coastal engineering projects; community-based sustainable use and monitoring of marine resources in traditional Indigenous territories; and enhancing knowledge about river and ocean protection laws, regulations, and competent authorities.

Chronologically the Xinshe ILSA was established 2 years prior to TEN. By the time the TEN was introduced, it had already had a clear vision of community-based connectivity needs and priorities. TEN identified the Xinshe SEPLS as a strategic connectivity point within the Northern Section of the Coastal Mountain Range conservation corridor. In the 1990s, with disaster risk reduction considerations in mind,

171 lateral structures (river weirs) were built across the entire length of the Dabuan/Jialang Stream, which blocked the spawning migration of 21 rare local shrimp varieties (Observer Ecological Consultant Co. 2021). Thus, addressing migratory obstruction and restoring freshwater–marine ecological corridors for migratory aquatic organisms has been the main objective of TEN here.

In 2018–2020, with the support from the MSP of the Xinshe ILSA, cross-sectoral and multi-stakeholder negotiations (including engagement of Hualien County government and ecological consultant companies), nine downstream river weirs were dismantled, including the largest one at the river mouth (Observer Ecological Consultant Co. 2021). To date, short-term ecological monitoring of shrimp populations has been showing a substantial increase in numbers (Hualien Natural Education and Ecology Consultant Ltd. 2022). The return of unique shrimp varieties has fostered a sense of pride and ownership among the locals. Shrimp as an indicator of the landscape–seascape ecological connectivity was also introduced into the place-based curriculum at Xinshe primary school. The success of this project has already become an exemplary case study for ecosystem restoration and conservation of aquatic migratory species for other SEPLS across the island.

10.3.3 Gangkou SEPLS

FANCA engagement in the Gangkou SEPLS dates to a terraced paddy restoration and biodiversity conservation project of 2009–2010 and biodiversity inventory project of 2016. Since 2018, TEN spatial planning has been the main strategy behind revitalization and sustainable management of the Gangkou SEPLS. Its strategic location at the intersection of the two conservation corridors has made it into a landscape–seascape biodiversity hotspot nested within the shallow mountains and coastal wetlands.

In 2021, RAWs were employed as a tool to assess community-based perspectives toward local connectivity needs and priorities in a systematic and comprehensive way. After a series of in-depth pre-RAWs consultations with community members, the Xinshe set of 20 indicators was further localized into the Gangkou version (Lin et al. 2023). RAWs results demonstrated that local connectivity priorities focused on the freshwater and marine issues within the Shitigang Stream and the estuary of the Xiuguluan River. They included (1) connectivity of natural waterways and irrigation channels; (2) water quality and availability for household use; (3) decrease in quantity and size (even disappearance) of freshwater and marine species (Chinese mitten-handed crab (*Platyeriocheir formosa*) and giant clam (*Tridacninae*)); (4) the need for poaching control and community-based river and coastal protection mechanism (for clams, coronate noon turban (*Lunella coronata*), silver-mouthed monodont (*Chlorostoma argyrostomum*)), common limpet (*Patella vulgata*), and seaweed); (5) control of invasive alien species (white leadtree (*Leucaena leucocephala*), parthenium weed (*Parthenium hysterophorus*), and American rope (*Mikania micrantha*)); and (6) enhancement of SEPLS-to-SEPLS connectivity alongside the Xiuguluan River conservation corridor (with Luoshan SEPL, Fengnan SEPL, and Qimei SEPL) with the monk goby (*Sicyopterus japonicus*) identified as the connectivity species.

Following post-RAWs analysis, the Gangkou Forest–River–Village–Ocean Eco-Agriculture and Aquaculture Initiative (the Gangkou Initiative) was established in March 2022. The main members of its MSP included the Gangkou locals, a government agency (FANCA-Hualien), and a facilitator (NDHU). Other relevant stakeholders—including county and local government, local school, and ecological consulting and creative industry companies—have also been attending the meetings based on the discussion topics (Figs. 10.8 and 10.9). Unlike the Fengnan ILA and the Xinshe ILSA, the Gangkou ILSA's action plan is structured around its landscape–seascape elements: forest, river, ocean (and intertidal zone), and village (income-generating activities, culture and education, infrastructure and livelihoods, and shared governance).

Connectivity issues identified through RAWs and reflecting TEN conservation objectives have been integrated into the action plan of the Gangkou ILSA. They include: working with Fengbin Township Office on addressing water shortage and waterways connectivity; installation of filtration systems to resolve issues of turbid water quality and availability of water for household use; promotion of ocean-friendly farming and restoration of coastal wetlands; biodiversity inventory and monitoring of land crabs and carnivorous plants such as Burmann's sundew (*Drosera burmannii*); restoration of spawning migration of fish and shrimp in the Shitigang Stream; working with Hualien County government to prevent littering and illegal fishing activities within the coastal zone (e.g., discharge of abandoned fishing nets); and community-based coastal patrol of intertidal flora and fauna.

Out of the four sites, the Gangkou SEPLS presents the strongest marine conservation and satoumi emphasis. Even though marine affairs are outside of TEN's scope and FANCA's area of jurisdiction, priority issues identified by RAWs have helped Gangkou community members to apply for satoumi restoration projects from other government agencies. Since 2022, they have been working with the Ocean Conservation Administration on community-based monitoring and restoration of the giant clam (*Tridacninae*) habitat and coastal patrol.

10.3.4 Luoshan SEPL

The Luoshan SEPL is one of the most recent "dots" within TEN's spatial planning strategy, having only been added as a strategic connectivity point within the Xiuguluan River conservation corridor in 2023. However, its efforts relevant to biodiversity conservation through eco-friendly production date back at least two decades. In addition to being the first organic village on the island (since 2002), in 2017, the Luoshan SEPL and Fuli Farmers' Association joined the International Partnership for the Satoyama Initiative (IPSI) and began actively promoting the Satoyama Initiative.

The Luoshan SEPL is in close vicinity and has close ties with the Fengnan SEPL. It has been long aware of the efforts by the Fengnan ILA and the Turtle River platforms. In October 2022, with support from the Ministry of Agriculture (former Council of Agriculture), FANCA, and NDHU, the Luoshan locals organized a three-day workshop entitled, "The Satoyama Initiative Outlook in the Luoshan SEPL". Community members of the Gangkou SEPLS and Fengnan SEPL were invited as participants and speakers. This was a turning point for realizing the unique role that the Luoshan SEPL can play in safeguarding biodiversity and enhancing connectivity in the upstream and midstream sections of the Xiuguluan River conservation corridor. The Luoshan locals, FANCA, and NDHU made a collective decision vvestablish the Luoshan ILA (Figs. 10.10 and 10.11).

In 2023, following a series of online dialogues with Luoshan community members, the Gangkou version of indicators of resilience in SEPLS was adjusted and localized into the 20 indicators for the Luoshan SEPL. Based on these detailed and place-sensitive indicators, Luoshan RAWs outlined the following connectivity priorities: (1) preventing construction of cemented buffer walls in Luoshan paddy fields; (2) improving water quantity and distribution for household and agricultural use; (3) enhancing connectivity between the Luozai and Jiu'An Streams and irrigation channels; (4) preventing flooding events during typhoons and heavy rains; (5) establishing a stream co-governance mechanism to prevent littering and poaching; (6) investigation of aquatic migratory species in the Luozai and Jiu'An Streams; (7) improving the environmental awareness of Luoshan residents; and (7) control of invasive plants such as parthenium weed (*Parthenium hysterophorus*) and American rope (*Mikania micrantha*).

Starting from 2024, these issues are being incorporated into the Luoshan ILA that is composed of the representatives of the Luoshan community, Fuli Farmers' Association, a government agency (FANCA—Hualien), and a facilitator (NDHU). More stakeholders are expected to join in the future. The action tasks will be closely aligned with TEN strategic action tasks, focusing on, among others, promotion of eco-friendly farming, conservation of rare plants such as the mangrove fern (*Acrostichum aureum*), and bamboo forest restoration and sustainable use.

10.4 Discussion

10.4.1 RAWs–ILSA–TEN: Works as a Bundle and in Various Directions

All four SEPLS in our study have applied the "tool–approach–strategy" bundle: RAWs as a place-sensitive, community-based tool for identification of local and connectivity priorities, ILSA as an approach to operationalize actions, and TEN as a biodiversity-focused spatial planning strategy. Due to local specifics of each SEPLS and their various development agendas and time of engagement (before or

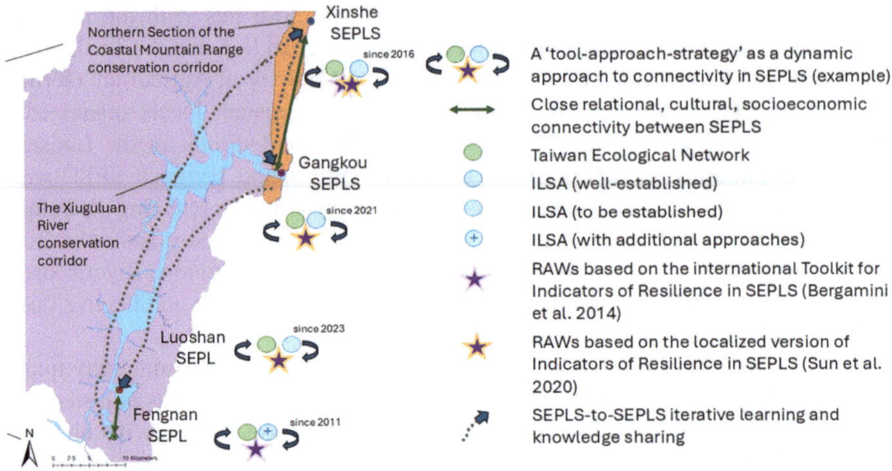

Fig. 10.12 Temporal and spatial application of the RAWs-ILSA-TEN "tool–approach–strategy" bundle within and across SEPLS. (Source: Prepared by the authors). Each bundle is unique in its combination of RAWs (international and/ or localized version) and ILSA (well-established, to be established, or with additional approaches), while TEN demonstrates a dual "top-down" (its objectives guiding local action) and "bottom-up" (local action feeding into its objectives) directionality. Double green arrows show that immediate connectivity synergies are more likely to occur between SEPLS located in closer proximity or possessing similar geo-ecological, sociocultural, and economic backgrounds (Xinshe and Gangkou, Fengnan and Luoshan). Over time (2011–2023), iterative learning and knowledge sharing have helped to enhance various types of connectivity across the four SEPLS (dotted line with arrows)

after 2018), the directionality of the bundle has differed. In each case, however, the combination of the three has helped to ensure synergies *within* each SEPLS as well as *across* SEPLS (Fig. 10.12).

Within SEPLS. In earlier study sites—the Fengnan SEPL and the Xinshe SEPLS—ILSA were established first (2011 and 2016, respectively), while RAWs as an evaluation and issue identification tool were applied later (2015–2016 and 2017–2018, respectively). With the introduction of TEN in 2018, the efforts of both platforms were organically aligned with TEN connectivity objectives. The Fengnan ILA itself was thematically repurposed to better fit local development aspirations in the establishment of the Turtle River Geopark. Thus, the direction of the bundle looked like "approach–tool–strategy." For the Gangkou SEPLS and Luoshan SEPL with chronologically more recent ILSA, the direction of the bundle looked like "strategy-tool-approach," where the biodiversity-focused spatial thinking behind TEN and institutional support from FANCA were the main stimulating factors. RAWs in these cases were applied from the start and their results became the foundation for ILSA action plans.

Across SEPLS. SEPLS do not exist in a vacuum. As complex socio-ecological systems, they are closely linked with each other through a network of ecological corridors and by relational, cultural, and socioeconomic ties (Butler et al. 2022).

Therefore, cross-pollination of "tool–approach–strategy" experiences between SEPLS is an inevitable and desirable phenomenon. One example is the gradual localization of the 20 indicators of resilience based on their earlier versions in other SEPLS (Karimova and Lee 2023). Another iterative learning example is the ILSA itself with the action tasks, MSP composition, and mechanism of regular meetings shared from Fengnan to Xinshe to Gangkou to the Luoshan SEPL (Fig. 10.12).

Whether within SEPLS or across SEPLS, TEN deserves a separate paragraph to highlight its dynamic, inclusive, and adaptive nature. Despite being a national biodiversity-focused spatial planning strategy (a priori a "top-down" arrangement), TEN has managed to effectively incorporate SEPLS processes, community-based bottom-up perspectives, and multi-stakeholder voices to continually re-adjust and refine itself. This is why in reality the "tool–approach–strategy" bundle works in various directions. Its dual directionality is shown with double arrows in Fig. 10.8. Over time, the "dots" on the TEN map also appear and become connected because of rigorous consultative processes and observations. The Luoshan SEPL is a good example of this, when in 2023, following "The Satoyama Initiative Outlook in the Luoshan SEPL" workshop and relevant dialogues, it was added as an important connectivity point within the Xiuguluan River conservation corridor.

From the RAWs–ILSA–TEN perspective, it is also important to have comprehensive knowledge of SEPLS currently located outside the conservation corridors (Fig. 10.1). Their proximity to already well-established "dots" on the map and availability of preliminary biodiversity data can determine how quickly and effectively they can join the connectivity process. Timely identification of such SEPLS and their inclusion within TEN spatial planning is one of the key tasks performed by each branch of FANCA.

10.4.2 Multi-Stakeholder and Cross-Sectoral Partnerships Are Key to Enhancing Connectivity

Engagement of multiple stakeholders and cross-sectoral nexus approaches are central to dealing with sustainability challenges at various scales (Hoff 2018). Connectivity issues in SEPLS are no exception. Our experience in the four SEPLS has elicited the following three characteristics of multi-stakeholder arrangements for effective, sustainable, and equitable partnerships.

The first characteristic is a *community-based and participatory approach* used across all elements of the "tool–approach–strategy" bundle. Continuous engagement of SEPLS communities throughout the process—during issue identification (RAWs), as a part of MSP for ILSA, or for the implementation of TEN strategic action tasks—has allowed for both the "digestion" of the national spatial planning and biodiversity agenda into SEPLS contexts and the incorporation of local perspectives into TEN. Weaving of diverse knowledge types with attention to IPLC engagement has been critical for such projects as payments for ecosystem services for endangered species and important habitats and removal of invasive alien species. SEPLS locals have fully embraced their role as environmental stewards by engaging in community-based monitoring, patrol, and reporting activities (FANCA 2024).

Second, *cross-sectoral coordination* has played a significant role in resource allocation, distribution of action tasks, and inclusion of professional expertise from various government sectors (Chen 2022). The partnership between FANCA and the Water Resources Agency within the Turtle River Species Conservation and Habitat Restoration Platform is a good example of a combination of forest and water, and conservation and eco-engineering areas of expertise. This synergistic approach also means building upon the efforts of other stakeholders involved in SEPLS. These include local schools (Xinshe, Gangkou, and Yongfeng primary schools), ecological consultant companies, arts collectives, Indigenous language promotion associations, and others.

Lastly, it is *science–policy partnership* that has been key for navigating the direction of change in SEPLS and realizing sustainability in practice (Saito 2017). In Hualien County, NDHU and FANCA have been working hand in hand for over 15 years, collectively sensing SEPLS needs and applying intellectual, administrative, and financial resources for the introduction and implementation of the "tool–approach–strategy" bundles. Here, we particularly underline the role of NDHU as a bridging stakeholder (boundary broker) and facilitator of connectivity processes both within and across SEPLS. Skilful—place- and time-sensitive—application of a wide variety of engagement and CEPA tools (local conferences and workshops, exchanges of RAWs experiences) has been crucial for enabling SEPLS-to-SEPLS knowledge sharing.

Among the observed and potential challenges, we point out dependence on the stakeholders' willingness to engage, cooperate, and sustain efforts. For example, the Qimei SEPL was identified by Gangkou locals as an essential connectivity point within the Xiuguluan River conservation corridor, but to date finding local partners willing to engage has been a challenge. Cooperation with private owners and private–public partnership in SEPLS is another important factor. For example, in the case of restoration works in the upper reaches of the Turtle River watershed, private landowners in the Fengnan SEPL agreed to allow the stream water to pass through their lands, which significantly increased the permeability of the freshwater corridor (Sinotech 2022). This is not always the case. In less fortunate situations, private land rights and public benefits such as biodiversity conservation and ecosystem connectivity can be viewed as contested agendas.

10.4.3 Connectivity Has Multiple Meanings

As we stated in the introduction to this study, in the island's context with a high overlap between wildlife biodiversity hotspots and satoyama areas, ecological connectivity goes hand in hand with socioeconomic and cultural connectivity across spatial and temporal dimensions (Karim et al. 2024; Lee 2017). Our experience with the Hualien SEPLS has also shown that connectivity has multiple meanings.

Ecological connectivity in the context of TEN determines species conservation, habitat restoration, and species migration capacity as its main objectives (Chen 2022). These objectives in the four SEPLS are demonstrated through terrestrial and aquatic biodiversity monitoring, restoration of freshwater corridors for aquatic migratory species, and improvement of native habitats for carnivorous plants and land crabs. Attention to other cross-cutting environmental issues, such as flood prevention, improvement of air quality, and climate mitigation and adaptation, has also been important for enhancing ecological connectivity.

It is also clear, however, that in SEPLS, ecological connectivity objectives can only be accomplished with inclusion of a *sociocultural* dimension (Hodgetts 2017). Eco-friendly production activities (B), weaving of Indigenous and local knowledge and modern science (C), community-based co-management systems (D), and nature-based income-generating activities and sustainable livelihoods (E)—all socio-ecological perspectives of the Satoyama Initiative have their indispensable role to play in ensuring landscape diversity and connectivity (A). Our previous research has shown that both RAWs (tool) and ILSA (approach) require all five perspectives to work in synergy (Karimova et al. 2022; Karimova and Lee 2023).

From *spatial* and *temporal* perspectives, connectivity reflects the bundled processes both within and across SEPLS (as discussed in Sect. 10.4.1) and further emphasizes the importance of multi-stakeholder interactions and networking in enhancing ecological connectivity over time. This way, for example, TPSI regional exchange workshops have played a particularly strong role in supporting the TEN agenda (Taiwan Satoyama News 2024). Like Butler et al. (2022), we have also observed that immediate connectivity synergies are more likely to occur across SEPLS located in closer proximity or possessing similar geo-ecological and sociocultural backgrounds (Fig. 10.12).

A comprehensive understanding of the multiple meanings of connectivity in SEPLS can help to address such potential trade-offs as prioritization of some local issues over others (e.g., placing ecological agenda before sociocultural development, and vice versa), or failure to understand the added value of a SEPLS' efforts within the scale of an entire ecological corridor, or balancing short-term development objectives with long-term conservation goals.

10.5 Conclusion

Our study demonstrated how application of community-based RAWs, operationalization of ILSA in SEPLS, and their alignment with biodiversity-focused spatial planning (TEN) were used as a "tool–approach–strategy" bundle in the four SEPLS to enhance connectivity *within* and *across* SEPLS over time. In conclusion, we summarize the main lessons learned and outline ways forward.

- Connectivity requires an open mind. From early on and throughout the process, SEPLS practitioners and all relevant stakeholders need to be mindful of and open to the multiple meanings of connectivity—ecological, sociocultural, spatial, temporal, and others.
- SEPLS management and spatial planning are a two-way street. There is a high value in aligning SEPLS management objectives with a biodiversity-focused spatial planning strategy, and vice versa—adjusting spatial planning objectives in response to SEPLS expectations and priorities.
- RAWs possess a high potential as a community-based tool for monitoring ecological connectivity in SEPLS. They can help to understand the precise meaning of ecological connectivity in each SEPLS context (e.g., as the integrity of all landscape elements or as the presence of indicator migratory species) and elicit priority issues and challenges at each stage of the adaptive co-management process.
- Multi-stakeholder and cross-sectoral partnerships are key for enhancing ecological connectivity within and across SEPLS over time. Inclusive and equitable arrangements that weave different values and knowledge types allow for more just and sustainable outcomes.
- Science–policy–community partnerships are the future of sustainability transformations. Facilitators, also known as bridging stakeholders or boundary brokers, play a key role in enabling lasting and effective connectivity processes. More capacity development, cross-site learning, and knowledge exchange in this regard is highly recommended.

In the future, during its third phase (2026–2029), TEN will continue to promote ecological connectivity. It will focus on improvement of community-based and scientific monitoring and measuring tools by combining RAWs results with long-term biodiversity monitoring surveys, such as CBMIS (community-based monitoring and information systems) . Timely adjustments based on the monitoring and measurement outcomes will be further aligned with the objectives of each conservation corridor, while appropriate cross-comparisons between sites (e.g., within one watershed) will be encouraged. Respect to the community ownership of collected data will be an important consideration throughout these processes. Moreover, closer partnership with the stakeholder networks (TPSI and Community Forestry Network (Community Forestry Network 2025)) and enhanced cross-sectoral collaboration will be imperative for promoting inclusive, participatory, and equitable approaches. Finally, TEN's efforts will be closely aligned with the 2030 targets of the Kunming-Montreal Global Biodiversity Framework. Within all these processes, connecting the "dots" into "lines" and "networks"—點-線-面 connectivity—will continue to play an ever-increasing role.

Acknowledgments We acknowledge with gratitude the continued support of our local partners from Fengnan, Xinshe, Gangkou, and Luoshan SEPLS. Without them, this study would not have been possible. We also deeply appreciate the opportunity to be a part of SITR-10 Volume, as it has allowed us to rethink and reevaluate the true meaning of connectivity in our past, present, and future efforts.

Appendix A: List of Abbreviations

ACM	adaptive co-management
ARDSWC	Hualien Branch, Agency of Rural Development and Soil and Water Conservation, Ministry of Agriculture
Eastern AFA	Eastern Region Branch, Agriculture and Food Agency, Ministry of Agriculture
FANCA	Forestry and Nature Conservation Agency, Ministry of Agriculture
FANCA Hualien	Hualien Branch, Forestry and Nature Conservation Agency, Ministry of Agriculture
HDARES	Hualien District Agricultural Research and Extension Station, Ministry of Agriculture
ILSA	integrated landscape–seascape approach
IPLCs	Indigenous Peoples and local communities
NbS	nature-based solutions
NDHU	National Dong Hwa University
RAWs	resilience assessment workshops
SEPLS	socio-ecological production landscapes and seascapes
TEN	Taiwan Ecological Network
Water Resources Agency, 9th Branch	Ninth River Management Branch of the Water Resources Agency

Appendix B: Basic Information About the Xiuguluan River Conservation Corridor and the Northern Section of the Coastal Mountain Range Conservation Corridor (Source: FANCA Official Database 2024; TaiCol 2025)

Name of the conservation corridor	Corridor type	Geographic scale	Administrative scale	Conservation objectives	Relevance to TEN strategic action tasks	Target eco-systems and habitats	Target conservation species
The Xiuguluan River conservation corridor (#37)	Stream	Xiuguluan River management zone (some areas extend outward including No. 2605 Protected Forest, Wuhe, Hegang, Dewu, Nan'an, and Luoshan) and Turtle Stream (from farmland surrounded by Fudong Highway and other roads to Shicuogou Stream and Zhonggou Stream)	Fengbin Township, Ruisui Township, Wanrong Township, Yuli Township, Zhuoxi Township, Fuli Township, Hualien County	Strengthen estuary environmental management and maintain habitats for rare plants and land crabs. Improve connections between stream waters and surrounding water channels and enhance corridors for migratory species. Promote eco-friendly production and provide breeding and habitat space for aquatic organisms such as Medaka (*Aphyocypris kikuchii*) Strengthen conservation of rare plants in streamside belts, connect streamside plant belts in series, create high beach ecological stepping stones functions, and provide animal movement or habitat use. Strengthen connection between riverside zone and terrestrial habitats to provide diverse habitats for animals.	Restoring native plants and removing invasive alien species. Connecting ecological corridors and constructing animal passages. Identifying high-risk areas and conserving endangered species. Fostering an eco-friendly production environment. Promoting the Satoyama Initiative and landscape conservation. Enhancing communication, education, and public awareness (CEPA).	Forests, streams, riverbanks, farmland, water channels	Chinese softshell turtle (*Pelodiscus sinensis*) Green pond frog (*Pelophylax fukienensis*) Land crabs Taitung river loach (*Hemimyzon taitungensis*) Medaka (*Aphyocypris kikuchii*) Tiny-spotted Rhinogoby (*Rhinogobius delicatus*) Monk goby (*Sicyopterus japonicus*) Rare carnivorous plants: Burmann's sundew (*Drosera burmannii*) Two-flower bladderwort (*Utricularia bifida*) *Utricularia heterosepala* Striped bladderwort (*Utricularia striatula*) Luoshan veronicastrum (*Veronicastrum loshanense*) Witchweed (*Striga lutea*) Oblong-leaf blumea (*Blumea oblongifolia*) Bogrush (*Schoenus falcatus*) Mangrove fern (*Acrostichum aureum*) Tashiro skullcap (*Scutellaria tashiroi*) Pouzolz's bush (*Gonostegia pentandra*) Ryukyu spikemosses (*Selaginella lutchuensis*) Ryukyu Macartney rose (*Rosa bracteata*)

Name of the conservation corridor	Corridor type	Geographic scale	Administrative scale	Conservation objectives	Relevance to TEN strategic action tasks	Target eco-systems and habitats	Target conservation species
The Northern Section of the Coastal Mountain Range conservation corridor (#38)	Coastal	From the east of the coastal mountain ridge to the coast and north of the Xiuguluan River estuary	Shoufeng Township, Fenglin Township, Guangfu Township, Ruisui Township, Yuli Township, Hualien County	Improve longitudinal corridors of independent streams to maintain corridors for migratory species. Maintain connection and habitat quality between streams and coastal forests/flats to maintain land crab diversity. Maintain forest habitat within the Northern and Southern Sections of the Coastal Mountain Range and preserve the pangolin habitat. Promote Satoyama Initiative to improve the coexistence of agricultural production and natural habitats.	Connecting ecological corridors and constructing animal passages. Identifying high-risk areas and conserving endangered species. Fostering an eco-friendly production environment. Promoting Satoyama Initiative and landscape conservation. Enhancing communication, education, and public awareness (CEPA).	Streams, paddy terraces, seaside meadows, and shrubs	Pangolin (*Manis pentadactyla pentadactyla*) Land crabs Migratory species (amphidromy): Kijimuna goby (*Lentipes kijimuna*) Monk goby (*Sicyopterus japonicus*) Red-tailed goby (*Sicyopterus lagocephalus*) Ornate goby (*Sicyopus zosterophorus*) Burny bean (*Mucuna gigantea subsp. tashiroi*) Jamaica swamp sawgrass (*Cladium jamaicense*) Ryukyu spikemosses (*Selaginella lutchuensis*) *Utricularia heterosepala* Purple spikerush (*Eleocharis atropurpurea*) Ornithopod finger grass (*Dimeria ornithopoda*) Witchweed (*Striga lutea*)

References

Bennett G, Mulongloy KJ (2006) Review of experience with ecological networks, corridors and buffer zones. In: CBD Technical Series No 23. Secretariat of the Convention on Biological Diversity, Montreal

Bergamini N, Dunbar W, Eyzaguirre P, Ichikawa K, Matsumoto I, Mijatovic D, Morimoto Y, Remple N, Salvemini D, Suzuki W, Vernooy R (2014) Toolkit for the indicators of resilience in socio-ecological production landscapes and seascapes. UNU-IAS, Biodiversity International, IGES, UNDP, Rome

Butler EP, Bliss-Ketchum LL, de Rivera CE, Dissanayake STM, Hardy CL, Horn DA, Huffine B, Temple AM, Vermeulen ME, Wallace H (2022) Habitat, geophysical, and eco-social connectivity: benefits of resilient socio-ecological landscapes. Landsc Ecol 37:1–29

Catalogue of Life in Taiwan (TaiCol) (2025) Viewed 6 January 2025. Retrieved from https://taicol.tw/zh hant/

Chen CT (2022) Chuanlian guotu baoyu kongjian kuayu tuidong guotu shengtai lvwang: linye yongxu sensen buxi (串聯國土保育空間跨域推動國土生態綠網:林業永續 森森不息 Cross-sectoral promotion of Taiwan Ecological Network to connect conservation spaces: sustainable forestry, forest longevity). Agric Policy Rev 366:12–17. (In Chinese)

Chen CT (2023) Guotu shengtai lvwang tuzi shangxian buju quantai kongjian baoyu celve ("國土生態綠網圖資"上線布局 全臺空間保育策略 Taiwan Ecological Network digital database goes online to lay out the Island's biodiversity-focused spatial planning strategy). Agric Policy Rev 373:86–91. (In Chinese)

Chung KF, Shao KT (2025) Catalogue of Life in Taiwan, viewed 11 January 2025. Retrieved from https://taicol.tw/zh-hant/policy

Community Forestry Network (2025) Shequ linye wangluo judian (社區林業網絡據點 Community Forestry Network), viewed 19 Februarty 2025. Retrieved from https://communityforestry.forest.gov.tw/

FANCA (2024) Binwei wuzhong ji zhongyao qidi shengtai fuwu jifu tuidong fangan (瀕危物種及重要棲地生態服務給付推動方案 Promotion plan for payments for ecosystem services of endangered species and important habitats) FANCA Official Website, viewed 6 March 2024. Retrieved from https://www.forest.gov.tw/0004614

FANCA (2025) Guotu shengtai lvwang (國土生態綠網 Taiwan Ecological Network). FANCA Official Website, viewed 15 January 2024. Retrieved from https://conservation.forest.gov.tw/TEN (In Chinese)

FlyingV (2023) Biexi mi – Nongmin shouhu hechuan muzi jihua (鱉溪米 - 農民守護河川募資計畫 The Turtle River rice brand – community-based fundraising project for river protection), viewed 5 March 2024. Retrieved from https://www.flyingv.cc/projects/32120. (In Chinese)

Hodgetts T (2017) Connectivity as a multiple: in, with and as 'nature'. Area 50(1):83–90

Hoff H (2018) Integrated SDG implementation—how a cross-scale (vertical) and cross-regional Nexus approach can complement cross-sectoral (horizontal) integration. In: Hülsmann S, Ardakanian R (eds) Managing water, soil and waste resources to achieve sustainable development goals. Springer, Cham, pp 149–163

Hualien Natural Education and Ecology Consultant Ltd. (2022) Hualien xian Fengbin xiang dulixi xiliu shengtai diaocha yu peili jihua (花蓮縣豐濱鄉獨流溪溪流生態調查與培力計畫 Independent stream ecological monitoring and capacity building project in Fengbin Township, Hualien County). FANCA Hualien project report. (In Chinese)

Karim PG, Lee KC, Liao RY, Chen MH (2024) You are my mountain, I am your community: rebuilding nature-culture connectivity in Taiwan's Lishan areas. In: Sarmiento FO, Gunya A (eds) Mountain Lexicon. Montology, vol 2. Springer, Cham, pp 223–232

Karimova PG, Lee KC (2022) An integrated landscape-seascape approach in the making: facilitating multi-stakeholder partnership for socio-ecological revitalisation in eastern coastal Taiwan (2016–2021). Sustainability 14(7):1–23

Karimova PG, Lee KC (2023) The many meanings of 'One': the concept and practice of the One Health approach in socio-ecological production landscapes and seascapes. CABI One Health. https://doi.org/10.1079/cabionehealth.2023.0026

Karimova PG, Yan SY, Lee KC (2022) SEPLS well-being as a vision: co-managing for diversity, connectivity and adaptive capacity in Xinshe Village, Hualien County, Chinese Taipei. In: Nishi M, Subramanian SM, Gupta H (eds) Biodiversity-health-sustainability Nexus in Socio-Ecological Production Landscapes and Seascapes (SEPLS). Satoyama Initiative Thematic review, vol 7. Springer, Cham, pp 61–88

Lee KC (2016) Sen-chuan-li-hai: yi Lishan changyi he shengtai nongye qianghua guotu lvse baoyu wangluo (森川里海:以里山倡議和生態農業強化國土綠色保育網絡 Forest-river-village-ocean: the Satoyama Initiative and eco-agriculture for strengthening Taiwan Ecological Network). Nature 134:26–29. (In Chinese)

Lee KC, Karimova PG (2021) From cultural landscape to aspiring geopark: 15 years of community-based landscape tourism in Fengnan Village, Hualien County, Taiwan (2006-2021). Geosciences 11(310):1–24

Lee KC, Yan SY (2019) Participatory planning and monitoring of protected landscapes: a case study of an indigenous rice paddy cultural landscape in Taiwan. Paddy Water Environ 17:539–548

Lee KC, Hsu T, Chiang SY, Kacaw L, Lee BS (2016) Conservation and management strategies for cultural landscapes: a case study of a continuing landscape in Taiwan. J Cult Heritage Conserv 34:66–92

Lee KC, Karimova PG, Yan SY (2019) Towards an integrated multi-stakeholder landscape approach to reconciling values and enhancing synergies: a case study in Taiwan. In: UNU-IAS, IGES (ed) Understanding the multiple values associated with sustainable use in socio-ecological production landscapes and seascapes (SEPLS), Satoyama Initiative Thematic review, vol 5. UNU-IAS, Tokyo, pp 118–135

Lee KC, Karimova PG, Yan SY, Li YS (2020a) Resilience assessment workshops: an instrument for enhancing community-based conservation and monitoring of rural landscapes. Sustainability 12(1):408–422

Lee KC, Karimova PG, Chiu YH, Lin HC (2020b) Taiwan Partnership for the Satoyama Initiative: think global, adapt national, act local. In: Lin KC, Lee KC (eds) Implementing the Satoyama Initiative for the benefit of biodiversity and human well-being. Food and Fertiliser Technology Centre, Taipei, Taiwan

Lin HC, Shia JS, Chen CJ, Wang CC (2018) Linye xinsiwei: Cong senlin dao hai'an - Guotu shengtai baoyu lvse wangluo jianzhi jihua zhi qidong (林業新思維:從森林到海岸 - 國土生態保育綠色網絡建置計畫之啟動 A new way of forestry thinking: from Forest to coast—launch of Taiwan Ecological Network). Taiwan Forest J 44(1):3–10. (In Chinese)

Lin HC, Lo YC, Shih CC, Chang WY, Hsiao CJ, Hsiao YF, Shia JS, Lin CY, Chang HY, Hsu YC, Lee CH, Chang CL, Chang T, Wang TW, Yang JF, Yang CY, Huang CT, Hsu FC, Wu CY, Hsu HC, Karimova PG, He HA, Lee KC, Chen MH, Huang HH, Hseuh ML (2023) International Partnership for the Satoyama Initiative (IPSI) case study platform, viewed 30 June 2023. Retrieved from https://satoyama-initiative.org/case_studies/resilience-assessment-workshops-and-beyond-fostering-adaptive-co-management-in-taiwans-sepls/

Observer Ecological Consultant Co. (2021) Hualien Shengtai baoyu lvse wangluo fazhan jihua (II) (花蓮生態保育綠色網絡發展計畫 II Development of Taiwan Ecological Network in Hualien Region,phase II). FANCA Hualien project report. (In Chinese)

Saito O (2017) Future science-policy agendas and partnerships for building a sustainable society in harmony with nature. Sustain Sci 12:895–899

Shih CC, Chen CT, Lo YC (2021) Taiwan de guotu shengtai lvwang: Shengtai, shenghuo, shengchan jiangu de baoyu jinxingshi (臺灣的國土生態綠網:生態、生活、生產兼顧的保育進行式 Taiwan Ecological Network—a progressive approach to inclusion of ecology, livelihoods, and production in conservation). Taiwan Forest J 47(2):6–8. (In Chinese)

Shiu HJ (2020) Guotu shengtai lvwang lantu guihua ji fazhan jihua (國土生態綠網藍圖規劃及發展計畫 Taiwan Ecological Network planning and development project. FANCA project report), viewed 5 March 2024. Retrieved from https://www.forest.gov.tw/research/0003819 (In Chinese)

Shiu HJ, Tsou MH, Wu TY (2022) Guotu shengtai lvwang lantu guihua: Miandui pingdi ji qianshan ziran baoyu de tiaozhan (國土生態綠網藍圖規劃:面對平地及淺山自然保育的挑戰 Blueprint planning for Taiwan Ecological Network—addressing conservation challenges in Taiwan's flatlands and shallow mountain areas). Taiwan Forest J 47(2):9–14. (In Chinese)

Sinotech Engineering Consultants, Ltd. (2022) Chunggou yexi jishuiqu fangsha gongcheng shengtai gongrong tiaozheng jihua (中溝野溪集水區防砂工程生態共融調整計畫 Chung Gou Stream watershed sediment control and ecological integration project) (In Chinese)

Sun XT, Yan SY, Lee KC (2020) Yingzao renxing de sen-chuan-li-hai: renxing zhibiao zaidihua bing rongru Hualian Xinshe cun shehui-shengtai-shengchan dijing yu haijing de tiaoshi jingying (營造韌性的森-川-里-海:韌性指標在地化並融入花蓮新社村社會-生態-生產地景與海景的調適經營 Localised resilience indicators for adaptive management: building up resilient SEPLS in Xinshe Village, Hualien County, Taiwan). Taiwan Forest J 46(6):58–80. (In Chinese)

Taiwan Satoyama News (2024) FANCA Official Website, viewed 5 March 2024. Retrieved from https://conservation.forest.gov.tw/EN/0002320

Teng LS, Lee CT, Tsai YB, Hsiao LY (2000) Slab breakoff as a mechanism for flipping of subduction polarity in Taiwan. Geology 28:155–158

Tse-Xin Organic Agriculture Foundation (TOAF) (2017) Tiandi jinghua yunyu qingjing zhi zui – Hala mi (天地精華孕育 清淨之最 - 哈拉米 The essence of Heaven and Earth nurtures the purest—Hala Rice), viewed 5 March 2024. Retrieved from https://toaf.org.tw/organic/g/610-2017-04-07-05-53-53 (In Chinese)

Chapter 11
Toward Harmony with Nature: The Sustainable Transformation of Rice Industry to Protect the Ecosystem Services

Chen-Yang Lee, Jung-Chun Chen, Ling-Tsen Chen, Chen-Fa Wu, Szu-Hung Chen, Zhao-Hong Zhang, Ching-Chien Huang, Hao-Yun Chuang, Chun-Hsien Lai, Luu Van Thong Trac, Hao-Wei Hsu, Yu-Chin Chen, and Da-Li Lin

Abstract The Wufeng District in Taiwan, China, has experienced significant infrastructural development, leading to challenges such as soil degradation and habitat loss in local agroecosystems. Farmers face issues like small, fragmented landholdings, resulting in low yields and incomes. Conventional farming practices, characterized by heavy pesticide use, have exacerbated these problems. To address these concerns, it is essential to balance agricultural production with ecosystem services, ensuring both environmental health and farmer livelihoods. The Wufeng Farmers' Association, alongside the Agency of Rural Development and Soil and Water Conservation, plays a critical role in promoting sustainable agricultural practices. By transitioning conventional rice cultivation to eco-friendly methods, as seen in

C.-Y. Lee · L.-T. Chen · H.-Y. Chuang
Agency of Rural Development and Soil and Water Conservation, Nantou City, Nantou County, Taiwan

J.-C. Chen · Z.-H. Zhang · Y.-C. Chen
Taichung Branch Agency of Rural Development and Soil and Water Conservation, Taichung City, Taiwan

C.-F. Wu (✉) · L. V. T. Trac (✉)
Department of Horticulture, National Chung Hsing University, Taichung City, Taiwan
e-mail: cfwu@dragon.nchu.edu.tw; thongtracluuvan@smail.nchu.edu.tw

S.-H. Chen
International Master Program of Agriculture, National Chung Hsing University, Taichung City, Taiwan

C.-C. Huang
Wufeng Farmers' Association, Taichung City, Taiwan

C.-H. Lai · H.-W. Hsu
Taiwan Landscape Environment Association, Kaohsiung City, Taiwan

D.-L. Lin
Taiwan Biodiversity Research Institute, Nantou, Taiwan

M. Nishi et al. (eds.), *Ensuring Ecological Connectivity in Socio-Ecological Production Landscapes and Seascapes (SEPLS)*, Satoyama Initiative Thematic Review, https://doi.org/10.1007/978-981-95-1474-8_11

the Wufeng aromatic rice brand, the association advocates organic farming through the Organic Agriculture Promotion Act, which includes subsidies and market incentives. One notable initiative involves utilizing the black-winged kite, a key species for biological pest control, reducing the need for chemical pesticides. The association collaborates with National Chung Hsing University to establish centers for organic agriculture promotion, certification, and technological investment, enhancing rice industry competitiveness. While the shift toward organic farming supports biodiversity and ecological connectivity, it may also reduce habitat heterogeneity for nontarget species and pose economic and social challenges related to market expansion and equity. By maintaining connectivity through landscape structure metrics such as patch size, edge density, and movement corridors, sustainable land management can be achieved. Raptors like the black-winged kite require home ranges of 5–25 km^2 and foraging distances of 1–5 km, highlighting the need for open landscapes. Ensuring corridors 50–500 m wide supports movement, while reducing pesticide use by 30–50% enhances ecological health and sustainable land management.

Keywords Farmland · Agroecosystems · Organic agriculture · Black-winged kite · Eco-friendly farming · Sustainable rice production · Biodiversity conservation

11.1 Introduction

11.1.1 Geographic and Socioeconomic Overview

The Wufeng District in Taichung City spans around 98 km^2 (Fig. 11.1), with its terrain varying from eastern slopes at elevations of 100–600 m to western floodplains along the Wu River, with elevations of 40–100 m (Table 11.1). The district benefits from a developed irrigation network fed by major rivers such as the Dali, Caohu, Beigou, Qian, and Wu. With a subtropical monsoon climate, Wufeng experiences average temperatures of 19.4 °C, with a peak from May to September and cooler periods in January and February. Rainfall, primarily concentrated in summer, ranges from 1400 to 1670 mm annually. The soil composition supports diverse agricultural activities. The eastern, weakly acidic yellow-brown sandy soil is ideal for lychee and longan cultivation, covering approximately 3100 ha. The western sandy shale alluvial soil is suited for rice, spanning around 1700 ha, including 500 ha dedicated to the aromatic rice variety Tainung 71 (Yihchuan aromatic rice). The Wufu community, located in the northwest, covers over 320 ha and represents 4.2% of the district's total area. Traditionally, the community's economy has been rooted in agriculture, particularly rice cultivation, with muskmelon taking prominence during fallow periods. In 2020, the district reported longan production of over 1170 ha

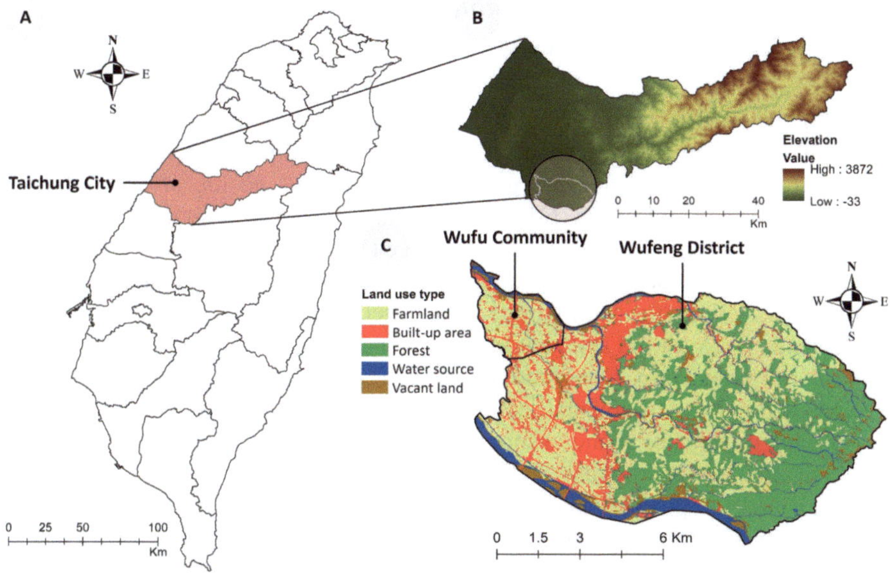

Fig. 11.1 Map of the case study area. (**a**) Administrative map of Taiwan Island; (**b**) Elevation map of Taichung City; (**c**) Map of land use in Wufeng District and the location of Wufu Community. (Source: Luu Van Thong Trac created the maps using the land use data obtained from National Land Surveying and Mapping Center (https://www.nlsc.gov.tw/))

Table 11.1 Basic information of the study area

Country/Region	Taiwan, China
Province	Taichung City
District	Wufeng District
Municipality	N/A
Size of geographical area (ha)	9808
Dominant ethnicity(ies), if appropriate	Han Chinese
Size of case study/project area (ha)	9808
Dominant ethnicity in the project area	Han Chinese
Number of direct beneficiaries (people)	6812
Number of indirect beneficiaries (people)	**65,094**
Geographic coordinates (latitude, longitude)	24.0454226, 120.7222266

yielding over 3500 metric tons, lychee of around 500 ha yielding about 1250 metric tons, and rice cultivation of over 2200 ha yielding over 11,700 metric tons (Taichung City Government 2021). The district also hosts industrial zones in the Wufeng urban area, contributing to commercial growth and population increases.

11.1.2 Agricultural Production and Organic Farming Practices

In 2020, the Wufeng District had 6812 registered farmers, most of whom operated on small landholdings. Accordingly, the average rice cultivation area per household was 0.96 ha, with each hectare generating an annual income of approximately 111,890 New Taiwan Dollars (NTD). Small landholdings necessitate improved economic viability for farmers, who are challenged by climate variability and growing competition from larger agricultural operations and international markets. To address these issues, the community has launched a rural regeneration plan focusing on aromatic rice varieties and natural farming methods. This initiative aims to boost agricultural production, foster community cohesion, and attract youth back to the region. Through environmental, cultural, and industrial revitalization, Wufu seeks to enhance its reputation and ensure sustainable development (Agriculture Bureau of Taichung City Government 2019).

The island's tropical climate drives frequent crop infestations, leading farmers to rely heavily on pesticides and chemical fertilizers. While these practices boost yields, they also result in pesticide residues that pose environmental and health risks, particularly affecting biodiversity and human well-being. Raptors are particularly vulnerable to secondary poisoning from anticoagulant rodenticides used in pest control (Hong et al. 2019). Studies have detected rodenticide residues in 47% of raptor liver samples, highlighting the need for a reassessment of pest control strategies to balance agricultural productivity with environmental conservation (Chen and Shyu 2019).

The Wufeng Farmers' Association plays a crucial role in addressing technical, financial, and production-marketing challenges faced by local farmers. With the island's accession to the World Trade Organization (WTO) in 2002, domestic rice prices have declined due to increased rice imports, impacting farmers' incomes. The government's "Water and Land Utilization Adjustment Plan" aims to stabilize rice production and prices. In this context, the Farmers' Association's efforts in counselling and innovative management are vital for maintaining stable incomes and supporting local agricultural development.

The Organic Agriculture Promotion Act, introduced in 2018, encourages eco-friendly farming practices by offering subsidies, technology, and educational support (Fig. 11.2). The Wufeng Farmers' Association has embraced this act by enhancing rice quality, promoting organic farming, and supporting land leasing for larger-scale operations. While these efforts aim to improve profitability through economies of scale, the association recognizes the potential trade-offs between agricultural intensification and biodiversity (Macchi et al. 2020). To mitigate these trade-offs, adaptive management strategies have been implemented, focusing on maintaining landscape heterogeneity and ecological connectivity. For instance, the association promotes land-sharing practices that integrate seminatural habitats within farmlands, providing critical environments for wildlife species (Ministry of Agriculture 2018). Additionally, woodland patches and grasslands are preserved to support farmland bird diversity, ensuring that agricultural expansion does not come

Fig. 11.2 Eco-friendly farming zone for sustainable rice production and marketing in Wufeng District. (Source: Wufeng Farmers' Association)

at the expense of biodiversity (Sasaki et al. 2020). Collaborating with National Chung Hsing University, the association monitors these ecological dynamics, aiming to balance agricultural growth with biodiversity conservation, particularly in the face of climate variability and landscape changes (Ministry of Agriculture 2018).

11.1.3 Role of Black-Winged Kite as an Indicator Species for Agroecosystems

The black-winged kite (*Elanus caeruleus*, BWK) is a medium-sized raptor with striking white plumage and distinctive black wing patches. It measures 30–37 cm in body length and has a wingspan ranging from 77 to 92 cm (Kemp et al. 2020; Wu et al. 2023). The island of Taiwan, China, as a distinct geographic unit along the East Asian Oceanic Flyway, plays a critical role as a stopover site for southward-migrating raptors from the Northern Hemisphere, especially in autumn (Chen et al. 2021). Some species, such as the BWK, have adapted their migratory strategies and established resident populations on the island (Lin and Su 2019). The BWK was first recorded on the west coast in 2001, and since then, their population has expanded rapidly across the island (Tseng et al. 2018; Wu et al. 2023). It is believed that by 2010, they began colonizing other regions, including Wufeng District. The nearest known breeding site prior to this was around 50–60 km away, along the western coastal areas.

Wufeng District, in particular, with its diverse agricultural landscape, became a key area for BWK establishment. Although specific data on the origin of the Wufeng individuals is limited, it is likely that their spread was driven by the availability of suitable seminatural habitats and prey as they moved inland. The district's agroecosystems are characterized by a mosaic of farmland types, as shown in Fig. 11.3a. To our best knowledge, the movement of BWKs from coastal to inland regions was likely facilitated by ecological corridors that connected suitable habitats across farmland areas. From a previous study by Wu et al. (2023), we identified the key factors supporting the niche establishment of the BWK population, including landscape ecological integrity and farmland habitat connectivity. Furthermore, the expansion of the BWK population is likely supported by land-use changes that favor open landscapes with scattered trees (Bed'Hom et al. 2003; Balbontín et al. 2008). Over the past decade, BWKs have been frequently observed foraging and breeding in agroecosystems (Howard et al. 2016). Their movements and breeding performance are closely tied to rodent abundance (Mendelsohn 1982; Mendelsohn and Jaksić 1989).

We believe that the BWK plays a key role in maintaining ecosystem health in the island's agroecosystems. Raptors like the BWK serve as umbrella species. As a high trophic consumer, the BWK is a specialist in preying on small mammals, such as *Bandicota bengalensis* and *Rattus nitidus* (Parejo et al. 2001; Manaa et al. 2013), which often damage crops (Chakma et al. 2019; Witmer 2022). The presence of birds of prey in agroecosystems helps regulate rodent populations, reducing the need for chemical pest controls like rodenticides (Yonas and Leirs 2019; Paz Luna et al. 2020; Zagorski and Swihart 2020), which can harm the environment and non-target species through secondary poisoning (Lohr and Davis 2018; Ravindran et al. 2022). Raptors serve as natural pest controllers, benefiting both agriculture and biodiversity by helping to reduce conflicts between food production and conservation efforts (Gómez-Ramírez et al. 2014).

Efforts to promote raptor conservation and sustainable agriculture have fostered collaboration among stakeholders. Eco-friendly farming practices, introduced in 2014, successfully addressed rodent control issues while supporting agricultural sustainability. The distribution of BWKs within the Wufu Community is primarily concentrated in the southern and central areas, where a variety of farmland types, including paddy fields, dry fields, and orchards, are present (Fig. 11.3b). The minimum convex polygon defines the area utilized by the BWK, with key observation points overlapping an eco-friendly farming zone. The eco-friendly farming zone, characterized by sustainable agricultural practices, is vital for enhancing habitat

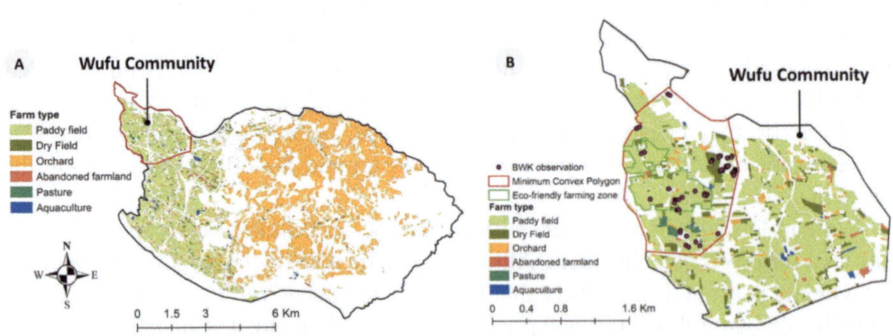

Fig. 11.3 Farmland areas in the study area. (**a**) Location of six farm types in Wufeng District; (**b**) distribution of BWKs in Wufu Community. (Source: Luu Van Thong Trac created the maps using the land use data obtained from National Land Surveying and Mapping Center (https://www.nlsc.gov.tw/) and bird observation data obtained from Taiwan Biodiversity Research Institute (https://www.tbri.gov.tw/))

Fig. 11.4 Farmers working together to set up perch-mounted camera traps and BWK preying on perch-mounted camera traps. (Source: Taichung Branch Agency of Rural Development and Soil and Water Conservation)

suitability for birds. Eco-friendly farming promotes landscape connectivity by providing corridors for the BWK to move between foraging and roosting sites. Moreover, the zone improves habitat quality by reducing harmful rodenticides and encouraging BWK in farmland ecosystems to control rodents. Those practices mutually benefit farmers by improving yields and promoting environmental sustainability. Further conservation measures, such as the installation of bamboo perches in 2017 (Fig. 11.4), increased available hunting perches for the kites, enhancing their pest control effectiveness. The use of infrared cameras and GPS tracking technology has enabled better monitoring of kite activity, revealing a considerable range of movement within the farmland landscape of Wufu Community. Although recorded predation rates show a significant impact on pest populations, field observations suggest that their contribution to pest regulation may be even greater than that the current data indicates.

11.2 Description of Activities

11.2.1 Wufeng Farmers' Association Provides Guidance to Farmers

Since the first crop of rice in 2002, the Wufeng Farmers' Association has collaborated with agricultural research institutions to promote local production of Tainung 71 (Yihchuan aromatic rice), a high-quality japonica rice variety known as "Wufeng fragrant rice". This collaboration, overseen by the Taiwan Agricultural Research Institute, Ministry of Agriculture, facilitated the transformation of the rice industry. To ensure rice quality, the Wufeng Farmers' Association actively assisted in establishing rice production and marketing classes, employing comprehensive contract cultivation methods. They conducted regular field inspections, strictly controlled pesticide and fertilizer usage, and invested significantly in computerized rice drying and low-temperature storage processes. Additionally, they imported Japanese four-view color-sorting machines to eliminate impurities and established the "Wufeng fragrant rice" brand. Through contract procurement, they stabilized producers' income, enhanced economic incentives, and ensured the production of high-quality rice. By 2020, the contracted rice cultivation area in Wufeng had expanded to approximately 267 ha, yielding around 1555 metric tons per period, accounting for approximately 23% of the total rice cultivation area in Wufeng, significantly contributing to the development of the local rice industry and stabilizing farmers' income (Huang et al. 2020).

11.2.2 Establishing an Organic Rice Production Verification Mechanism Through Industry–Academia Cooperation

The Ministry of Agriculture (MOA) collaborates with National Chung Hsing University (NCHU) to establish centers for promoting organic agriculture, pesticide residue testing, agricultural product verification, NCHU mart, and organic farmers' markets. These centres aid county governments and the Wufeng Farmers' Association in fostering organic farming practices and resolving production challenges. They provide research support and professional education services and assist in developing organic rice cultivation plans. The NCHU Agricultural Products Verification Center ensures compliance with regulations and offers training in organic crop farming and for auditing staff. The NCHU Pesticide Residue Analysis Center conducts safety checks to uphold food hygiene, enhance market competitiveness, and safeguard the rights of farmers and consumers, especially in the Wufeng District, ensuring rice quality and compliance with pesticide regulations.

Establishing the NCHU Mart at National Chung Hsing University (https://shop.nchu.edu.tw) aims to integrate research, teaching, and service, developing innovative agricultural products distinctive to the college. It aims to create a learning, internship, production, quality control, and marketing-integrated exhibition platform, collaborating with the Wufeng Farmers' Association to set up a Taichung City agricultural specialty counter in the NCHU mart, showcasing organic, eco-friendly, traceable, or top 100 agricultural products of Taichung's various farmers' associations. The Organic Farmers' Market has 30 booths, operating every Saturday from 08:00 to 12:00, intending to assist disadvantaged small farmers through local production and consumption, building consumer confidence through university verification and supervision, and promoting organic farming methods further. All market farm products must obtain organic verification qualifications approved by the verification unit to be sold, with many farms currently verified by institutions such as the MOA, the Tse-Xin Organic Agriculture Foundation (TOAF/TOC), NCHU, and the Taiwan Organic Production Association (TOPA). To safeguard consumer rights, the market periodically commissions inspection agencies such as the Taipei

Liu-Kung Agriculture Foundation to inspect pesticide residues in market-sold agricultural products, ensuring food safety. Organic farmers from the Wufeng District practicing organic farming can also join to sell organic agricultural products.

11.2.3 Enterprises' Support Through Adopting Organic Rice Fields

The Wufeng Farmers' Association promotes a specialized rice production and sales zone through contractual arrangements. Local farmers employ eco-friendly and organic cultivation methods, eschewing pesticides and fertilizers, resulting in notable success. Expanding the promotion of eco-friendly cultivation, the Wufeng Farmers' Association initiated the Hundred Acres Revitalization Project in 2017. Through the direct involvement of local enterprises in land adoption, trust and cooperation between farmers and enterprises are fostered. Farmers, supported by the assurance of enterprise backing, provide safe and healthy rice to adopting enterprises. Through rice field adoption, enterprises fulfill social responsibility, enhance corporate image, and contribute to ecological improvement, preserving the authentic and pure rural environment for future generations. Enterprises may adopt rice fields for current or upcoming harvest seasons, with adoption based on a 0.1 ha unit, totalling 30 units at 38,000 NTD per unit.

11.2.4 Developing "Black-Winged Kite" Brand Rice

In August 2019, the Wufeng Farmers' Association introduced the "Black-winged Kite" brand rice made from eco-friendly field-cultivated Yihchuan aromatic rice (Fig. 11.5). With an annual sales volume of approximately 10,000 bags, the initiative aimed to promote the concept of black-winged kites in rodent control throughout the island. One hundred bags of "Black-winged Kite" rice were donated for charity sales, with proceeds allocated to bird-related research at the National Pingtung University of Science and Technology. The charity sale concluded within 3 days, indicating widespread societal recognition and support for eco-friendly products. Subsequently, "Black-winged Kite" rice was distributed nationwide through agricultural cooperative marketing channels and Taichung Branch Agency of Rural Development and Soil and Water Conservation (ARDSWC) service outlets. It also gained popularity through direct sales at the NCHU mart, the direct sales at stores established by the Wufeng Farmers' Association, and sales at farmers' markets.

Fig. 11.5 Local agricultural products with the "Black-winged Kite" rice brand. (Source: Wufeng Farmers' Association)

11.2.5 Establishing a Winery for the Production of Rice Wine

To establish a winery for the production of rice wine aimed at addressing potential rice surpluses, prolonged storage challenges, and enhancing the economic value of rice products, the Wufeng Farmers' Association Winery (Agriculture Bureau of Taichung City Government 2023), since 2006, initiated the learning of sake brewing in Niigata Prefecture, Japan, by its first-generation brewmaster. It commenced utilizing the Tainung 71 rice variety from the Wufeng District for sake production. Crafted from the Yihchuan aromatic rice, the exquisite "Hatsugiri Shochu" garnered numerous accolades, including the gold and silver awards at the Golden and Silver Award of German International Spirits Award from 2010 to 2014. Similarly, the "Hatsugiri Junmai Ginjo" received the Silver Medal Award at the International Liquor Awards in Paris, France in 2011 and 2015. In addition to fragrant rice, lychee and honey are significant agricultural products in the Wufeng region. The research-developed Lychee Honey Wine attained the Gold Medal at the Concours Mondial de Bruxelles in Belgium in 2014 (Chen 2018).

11.3 Results

11.3.1 Transforming Farmers' Behavior Toward Organic Farming and Biodiversity Conservation

The rapid decline in raptor habitats, exacerbated by human activities such as pesticide use and pollution, poses a significant threat to raptor populations. This is particularly true for species like the BWK, whose numbers are dwindling due to habitat fragmentation and uniformity (Wu et al. 2023). Initially, chemical pesticides were introduced in agriculture to control pests, but their impact on raptors at the top of the food chain has been detrimental. As a result, conflicts between raptors and farmers have intensified, further endangering these birds. While some raptor species have adapted to these environmental changes, many like the BWK may face the risk of local extirpation.

In response, a collaboration between the Taichung Branch ARDSWC and Pingtung University of Science and Technology has been established to monitor BWK breeding in Wufeng's ecosystems. The discovery of successful breeding in October 2018, with three chicks raised and migrating to other fields by February 2019, highlights the importance of conserving their habitats. Farmers have played a crucial role in this effort by monitoring the breeding process and protecting nest areas from human interference.

From 2017 to 2020, a shift in farmers' attitudes toward the BWK was observed as they engaged in ecological observation and conservation practices. By installing artificial perches to attract these raptors for natural pest control, farmers not only protected their crops but also contributed to raptor conservation. This initiative was further supported by the Taichung Branch ARDSWC, which provided training on effective perch installation, inspiring neighboring communities to adopt similar practices. The success in Wufeng even led to increased bird activity in Longjing, where Eurasian scops owls began using the perches for nocturnal hunting. The transformation in farming practices underscores the potential for integration of organic farming with biodiversity conservation. By aligning agricultural activities with ecological goals, farmers can contribute to the preservation of raptor species like the BWK, while also enhancing their agricultural productivity and sustainability (Liu 1999).

11.3.2 Trade-Offs and Ecological Connectivity in Transforming Wufeng's Rice Industry

The transformation of the rice industry in the Wufeng region presents a complex interplay between ecological conservation, economic viability, and social dynamics, which directly impacts ecological connectivity. As the region shifts toward organic farming and biodiversity conservation, particularly focusing on the BWK species, several trade-offs emerge that need careful consideration.

Ecologically, the reduction in pesticide use and the creation of habitats conducive to raptors enhance the survival and reproductive success of wildlife species like the BWK (Cech et al. 2022; Moreau et al. 2022). These changes support ecological connectivity by maintaining landscape-level habitat networks, which are crucial for species movement and gene flow (LaPoint et al. 2015; Préau et al. 2022). However, the uniformity of agricultural landscapes, driven by agricultural intensification, might reduce habitat heterogeneity, potentially limiting biodiversity and disrupting the connectivity of nontarget species (Burel et al. 2004; Emmerson et al. 2016). While raptors benefit from these modifications, other species may face habitat loss, leading to a complex dynamic where connectivity is selectively improved (Burel et al. 2004; Firbank et al. 2007).

Economically, the introduction of ecologically branded products, such as "Black-winged Kite" rice, significantly increases the value of agricultural products, raising farmer incomes. The premium pricing strategy creates a niche market for eco-friendly products, encouraging further adoption of sustainable practices (Kumar and Christodoulopoulou 2014; Barge et al. 2015). However, this economic model might alienate lower-income consumers and pose a risk if the market for high-priced ecological products fails to expand (Magnani 2000; Mantovani et al. 2017). Additionally, the financial burden on farmers to invest in these practices without immediate returns could create disparities, potentially leading to a reversion to conventional methods if the economic benefits are not equitably distributed (Kerselaers et al. 2007; Łuczka and Kalinowski 2020; Michalke et al. 2023).

Socially, the collective action facilitated by the Wufeng Farmers' Association strengthens community engagement in conservation efforts. Farmers' involvement in monitoring raptor populations and managing habitats fosters a sense of shared responsibility, which is crucial for the long-term success of ecological initiatives (Greiner 2015; Rizzardi Orlandi et al. 2024). However, this collective approach may also limit individual autonomy and lead to social conflicts, particularly if the distribution of resources and benefits is perceived as unequal (Hauck et al. 2013; Skrimizea et al. 2020).

11.4 Discussion

The initiative exemplifies the operationalization of socio-ecological production landscapes and seascapes (SEPLS) in spatial planning. Through collaborative conservation efforts and innovative pest control methods, we ensured ecological connectivity, mitigated rodent issues, and increased farmers' incomes. By emphasizing the interplay between environmental, economic, and agricultural elements, Wufeng's model offers insights into sustainable development. SEPLS management efforts in spatial planning can enhance ecological connectivity by promoting practices that harmonize agricultural productivity with biodiversity preservation and economic viability, ensuring balanced and resilient ecosystems.

11.4.1 Farmers' Shift to Eco-Friendly Practices

In Wufeng District, farmers have embraced organic farming methods, fostering a harmonious relationship among agricultural practices, farmland, and biodiversity. This transformation has been driven by initiatives from agricultural experimental stations and farmers' associations, which have shifted perceptions from conventional to eco-friendly practices. Initially, the transition faced challenges such as pest outbreaks and reduced rice yields. However, these were mitigated with support from the Wufeng Farmers' Association and institutions like National Chung Hsing University. The introduction of BWK for rodent control has shifted farmers' attitudes from viewing raptors as pests to recognizing their role in conservation, enhancing both ecological balance and rice production (Gómez-Ramírez et al. 2014; Donázar et al. 2016).

11.4.2 Ecological Industry and Governmental Support

Development of an ecological industry across the island, bolstered by increasing consumer support for organic products, illustrates the connection between regional ecological conservation and agricultural economics (Lu and Cheng 2023). Since 1999, the area under organic farming on the island has expanded significantly, reflecting growing consumer trust in organic agriculture (Hsieh 2005; Tsai et al. 2021). The Organic Agriculture Promotion Act of 2018 further accelerated this trend, increasing organic farmland and fostering ecological habitats that support species like raptors and mammals (Petway et al. 2019; Lu and Cheng 2023). The Wufeng Farmers' Association has leveraged this consumer acceptance to drive large-scale organic farming, which not only protects BWK habitats but also boosts farmers' incomes through economies of scale and value-added products like premium sake. The association's efforts have created a sustainable economic model that integrates ecological conservation with agricultural productivity.

11.4.3 Cross-Domain Collaboration and Industry Innovation

Rice, a staple in the island's agriculture, faced market saturation and competition with imports following entry into the WTO (Chiang et al. 2004). The Wufeng Farmers' Association responded with innovative strategies, including the adoption of resilient rice varieties like Tainung 71, and collaborations with the Taiwan Agricultural Research Institute to advance breeding techniques. Eco-friendly cultivation practices have reshaped the rice industry, with initiatives like the Baijia Liangtian Restoration Project underscoring a commitment to sustainability. This project was initiated by the Wufeng Farmers' Association to encourage local businesses to collaborate with the association. With the support of the adopting businesses, the farmers can provide safe, healthy, and pesticide-free rice. On the other hand, businesses can fulfill their social responsibility while contributing to the conservation of rice paddies and BWK habitats. Legislative support, such as the Organic Agriculture Promotion Act, has facilitated this transition, promoting a robust ecosystem for organic rice production. Successful cross-domain collaboration has been crucial to this transformation. Partnerships with local organizations and investments in advanced equipment have enabled market expansion and competitiveness. The integration of ecological balance, exemplified by the BWK, has strengthened product marketing and consumer confidence, highlighting the role of environmental stewardship in agricultural success (Fig. 11.6).

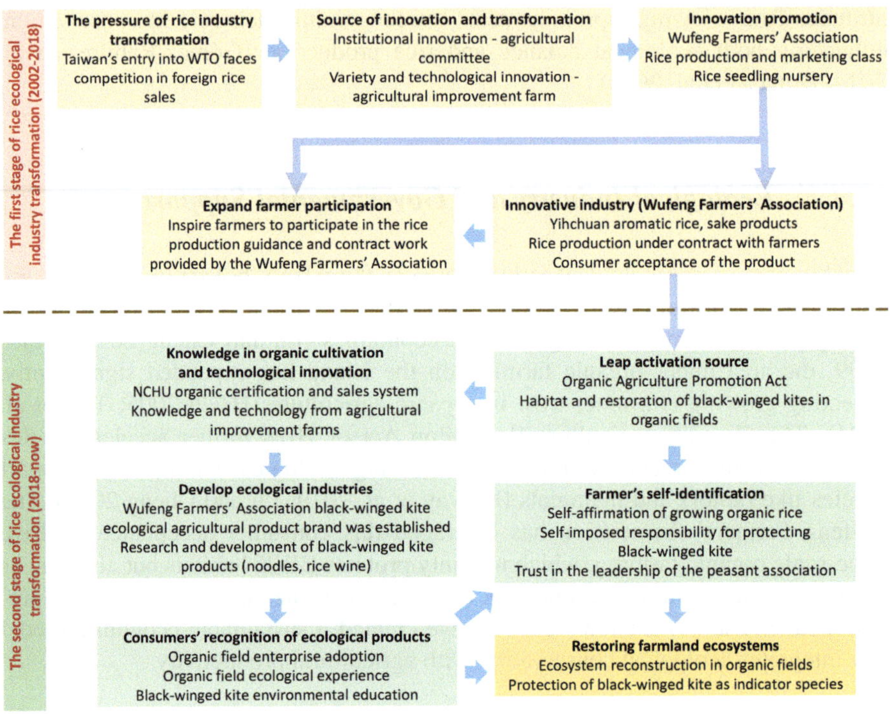

Fig. 11.6 Transformation mechanism of the ecological rice industry in Wufeng District. (Source: Luu Van Thong Trac)

11.4.4 Carbon Reduction in Organic Rice Production

The environmental impact of rice production, specifically its carbon footprint, has been a focus of recent initiatives (Ferng 2011; Hu et al. 2019; Zhang et al. 2021). We acknowledge that Wufeng District's rice production potentially emits significant carbon dioxide (CO_2). The implementation of eco-friendly practices has been proven to result in notable reductions in emissions globally (Venkat 2012; Lin and Fukushima 2016; Squalli and Adamkiewicz 2018; Holka et al. 2022). These practices include the adoption of organic farming techniques, reduced reliance on chemical fertilizers, and improved water management methods, which help minimize methane emissions—a major contributor to the carbon footprint in paddy fields (Lin and Fukushima 2016; Lu et al. 2020). Additionally, energy-efficient machinery and reduced tillage have further lowered CO_2 emissions (Smith et al. 2015; Sun et al. 2022). Our efforts in agricultural transition are reflected in the carbon labeling of products like Tainung 71 rice, the first to receive a Ministry of Environment-certified carbon label. Finally, studies on Taiwanese consumer behavior reveal strong support for carbon labeling, with a majority willing to pay a premium for products that reflect energy-efficiency improvements and reduced carbon emissions (Ferng 2011; Liu et al. 2016; Liang et al. 2020).

11.5 Conclusions: Key Messages

The sustainability of socio-ecological production landscapes and seascapes (SEPLS) necessitates a comprehensive approach and cross-sector collaboration involving communities, governments, and academia. This case underscores several vital points. First, the successful breeding of black-winged kites has prompted farmers to adopt organic farming methods and prioritize biodiversity conservation. Second, the emergence of consumer-driven ecological industries has facilitated the integration of regional ecological conservation with agricultural economic interests. Third, collaborative efforts across different domains have enabled the transition of the rice industry, leading to the establishment of regional rice ecological production chains. Lastly, organic rice cultivation has demonstrated a significant contribution to carbon reduction initiatives. Consequently, the experiences of the Wufu Community offer valuable insights for the practical implementation of the International Partnership for the Satoyama Initiative (IPSI) in hillside agricultural landscapes across East Asia.

References

Agriculture Bureau of Taichung City Government (2019) Wufu Community, Wufeng District, viewed 1 March 2024. Retrieved from https://www.agriculture.taichung.gov.tw/1042359/post/

Agriculture Bureau of Taichung City Government (2023) Wufeng Farmers' Association Winery, viewed 4 March 2024. Retrieved from https://www.agriculture.taichung.gov.tw/2364536/post

Balbontín J, Negro JJ, Sarasola JH, Ferrero JJ, Rivera D (2008) Land-use changes may explain the recent range expansion of the Black-shouldered Kite Elanus caeruleus in southern Europe. Ibis 150(4):707–716

Barge DS, More DK, Bhola SS (2015) Eco friendly products attitude towards pricing. Pravara Manag Rev 13(2):29–35

Bed'Hom B, Coullin P, Guillier-Gencik Z, Moulin S, Bernheim A, Volobouev V (2003) Characterization of the atypical karyotype of the black-winged kite Elanus caeruleus (Falconiformes: Accipitridae) by means of classical and molecular cytogenetic techniques. Chromosom Res 11(4):335–343

Burel F, Butet A, Delettre YR, Millàn de la Peña N (2004) Differential response of selected taxa to landscape context and agricultural intensification. Landsc Urban Plan 67(1):195–204

Cech RM, Jovanovic S, Kegley S, Hertoge K, Leisch F, Zaller JG (2022) Reducing overall herbicide use may reduce risks to humans but increase toxic loads to honeybees, earthworms and birds. Environ Sci Eur 34(1):44

Chakma N, Sarker NJ, Sarker SU, Sarker SK, Shafali RB, Belmain SR (2019) Impact of trap barrier systems on rodent damage to upland rice cropping systems during bamboo masting events. Crop Prot 126:104939

Chen C-W (2018) Winemaking Plant Tour. Country road 44(8):82–83

Chen H-S, Shyu T-H (2019) Analytical method for the determination of residual anticoagulant Rodenticides in Raptorial bird livers. Taiwan Pestic Sci 6(6):105–117

Chen C-C, Lin J-G, Cheng Y-J, Chen T-C, Tsai J-S, Sun Y-H (2021) Autumn migration of Diurnal Raptors across Kenting National Park, Taiwan, along the East-Asian Oceanic flyway. J Raptor Res 56(1):95–100

Chiang F-S, Sun C-H, Lin C-H (2004) The impact of Taiwan's WTO entry on its domestic agriculture sector. Rev Urban Region Dev Stud 16(1):1–13

Donázar JA, Cortés-Avizanda A, Fargallo JA, Margalida A, Moleón M, Morales-Reyes Z, Moreno-Opo R, Pérez-García JM, Sánchez-Zapata JA, Zuberogoitia I, Serrano D (2016) Roles of raptors in a changing world: from flagships to providers of key ecosystem services. Ardeola 63(1):181–234

Emmerson M, Morales MB, Oñate JJ, Batáry P, Berendse F, Liira J, Aavik T, Guerrero I, Bommarco R, Eggers S, Pärt T, Tscharntke T, Weisser W, Clement L, Bengtsson J (2016) How agricultural intensification affects biodiversity and ecosystem services. In: Dumbrell AJ, Kordas RL, Woodward G (eds) Advances in ecological research, vol 55. Academic Press, pp 43–97

Ferng J J (2011) Measuring and locating footprints: a case study of Taiwan's rice and wheat consumption footprint. Ecol Econ 71:191–201

Firbank LG, Petit S, Smart S, Blain A, Fuller RJ (2007) Assessing the impacts of agricultural intensification on biodiversity: a British perspective. Philos Trans R Soc B Biol Sci 363(1492):777–787

Gómez-Ramírez P, Shore RF, van den Brink NW, van Hattum B, Bustnes JO, Duke G, Fritsch C, García-Fernández AJ, Helander BO, Jaspers V, Krone O, Martínez-López E, Mateo R, Movalli P, Sonne C (2014) An overview of existing raptor contaminant monitoring activities in Europe. Environ Int 67:12–21

Greiner R (2015) Motivations and attitudes influence farmers' willingness to participate in biodiversity conservation contracts. Agric Syst 137:154–165

Hauck J, Görg C, Varjopuro R, Ratamäki O, Jax K (2013) Benefits and limitations of the ecosystem services concept in environmental policy and decision making: some stakeholder perspectives. Environ Sci Pol 25:13–21

Holka M, Kowalska J, Jakubowska M (2022) Reducing carbon footprint of agriculture—can organic farming help to mitigate climate change? Agriculture 12(9):1383

Hong S-Y, Morrissey C, Lin H-S, Lin K-S, Lin W-L, Yao C-T, Lin T-E, Chan F-T, Sun Y-H (2019) Frequent detection of anticoagulant rodenticides in raptors sampled in Taiwan reflects government rodent control policy. Sci Total Environ 691:1051–1058

Howard KA, Schwertner TW, Muir JP, Monadjem A (2016) Habitat selection of the Black-shouldered Kite Elanus caeruleus in agroecosystems of Swaziland. Ostrich 87(2):155–159

Hsieh S-C (2005) Organic farming for sustainable agriculture in Asia with special reference to Taiwan experience, viewed 10 October 2024. Retrieved from https://km.fftc.org.tw/article/874

Hu AH, Chen C-H, Huang LH, Chung M-H, Lan Y-C, Chen Z (2019) Environmental impact and carbon footprint assessment of Taiwanese agricultural products: a case study on Taiwanese Dongshan tea. Energies 12(1):138

Huang C-C, Chi S-Y, Chien L-H (2020) An analysis of the intention of farmers' participation in organizing contracts – a case of yihchuan aromatic rice farmingcontract of farmers' association in Wufeng district. Journal of Agriculture and Forestry 67(3):195–205

Kemp AC, Kirwan GM, Marks JS, Motis A, Garcia E (2020) Black-winged Kite (*Elanus caeruleus*), version 1.0. In: del Hoyo J, Elliott A, Sargatal J, Christie DA, de Juana E (eds) Birds of the world. Cornell Lab of Ornithology, Ithaca. https://doi.org/10.2173/bow.bkskit1.01

Kerselaers E, De Cock L, Lauwers L, Van Huylenbroeck G (2007) Modelling farm-level economic potential for conversion to organic farming. Agric Syst 94(3):671–682

Kumar V, Christodoulopoulou A (2014) Sustainability and branding: an integrated perspective. Ind Mark Manag 43(1):6–15

LaPoint S, Balkenhol N, Hale J, Sadler J, van der Ree R (2015) Ecological connectivity research in urban areas. Funct Ecol 29(7):868–878

Liang T-C, Situmorang ROP, Liao M-C, Chang S-C (2020) The relationship of perceived consumer effectiveness, subjective knowledge, and purchase intention on carbon label products—a case study of carbon-labeled packaged tea products in Taiwan. Sustainability 12(19):7892

Lin H-C, Fukushima Y (2016) Rice cultivation methods and their sustainability aspects: organic and conventional rice production in industrialized tropical monsoon Asia with a dual cropping system. Sustainability 8(6):529

Lin J-C, Su S-J (2019) Taiwan's Geoparks. In: Lin J-C, Su S-J (eds) Geoparks of Taiwan: their development and prospects for a sustainable future. Springer International Publishing, pp 1–53

Liu C-Y (1999) Compact service activities of farmers' associations in Taiwan. Rev Agric Exten Sci 16:45–65

Liu T, Wang Q, Su B (2016) A review of carbon labeling: standards, implementation, and impact. Renew Sust Energ Rev 53:68–79

Lohr MT, Davis RA (2018) Anticoagulant rodenticide use, non-target impacts and regulation: a case study from Australia. Sci Total Environ 634:1372–1384

Lu C-F, Cheng C-Y (2023) Exploring the distribution of organic farming: findings from certified rice in Taiwan. Ecol Econ 212:107915

Lu H-l, Chang Y-H, Wu B-Y (2020) The compare organic farm and conventional farm to improve sustainable agriculture, ecosystems, and environment. Org Agric 10(4):409–418

Łuczka W, Kalinowski S (2020) Barriers to the development of organic farming: a polish case study. Agriculture 10(11):536

Macchi L, Decarre J, Goijman AP, Mastrangelo M, Blendinger PG, Gavier-Pizarro GI, Murray F, Piquer-Rodríquez M, Semper-Pascual A, Kuemmerle T (2020) Trade-offs between biodiversity and agriculture are moving targets in dynamic landscapes. J Appl Ecol 57(10):2054–2063

Magnani E (2000) The Environmental Kuznets Curve, environmental protection policy and income distribution. Ecol Econ 32(3):431–443

Manaa A, Souttou K, Sekour M, Bendjoudi D, Guezoul O, Baziz-Neffah F, Doumandji S, Stoetzel E, Denys C (2013) Diet of Black-shouldered Kite Elanus caeruleus in a farmland area near Algiers, Algeria. Ostrich 84(2):113–117

Mantovani A, Tarola O, Vergari C (2017) End-of-pipe or cleaner production? How to go green in presence of income inequality and pro-environmental behavior. J Clean Prod 160:71–82

Mendelsohn JM (1982) The feeding ecology of the Black shouldered Kite Elanus caeruleus Aves: Accipitridae. Durban Mus Novit 13:75–116

Mendelsohn JM, Jaksić FM (1989) Hunting behaviour of Blackshouldered Kites in the Americas, Europe, Africa and Australia. Ostrich 60(1):1–12

Michalke A, Köhler S, Messmann L, Thorenz A, Tuma A, Gaugler T (2023) True cost accounting of organic and conventional food production. J Clean Prod 408:137134

Ministry of Agriculture (2018) Organic Agriculture Promotion Act, viewed 31 August 2022. Retrieved from https://law.moj.gov.tw/ENG/LawClass/LawAll.aspx?pcode=M0030093

Moreau J, Rabdeau J, Badenhausser I, Giraudeau M, Sepp T, Crépin M, Gaffard A, Bretagnolle V, Monceau K (2022) Pesticide impacts on avian species with special reference to farmland birds: a review. Environ Monit Assess 194(11):790

Parejo D, Aviles J, Ferrero J, Rivera D, Casas J (2001) Communal roosting and diet of Black-shouldered Kites (Elanus caeruleus) wintering in southwestern Spain. J Raptor Res 35(2):162–164

Paz Luna A, Bintanel H, Viñuela J, Villanúa D (2020) Nest-boxes for raptors as a biological control system of vole pests: high local success with moderate negative consequences for non-target species. Biol Control 146:104267

Petway JR, Lin Y-P, Wunderlich RF (2019) Analyzing opinions on sustainable agriculture: toward increasing farmer knowledge of organic practices in Taiwan-Yuanli township. Sustainability 11(14):3843

Préau C, Tournebize J, Lenormand M, Alleaume S, Boussada VG, Luque S (2022) Habitat connectivity in agricultural landscapes improving multi-functionality of constructed wetlands as nature-based solutions. Ecol Eng 182:106725

Ravindran S, Noor HM, Salim H (2022) Anticoagulant rodenticide use in oil palm plantations in Southeast Asia and hazard assessment to non-target animals. Ecotoxicology 31(6):976–997

Rizzardi Orlandi A, Ambrosini R, Rubolini D, Romano A, Orsi L, Brambilla M, Costanzo A (2024) Long-term monitoring of a flagship bird species in agroecosystems enhances stakeholder engagement in biodiversity conservation. Glob Ecol Conserv 54:e03090

Sasaki K, Hotes S, Kadoya T, Yoshioka A, Wolters V (2020) Landscape associations of farmland bird diversity in Germany and Japan. Glob Ecol Conserv 21:e00891

Skrimizea E, Lecuyer L, Bunnefeld N, Butler JRA, Fickel T, Hodgson I, Holtcamp C, Marzano M, Parra C, Pereira L, Petit S, Pound D, Rodríguez I, Ryan P, Staffler J, Vanbergen AJ, Van den Broeck P, Wittmer H, Young JC (2020) Sustainable agriculture: recognizing the potential of conflict as a positive driver for transformative change. In: Bohan DA, Vanbergen AJ (eds) Advances in ecological research, vol 63. Academic Press, pp 255–311

Smith LG, Williams AG, Pearce BD (2015) The energy efficiency of organic agriculture: a review. Renew Agric Food Syst 30(3):280–301

Squalli J, Adamkiewicz G (2018) Organic farming and greenhouse gas emissions: a longitudinal U.S. state-level study. J Clean Prod 192:30–42
Sun J, Wang Z, Du Y, Zhang E, Gan H, Sun D, Niu W (2022) Optimized tillage improves yield and energy efficiency while reducing carbon footprint in winter wheat-summer maize rotation systems. Sci Total Environ 820:153278
Taichung City Government (2021) Annual Statistic Report. 463–415
Tsai M-H, Chang Y-C, Yang T-Y, Luh Y-H (2021) Factors determining rice farm households' adoption of organic farming in Taiwan. Agronomy 11(11):2195
Tseng C-W, Huang C-Y, Yao M-C, Tsai Y-H, Yao C-T (2018) Observation of Black-winged Kite breeding in Nan'ao. Nat Conserv Quarter 103:74–79
Venkat K (2012) Comparison of twelve organic and conventional farming systems: a life cycle greenhouse gas emissions perspective. J Sustain Agric 36(6):620–649
Witmer G (2022) Rodents in agriculture: a broad perspective. Agronomy 12(6):1458
Wu C-F, Lai J-H, Chen S-H, Trac LVT (2023) Key factors promoting the niche establishment of black-winged kite Elanus caeruleus in farmland ecosystems. Ecol Indic 149:110162
Yonas M, Leirs H (2019) Raptor Perch Sites for biological control of agricultural Pest Rodents. In: Nyssen J, Jacob M, Frankl A (eds) Geo-trekking in Ethiopia's Tropical Mountains: the Dogu'a Tembien District. Springer International Publishing, pp 279–292
Zagorski ME, Swihart RK (2020) Killing time in cover crops? Artificial perches promote field use by raptors. Ann Appl Biol 177(3):358–366
Zhang L, Ruiz-Menjivar J, Tong Q, Zhang J, Yue M (2021) Examining the carbon footprint of rice production and consumption in Hubei, China: a life cycle assessment and uncertainty analysis approach. J Environ Manag 300:113698

Chapter 12
Green Space and Habitat Connectivity in Peri-Urban Delhi, India: Spatio-Temporal Dynamics, Drivers, and Implications

Somajita Paul, Krishna Gopal Saxena, and S. Sreekesh

Abstract The present study aimed to assess trends, driving factors, and implications of forest habitat extent and connectivity over the 1973–2017 period and to identify policy actions for enhancing landscape multifunctionality in peri-urban Delhi (India). Land cover/use in 1973, 1986, 1996, 2005, and 2017 was mapped by supervised classification of satellite data, and forest habitat connectivity was assessed by Morphological Spatial Pattern Analysis. Tree species diversity and carbon stocks were assessed by quadrat sampling and people's perceptions from participatory discussions with individuals/focus groups and responses to questionnaires. Urban area expanded and cultivated area diminished at annual rates of 29% and 2%, respectively. Urbanization occurred mostly in scrublands. Croplands declined largely because of the development of farm houses, which are the country houses of urban elites for retreats. Afforestation/reforestation compensated for deforestation/forest degradation such that spatial connectivity measured as "branch," "bridge," and "loop" was reduced, while "core dense–open forest mosaics" were enlarged. Yet, management was effective in enhancing climate change mitigation function but not biodiversity conservation and livelihood functions of forests. Plant/animal species literacy was quite poor. People were ignorant of carbon sequestration and species classification based on global conservation values. Gender, age, education, and occupation impacted perceptions of forest ecosystem services/disservices. Many agencies managed green spaces. None of them explicitly addressed ecological or social connectivity and feedback loops between forest and nonforest land uses and other socioeconomic activities in the landscape. Management plans lacked an adaptive co-management approach. There is a need for policies supporting collaborative action research turning concepts into practices effective in enhancing landscape multifunctionality and ensuring their incorporation in master development plans.

S. Paul · K. G. Saxena (✉)
School of Environmental Sciences, Jawaharlal Nehru University, New Delhi, India

S. Sreekesh
Centre for the Study of Regional Development, School of Social Sciences, Jawaharlal Nehru University, New Delhi, India

M. Nishi et al. (eds.), *Ensuring Ecological Connectivity in Socio-Ecological Production Landscapes and Seascapes (SEPLS)*, Satoyama Initiative Thematic Review, https://doi.org/10.1007/978-981-95-1474-8_12

Keywords Aravali Mountain · Biogeography · Habitat loss · Fragmentation and degradation · Invasive alien species · Keystone species · Migrants · Novel ecosystems · Adaptive co-management

12.1 Introduction

Connectivity is a plural and value-laden concept (Hodgetts 2018; D'Aloia et al. 2019; Beger et al. 2022) centered around material, energy, and gene flows and organismal movements in ecological sciences (MacArthur and Wilson 1967; Forman and Gordon 1986; Taylor et al.1993; LaPoint et al. 2015) and all material and nonmaterial flows among people in biomedical and social sciences (VanderWeele 2017; Bailey et al. 2018). An integration of ecological and social dimensions of connectedness is required when a landscape is adopted as a spatial unit for addressing intertwined biophysical and socioeconomic problems (Collins et al. 2011; Saito et al. 2019; Rao et al. 2022a, b).

Interventions altering connectivity and its outcomes may enormously differ in space and time (Egerer et al. 2020; Maciejewski et al. 2021). Improving connectivity between protected areas is a significantly more cost-effective approach to conserve biodiversity than expanding them. Infrastructure development improves connectivity between producers, business enterprises, and consumers but may counter the processes favoring biodiversity conservation. Enhancement/maintenance of ecological connectivity mitigates problems like species extinction, decline in yields of cross-pollinated crops, depleted soil seed banks, and soil fertility loss in alluvial plains/deltas but aggravates problems like communicable diseases, biological invasion, and depredation (Prist et al. 2022). Development of corridors between protected areas favors conservation of charismatic large-top carnivores and megaherbivores but may also create livelihood problems to the displaced people in thickly populated developing countries. Ignorance or inadequate knowledge of feedback loops between ecological and social systems may create new or aggravate existing problems (Saxena 1986; Brockington et al. 2008; Laishram et al. 2012; Beger et al. 2022). Translating theories underlying connectivity into effective plans and practices is a global challenge (Webster et al. 2017; Bailey et al. 2018; IPBES 2019; Kastner et al. 2021) as is capturing its multiple dimensions in simple metrics (Keeley et al. 2021; Lookingbill et al. 2022; Kirk et al. 2023).

With more than half of the world's population living in cities and continuation of urbanization in the near future, biodiversity and ecosystem services in urban and peri-urban landscapes are becoming important. The biodiversity-inclusive spatial planning (Fastré et al. 2021) envisaged explicitly in the Kunming-Montreal Global Biodiversity Framework, within the context of India and many other developing countries, would seek alignment of the primary interest of poor migrants in income generation with that of urban elites disgusted from pollution and poor aesthetics in "bringing nature back into cities" and "nature-based solutions." Greening of urban areas creates novel urban ecosystems (Kowarik 2011; Caryl et al. 2013; Nor et al.

2017) and thus contributes to the functions of protected areas and the livelihoods of both urban and rural masses (Apfelbeck et al. 2019; Mata et al. 2020).

Spread around the Aravali Range, one of the oldest geological features on the planet, the Central National Capital Region of Delhi is likely to overtake Tokyo as the largest city in the world by 2030 (United Nations 2018). Peri-urban Delhi is thus experiencing the enormous pressure of physical and economic development of the city core. Forest cover monitoring has been continued since 1987 but largely in terms of net changes in forest cover (Forest Survey of India 2021). Building on our previous studies (Paul et al. 2021), the present study in peri-urban Delhi aimed to assess (i) spatial distribution and connectivity of forests over the 1973–2017 period, (ii) outcomes of forest cover/habitat changes on tree biodiversity and carbon sequestration, and (iii) sociodemographic attributes impacting/impacted by forest quality and connectivity.

12.2 Materials and Methods

12.2.1 Study Area

The study area covered a cluster of three peri-urban census towns (viz. Rangpuri, Rajokri, and Ghitorni) with a 2 km buffer (Fig. 12.1). The terrain is undulating (170–240 m above mean sea level) with tropical dry thorn forest as climax (potential) vegetation (Champion and Seth 1968). The area is a preferred destination of both poor and rich migrants from all over the country (Table 12.1). Better education, medical facilities, living standards, and employment opportunities in the national capital became key attractors after 1950 (Kar et al. 2018). With a scarcity of affordable living places in the core urban area, the poor migrants bear the brunt of commuting long distances between workplaces in the core urban area and living places in peri-urban areas. Moreover, migrants increase pressure on forests by harvesting freely available products (Paul et al. 2021). Migration of urban people to rural areas induced by innate desire or by dedicated government support programs (e.g., in South Korea) (Paniagua 2002; Popescu 2020; Takahashi et al. 2024) is altogether lacking.

All greening initiatives started after proclamation of the National Forest Policy 1988 and gathered momentum following national commitments to the Convention on Biological Diversity (CBD) and United Nations Framework Convention on Climate Change (UNFCCC). The Forest (Conservation) Act (1980) checked deforestation and ensured compensatory afforestation in lieu of deforestation, which is inescapable for social development. Government agencies distribute saplings free of cost to the general public toward greening. The Indian Forest Act (1927), Wildlife Protection Act (1972), and Delhi Preservation of Trees Act (1994) check deforestation, forest degradation, poaching, and tree mortality due to human disturbances. By the turn of the century, the government started encouraging vermicomposting and made rainwater harvesting and horticultural wastewater recycling mandatory for

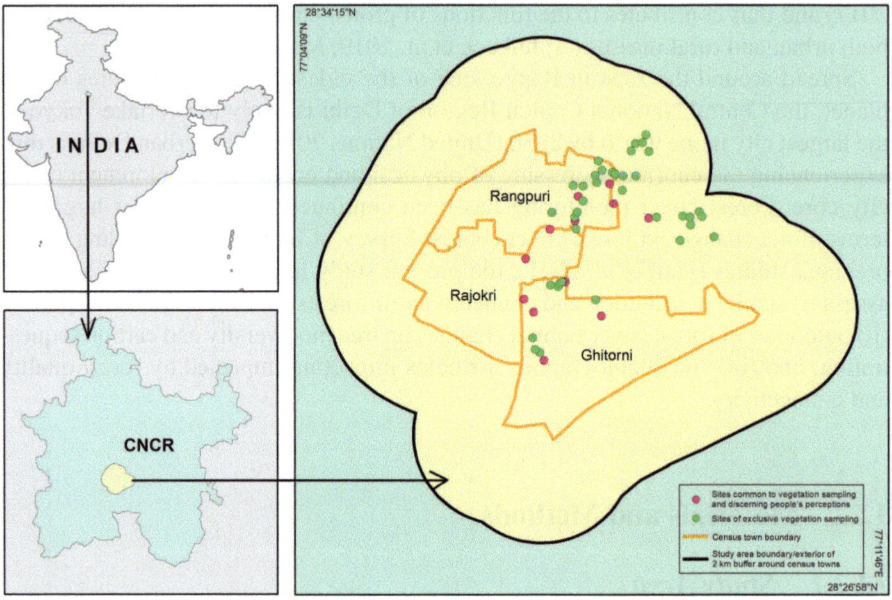

Fig. 12.1 Location of the study area in Central National Capital Region (CNCR) of Delhi. (Source: Prepared by authors combining information available in maps prepared by the Survey of India and archival records of Sub-Division Office, Vasant Vihar, New Delhi)

Table 12.1 Basic information of the study area

Country/region	India
Province	Central National Capital Region of Delhi
District	South West Delhi
Municipality	South Delhi Municipal Corporation
Size of geographical area (ha)	16,326 ha
Dominant ethnicity(ies), if appropriate	
Size of case study/project area (ha)	8486 ha
Dominant ethnicity in the project area	
Number of direct beneficiaries (people in the study area)	82,000
Number of indirect beneficiaries (people in South Delhi Municipal Corporation)	5,600,000
Geographic coordinates (latitude, longitude)	28°30′53.848″N, 77°7′57.605″E

Source: Authors, based on Google Earth Pro (2024)

urban units covering >100 m^2 area and releasing >10,000 L of wastewater per day. The National Green Tribunal was established in 2010 for quick disposal of environment-related litigations as well as for Suo Moto cognizance of environmental problems. Yet, Delhi needs to expand its forest cover by 43% to achieve the target of 33% of geographical area set in 1952 (Forest Survey of India 2021).

12.2.2 Land Cover/Use Mapping

Land cover/use in 1973, 1986, 1996, and 2017 was mapped from Landsat multispectral data and in 2005 from IRS (Indian Remote Sensing)-LISS III data. Imageries were registered onto UTM projection (zone 43 North; datum WGS 1984) using the nearest neighbor resampling algorithm. Error of registration was <15 m. Radiometric and atmospheric corrections were made by the one-step dark object subtraction method. Maximum Likelihood Classification was used to distinguish: (i) scrub (open ground, bare exposed rocks, grassy patches, and abandoned croplands with tree canopy cover <10%), (ii) open forests (canopy cover <60%), (iii) dense forests (canopy cover >60%), (iv) cultivated land, and (v) built-up area. Pooled information from Google Earth imageries, topographical sheets, special series maps, city guide maps, Master Plans for Delhi (1962, 2001, and 2021), structured interviews with 70 individuals followed by open discussions with nine focus groups (comprising 3–5 individuals selected based on their deep knowledge/interest in the subject), and own observations over 90 days in field were used to demarcate training sets (varying in the range of 200–250 points) across the 5 years selected for the study to train the classifier and assess accuracy. Training sets were increased unless accuracy of each land use/cover class was >80% and Kappa accuracy coefficient >0.8. Classified outputs were smoothed by a 3 × 3 majority filter and resampled to a common resolution of 30 m (details in Saxena et al. 1992; Paul et al. 2021) (Fig. 12.2).

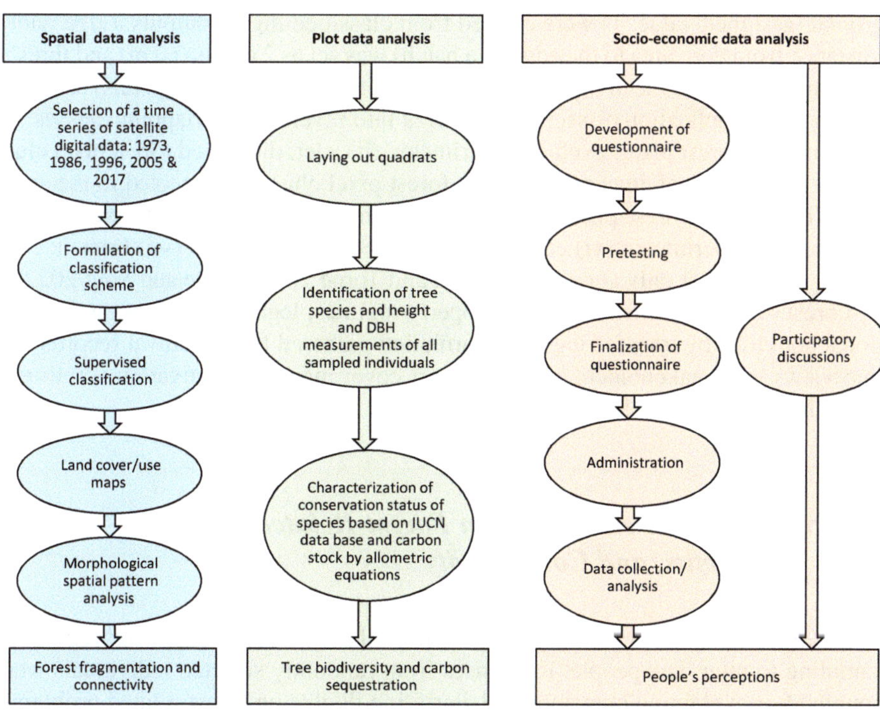

Fig. 12.2 Flowchart of the methodology. (Source: Prepared by authors)

12.2.3 Carbon Stock and Tree Species Diversity

Of 236 locations demarcated for feeding training sets of 2017, 49 were randomly selected for assessing tree diversity and carbon stocks. Tree height and diameter at breast (DBH) were measured in 100 m^2 circular plots, one each in 26 locations in open forests and 23 in dense forests, using a clinometer and tape, respectively (Higgins et al. 1996). Aboveground carbon pool of trees was computed from available allometric equations, wood density, biomass expansion factors, and IPCC default carbon fraction values (Brown 1997; Kishwan et al. 2009) and belowground pool from conversion factors determined by Cairns et al. (1997). Carbon stock in vascular plants other than trees was obtained using a conversion factor of 0.015 based on our measurements in comparable nearby sites. Species richness and diversity index were determined for each plot (Shannon and Weaver 1963). Species status in the Red List was ascertained from the IUCN database (https://www.iucnredlist.org/). Measurements made in plots in 2021 were assumed to represent the states in 2017.

12.2.4 Morphological Spatial Pattern Analysis

Forest habitat connectivity/fragmentation was assessed by a Morphological Spatial Pattern Analysis (MSPA) program in Guidos Toolbox 2.8. Binary maps showing forest pixels as foreground (assigned value: 2) and other land cover/uses as background (assigned value: 1) were created from classified digital outputs. Edge width (distance from core area to the edge of a patch) was set as 2 pixels (60 m) and threshold connectivity (connectivity of forest pixels with the surrounding pixels) as 8 pixels. Program application dissected forest area into seven nonoverlapping classes, (i) core: interior forest pixels excluding perimeter, (ii) islet: disjointed forest pixel clusters too small to contain core, (iii) loop: forest pixel clusters connected to the same core, (iv) bridge: forest pixel clusters connecting different cores, (v) perforation: internal forest perimeter, (vi) edge: external forest perimeter, and (vii) branch: connected to the forest only at one end (Saura and Torné 2009; Soille and Vogt 2022).

Forests were classified based on property regimes, legal obligations, and operational activities by transferring the information gathered from archival records and interviews with stakeholders onto the forest cover map to investigate institutional/social barriers to forest habitat connectivity.

12.2.5 People's Perceptions on Forest-Related Problems, Benefits, and Connectivity

Twelve locations were randomly selected from the set of 49 selected for vegetation sampling to interview people. Responses from randomly selected individuals were sought after explaining questions and clarifying doubts on forest-related problems, benefits, knowledge, management practices, and connectivity. Free and reflexive discussions followed the open-ended questions. Information gained from individuals was cross-checked in focus group discussions. With 25 respondents opting out during pilot interviews, information could be sought from 239 individuals. Significance of differences in responses by sociodemographic categories was assessed by Chi-square test.

12.3 Results

12.3.1 Land Cover/Use Changes

Both cultivated and scrub areas declined, more so after 1986 and 1996, respectively. The 1973–1986 period saw the highest increase in open forest and 1986–1996 in dense forest. Dominance of land cover/use shifted to forests and built-up areas in 2017 from scrubs and cultivation in 1973 (Fig. 12.3).

Gain and loss of scrub, cultivated land, and forest occurred all through the 44-year period, but the rates varied. After 1996, interconversion of scrub and cultivated areas became negligible as did agricultural expansion in forest land. Forest degradation, deforestation, and afforestation/reforestation substantially reduced after 2005. Built-up area increased throughout, but change was more pronounced in scrub and cultivated lands than forests (Table 12.2).

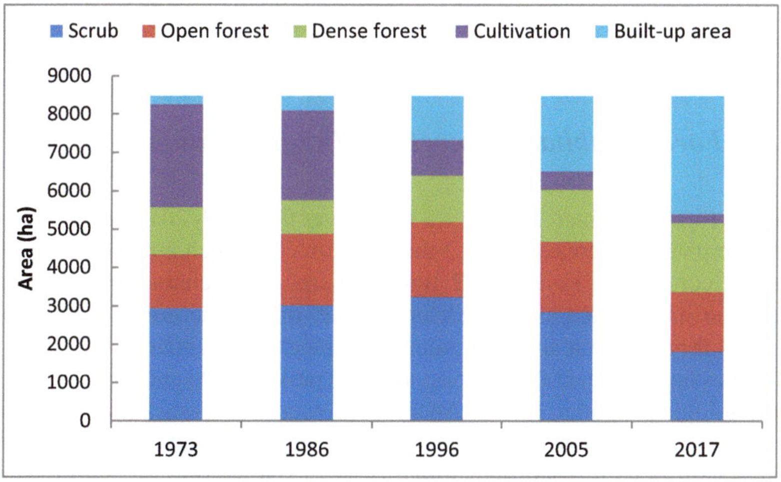

Fig. 12.3 Aggregate area of different land cover/use types during the 1973–2017 period. (Source: Prepared by authors)

Table 12.2 Land cover/use changes over the 1973–2017 period in the study area

Land cover/use change (hectares/year)	1973–1986	1986–1996	1996–2005	2005–2017
Conversion of scrub area				
to open forests	50.65	70.67	91.05	40.65
to dense forests	14.42	26.38	21.85	40.47
to cultivated land	47.08	28.74	19.90	4.52
to built-up area	6.38	35.06	48.34	65.53
Expansion of scrub area				
in open forests	35.76	73.50	71.10	37.25
in dense forests	26.45	16.73	19.40	13.32
in cultivated land	59.34	88.97	40.61	9.59
Conversion of cultivated area				
to scrub areas	59.34	88.97	40.61	9.59
to open forests	31.42	49.97	19.55	4.67
to dense forests	13.81	30.37	17.29	6.09
to built-up area	6.54	17.09	9.08	10.22
Expansion of cultivated area				
in scrub areas	47.08	28.74	19.90	4.54
in open forests	19.05	12.64	13.95	3.89
in dense forests	17.66	3.64	3.31	1.39
Afforestation/reforestation				
in scrub areas	65.07	97.05	112.90	81.03
in cultivated land	45.24	80.33	36.84	10.76
in open forests	14.33	25.09	32.05	23.54
Forest degradation (dense to open forests)	23.89	20.70	15.50	9.93
Deforestation	106.14	141.17	155.72	85.28

Source: Prepared by authors

12.3.2 Forest Habitat Connectivity/Fragmentation

Aggregate forest area increased by 1.3 times over the 1973–2017 period. The number of core patches increased by 1.2 times, mean core patch size by 2.4 times, and maximum core patch size by 3.2 times. These patch characteristics, however, showed one intermittent short spell of decline, unlike the consistent increase in aggregated forest habitat area. Minimum core patch size remained the same, while core patches larger than 100 ha in size occurred only in 1996 and 2017 (Table 12.3).

Morphological Spatial Pattern Analysis showed that branch, edge, islet, core, and bridge covered almost equal areas in 1973, while core and edge turned far more extensive than branch, islet, and bridge in 2017. New cores were formed in the western half, but most of them did not persist as opposed to the eastern half where cores existing in 1973 expanded and new cores outnumbered the vanished ones. Perforation surfaced from 1986 onward. All through the 44-year period, perforation and loops covered negligible areas (Fig. 12.4).

Table 12.3 Changes in forest area, core patch density, and core patch size over the 1973–2017 period

		1973	1986	1996	2005	2017
Total forest habitat area (ha)		2634.75	2741.4	3164.85	3188.79	3353.13
Number of core patches						
	>100 ha	0	0	1	0	3
	10–100 ha	10	15	14	21	77
	<10 ha	136	158	161	140	89
	All	146	173	176	161	169
Minimum core patch size (ha)		0.06	0.06	0.06	0.06	0.06
Maximum core patch size (ha)		64.87	69.01	99.57	64.22	206.02
Mean core patch size (ha)		2.54	3.61	3.08	4.64	6.15

Source: Prepared by authors

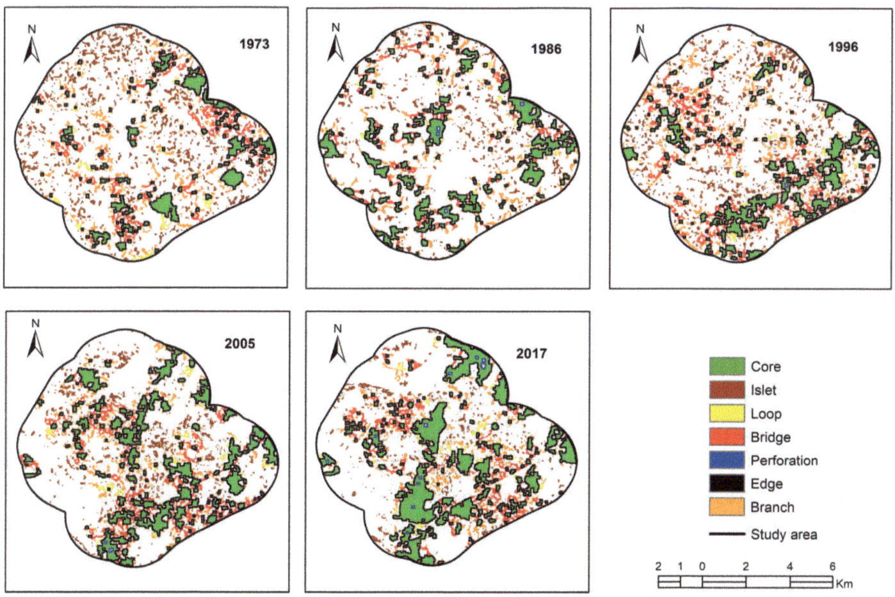

Fig. 12.4 Morphological Spatial Pattern Analysis of forests over the 1973–2017 period. (Source: Prepared by authors)

None of the morphological types displayed monotonous temporal trends. The core area increased throughout except for a marginal decline over the 1986–1996 period. Core area in 2017 was threefold larger than the area in 1973. On the other hand, spatial coverage of bridges decreased over the 1973–1986 period and rose during the 1986–1996 period with nominal changes thereafter. Edge area was almost stable after an initial increase during 1973–1986. Islet coverage fluctuated with a decline over initial (1973–1986) and terminal (1996–2017) periods and an increase over the intermittent (1986–1996) period (Fig. 12.5).

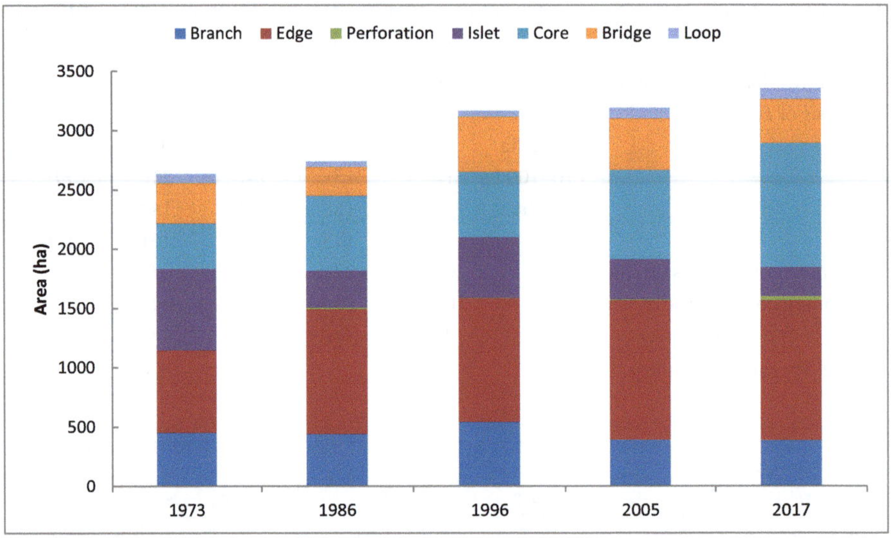

Fig. 12.5 Areas of different morphological forest types during the 1973–2017 period. (Source: Prepared by authors)

Figure 12.6 shows spatial distribution of forests differing in land/resource ownership regimes in 2017. Farm houses (residential retreats for urban elite) constituted the most dominant category. Protected forests were under the control of the Forest Department where activities other than tree cutting, planting nonforest species, and nonforest land uses were permitted. People have free access to public/community parks/gardens and city forests/forests on public land under the control of the Municipal Corporation (decision-making by elected representatives) and the Delhi Development Authority (decision-making by government officials) only for nonconsumptive resource uses. Farm houses were private properties with legal binding of maintaining green space over at least half the area. These units were not accessible to common people and were invariably fenced by cement/concrete walls, unlike other green spaces. The level of access for common people to defence, airport, and educational institutions was better than farm houses but worse than other green spaces. Each category had both dense and open forests. Forest habitat loss/degradation/expansion/restoration was not correlated with the legal status of the land. In all types, people were engaged merely as wage workers to implement top-down plans.

12.3.3 Tree Community Structure, Species Richness, and Diversity

Twenty species were confined to dense forests and only six to open forests. Species with high global conservation values native of Asia (*Aegle marmelos, Pterospermum acerifolium*, and *Tectona grandis*) or exotic (*Eucalyptus camaldulensis* native of

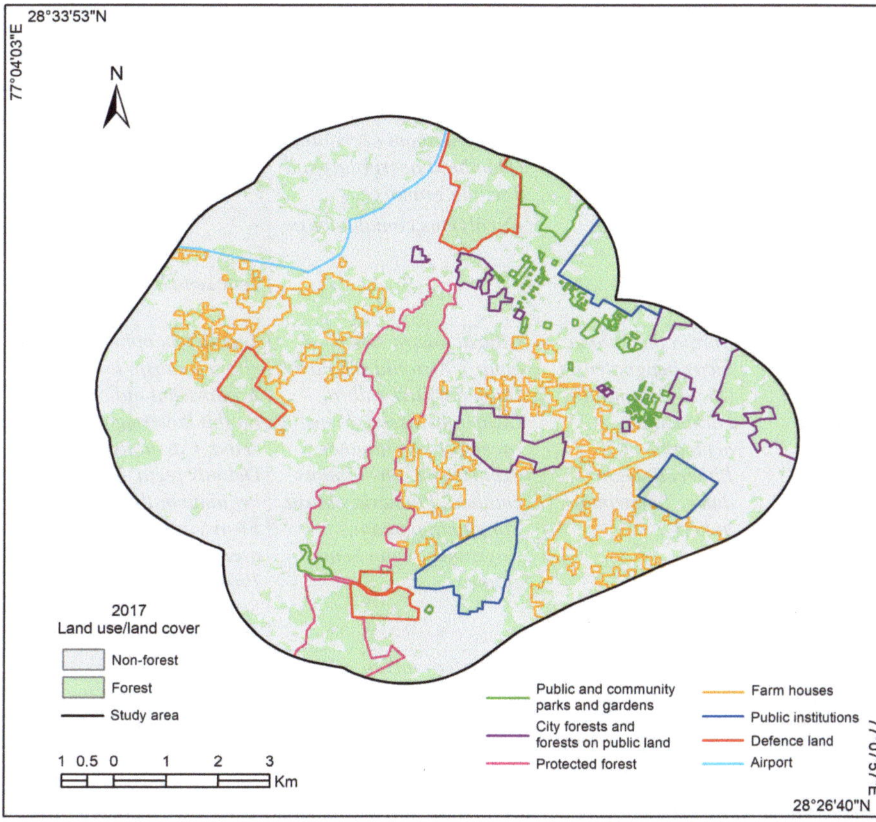

Fig. 12.6 Classification of forests based on land/resource ownership regimes in 2017. (Source: Prepared by authors by overlaying land/resource ownership information obtained from archival records)

Australia and *Jacaranda mimosifolia* of Latin America) occurred only in dense forests (Table 12.4).

Differences between dense and open forests were more pronounced in carbon density, species richness, and diversity index than tree density and canopy cover (Table 12.5).

12.3.4 People's Perceptions

Some perceptions about forests varied by gender (females and males: 38% and 62%, respectively, in the sample), age (<25 years, 25–55 years, and >55 years: 18%, 66%, and 16%, respectively), educational attainment (illiterates, undergraduates, graduates, and postgraduates: 24%, 48%, 22%, and 6%, respectively), and occupation (student, non-office job, office job, and at home: 9%, 19%, 51%, and 21%,

Table 12.4 Tree species captured in sample plots

	Species found only in open forest	Species found only in dense forest	Species common to open and dense forest
Vulnerable	–	*Jacaranda mimosifolia*	–
Endangered	–	*Pterospermum acerifolium (Dombeya acerifolia) and Tectona grandis*	–
Near-threatened	*Aegle marmelos*	*Eucalyptus camaldulensis*	–
Conservation-dependent	–	–	*Leucaena leucocephala*
Least concerned	*Cordia dichotoma, Ficus benghalensis (Ficus cotinifolia), Parkinsonia aculeata, Pithecellobium dulce, Tamarindus indica*	*Albizia amara subsp. amara, Alstonia scholaris, Bougainvillea glabra, Dalbergia sissoo, Ficus racemosa, Nyctanthes arbortristis, Phyllanthus emblica, Plumeria obtusa, Pongamia pinnata, Roystonea regia, Senna siamea*	*Acacia leucophloea, Acacia nilotica subsp. indica, Azadirachta indica, Bauhinia variegata, Bombax ceiba, Cassia fistula, Delonix regia, Ficus benjamina, Ficus religiosa, Ficus virens, Melia azedarach, Morus alba, Prosopis juliflora, Schleichera oleosa, Ziziphus mauritiana*
Data-deficient	–	*Mangifera indica*	–
Not evaluated	–	*Mitragyna parvifolia, Syzygium cumini, Polyalthia longifolia, Tabebuia aurea*	*Acacia senegal, Ailanthus excelsa, Holoptelea integrifolia*

Source: Prepared by authors using the information in IUCN database https://www.iucnredlist.org/

Table 12.5 Selected attributes of dense and open forests in peri-urban Delhi

	Dense forests ($n = 23$)	Open forests ($n = 26$)
Canopy cover (%)	60–80	30–60
Canopy stratification	Only partial	Absent
Tree height (m)	11.7 ± 4.3	10.05 ± 3.7
Tree density (mature trees/ha)	252 ± 122	168 ± 65
Total species richness	40	25
Species richness (mean number of species/plot)	4.74 ± 3.7	3.27 ± 2.4
Shannon Diversity Index	2.58 ± 0.07	1.62 ± 0.08
Carbon stock (t carbon/ha)	34.6 ± 27.85	16.67 ± 16.2

Source: Prepared by authors

respectively). All respondents referred to forests as the areas that hosted wild or wild-domesticated mixed tree associations irrespective of area, crown density, or tree height. While all respondents listed benefits, only 55% listed problems from forests. All respondents desired afforestation/reforestation on a larger scale though only 61% participated in it. Scarcity of time and space (dwelling) were the constraints behind nonparticipation. All respondents expressed displeasure on their exclusion from planning, monitoring, and evaluation, and neglect of their genuine needs and development aspirations. They substantiated this assertion by the fact that 47% of standing trees belonged to the invasive alien species (native of Latin America) *Prosopis juliflora*, lacking marketable products, or species yielding marketable fruits (gooseberry, Indian blackberry, mango, and wood apple) serving their interests were neglected in organized management. Older people (>55 years) were more intensively involved than the younger ones ($P < 0.05$) without any gender, education, and occupation bias (Table 12.6).

People recognized environmental benefits most frequently, followed by health benefits, recreational benefits, social benefits, and material/income benefits. Educational attainment was a significant ($P < 0.01$) determinant of these perceptions. Graduates/postgraduates were concerned about health and recreational benefits more than school children and illiterates. Only illiterates herded livestock, collected forest products, and earned wages from forestry activities. Insecurity from anti-social elements was the most common problem (forest ecosystem disservice), followed by health problems and damage to property. Threats from anti-social elements were reported more by women than men, people at home than employed people, and more by students than the illiterates. Women were disgusted by waste dumping and open defecation more than men (Table 12.6).

Women could correctly identify more species than men, postgraduates more than illiterates/school children/graduates, and older people (>55 years) more than the younger ones ($P < 0.05$). None of the respondents were aware of species classification based on probability of extinction by IUCN or the carbon sequestration/climate change mitigation connection.

People elaborated biophysical connectivity in terms of visible aboveground structures and social connectivity in terms of relationships within and between sociodemographic classes and institutions. Knowledge of connectivity between belowground and aboveground biodiversity was missing. People understood positive impacts of forests on agriculture but not the underlying mechanisms: forests hosting predators/parasitoids and checking pathogens/insect pests damaging crops and their pollinators (Table 12.7).

12.4 Discussion

12.4.1 Forest Habitat: The Extent

The lack of novel forest ecosystems, such as community gardens (Egerer et al. 2020) and forest gardens (Wiersum 2004), which combine the primary interests of urban elites in pollution mitigation/aesthetics and of poor migrants in commodities/income reflects a legacy of conventional forestry treating agriculture and horticulture as threats to forests. The existence of novel urban ecosystems like dwellings/buildings with rooftop water harvesting, wastewater recycling, and solar heating/

Table 12.6 Benefits and problems from forests reported by people in peri-urban Delhi

Benefits (ecosystem services)	% of respondents reporting the benefit[a]	Problems (ecosystem disservices)	% of respondents reporting the problem[b]
Environmental benefits: pollution mitigation, protection from harsh weather	92	Safety issues: sheltering anti-social elements	38
Health benefits: discourses and exercises in a less-polluted environment and information-rich system	26	Health problems: open defecation and waste dumps favoring allergens (*Parthenium hysterophorus*), pathogens, and harmful insects	21
Recreational benefits: interactions with kins and friends	21	Damage to property: tree roots damaging buildings and uprooting/branch falls both buildings and vehicles	8
Social benefits: safety from anti-social elements from vigilant forest guards	11	Travel/transport problems: longer travel time/expenses from closure of roads blocked by uprooted trees and closed for protection of plantations	6
Material/income benefits: wage employment, fruits, fuelwood, medicinal herbs, fodder, grazing, avoidance of expenditure on modern toilets[c]	5	Other problems: encroachment, invasive alien species (*Lantana camera*), rodents, and snakes	4

Source: Prepared by authors
[a]Statistical treatment of responses by gender, age group, education level, and occupation (df = 4, 8, 12, and 12, respectively; $\chi 2$ = 8.7, 13.3, and 23.9, and 13.1, respectively; P = 0.07, 0.10, 0.02, and 0.30, respectively);
[b]df = 5, 10, 18, and 15, respectively; $\chi 2$ = 35.92, 14.51, 50.82, and 35.39, respectively; P = <0.001, 0.15, <0.001, and 0.002, respectively);
[c](df = 1, 2, 3, and 3, respectively; $\chi 2$ = 0.73, 0.91, 7.15, and 17.42, respectively, P = 0.93, 0.63, 0.06, and 0.0005, respectively), collection of forest products (df = 1, 2, 3, and 3, respectively, $\chi 2$ = 0.08, 1.51, 34.07, and 37.84, respectively, P = 0.776, 0.46, <0.001, and <0.0001, respectively), and grazing (df = 1, 2, 3, and 3, respectively, $\chi 2$ = 0.05, 1.67, 7.0, and 8.58, respectively, P = 0.82, 0.34, 0.07, and 0.03, respectively)

lighting systems, but the absence of ones with green roofs (Labuschagne and Zulch 2016) or vertical gardens/farms (Beacham et al. 2019) seems to be rooted in regulation enforcement (mandatory water harvesting/waste recycling for buildings >100 m^2 and generating >10,000 litres wastewater/day) as well as subsidies only for the former.

Urbanization along with forest cover expansion in the study area (Dutta and Rahman 2017; Tian et al. 2014) is a combined outcome of several policies and processes. First, deforestation for crop cultivation declined as a result of improvement

Table 12.7 People's articulation of connectivity discerned from participatory discussions and questionnaires

Type of connectivity	People's articulation of connectivity
Biophysical connectivity	Special structures formed by one species: termite mounds, ant hills, and earthworm casts
	Products/structures formed by two or more species: honey and nests of birds
	Wildlife connecting forests and dwellings: ants, termites, insects, wasps, rodents, snakes, lizards, and monkeys
	Wildlife not connecting forests and dwellings: porcupines, jackals, foxes
	Streams connecting highlands and lowlands
Social connectivity (man-to-man)	Interdependence of people differing in sociodemographic profiles
Social connectivity (economic/ material benefits)	Utilization of fodder, fuelwood, fruits, green vegetables, medicinal herbs, and small timber
Social connectivity (nonmaterial benefits appropriated in groups)	Feeding wheat flour to ants around forest openings
	Feeding fruits to monkeys around dense forests
	Emotional connections manifested in voluntary tree planting
	Collective utilization of nontimber products in dense forests
	Discourses on linkage between culture and human well-being: (i) forest classification: *Aranyavan* (free from war), *Tapovan* (for penance and self-learning), *Mahavana* (dwelling of supergods), *Devavana* (dwelling of deities), and *Srivana* (enabling wealth creation); (ii) descriptions of tree species: *Ficus religiosa* (The God himself), *Ficus benghalensis* (Root as the creator of the earth system and branches/leaves as abodes of gods/ goddesses), *Azadirachta indica* (the only wood used for making idols of Gods/Goddesses), *Aegle marmelos* (the sole tree whose leaves offered to Lord Shiva); and (iii) descriptions of wildlife: Gods/Goddesses learning lessons from pigeon, python, moth, honeybee, elephant, deer, fish, spider, and beetle, riding mouse, peacock, eagle, bull, horse, donkey, lion, buffalo, dog and elephant, using cobra as ornament/bedding for meditation, soliciting support of squirrels, parrot, crow, monkey, eagle, dog and bear, and incarnating only as fish, tortoise, boar, man-lion and man.
Social connectivity (nonmaterial benefits availed individually)	Individuals meditating, walking, and jogging around forests (induction of physiological processes enhancing physical and mental health and pro-nature thoughts and actions)

Source: Prepared by authors

in policy implementation (supply of subsidized food grains from public distribution system) and distraction from agriculture (yield stagnation, smallholdings, and higher social status of people associated with secondary/tertiary development than farmers) (Semwal et al. 2013). Second, deforestation and forest degradation from fuelwood, fodder, and timber harvests diminished as a result of supply of their alternatives at subsidized prices by government agencies and proliferation of business (ecotourism, nature-cure, and farm houses) necessitating forest protection and restoration (Vij et al. 2018; Master Plan for Delhi 2021). Third, prohibition of quarrying in forests and restrictions on drawing groundwater reduced the scale of industrial/urban expansion (Sinha 2014). Lastly, tree planting was boosted along with creation of an urban greenspace network on the lines of a protected area network of residual intact forests (Mennis 2006; de la Barrera et al. 2016; Paul et al. 2021). However, urbanization progressed at the expense of deforestation in many other census towns (Das and Das 2017) resulting in stagnation in regional forest cover (Forest Survey of India 2021). The Green Credit Programme set out in 2023 is likely to bolster forest tree planting throughout the country.

Lack of uniformity in time frames and classification/change detection methodology limited the scope of comparing data and conclusions from different individuals/institutions (Rao et al. 2016; Paul et al. 2021). Divergence in scientific (definition of forests by FAO: average tree height >5 m and crown density >10% in a minimum area of 1 ha; Government of India: crown density >10% in a minimum area of 0.5 ha without horticultural trees) and people's perspectives (in the present study area: a land parcel with wild trees irrespective of spatial coverage, tree height, and crown density) often results in people–policy conflicts (Rao et al. 2016, 2022a, b).

12.4.2 People–Forest Nexus

Nature, drivers, causes, and outcomes of people–people or people–forest connectivity may vary by sociodemographic conditions, knowledge, and concerns of people. Emotional connectivity and a deep sense of "we-ness" could have resulted in more intensive engagement of older people in forest protection and expansion than the youth (Frantz and Mayer 2014; Budruk et al. 2009). All understood forests as ideal places of relieving mental stress (Maas et al. 2006; Paul and Nagendra 2017), but the poor could not appropriate this benefit due to lack of time for the needed exercises (Sugiyama and Thompson 2007; Barton et al. 2009). Limitations of Indigenous knowledge and lack of policies to overcome them underpinned lack of appreciation of forests for their contributions to climate change mitigation, sustainable food production, air purification, and water conservation (Manavvi and Rajasekar 2020; Buchel and Frantzeskaki 2015). Low levels of species literacy were a reflection of the limitations of Indigenous knowledge as well as weak connectedness between different sociodemographic classes (Jimenez and Lindemann-Matthies 2015; Saxena and Rao 2016; Randler and Heil 2021; Hooykaas et al. 2019).

Consideration of open defecation as an ecosystem disservice only by the rich and increase in travel/transport costs from uprooted trees blocking public transport routes only by the poor underscore socioeconomic conditions as determinants of people's perceptions and attitudes (Elledge and McClatchey 2013; Voumard et al. 2018). All sections of the society are unlikely to appreciate forest conservation, afforestation, or reforestation if it shelters anti-social elements (Rae et al. 2011; Paul and Nagendra 2017; Rani et al. 2019) and depredators (Lyytimäki et al. 2008).

Contribution of green spaces to social bonding and bridging (Enssle and Kabisch 2020; Sen and Guchhait 2021) has been declining with the expansion of online interactions (Penglee et al. 2019). Pro-environment attitudes of people can be boosted by enriching Indigenous knowledge with scientific knowledge and by policies balancing people-to-people and people-to-environment connectedness (Rao et al. 2022a, b).

12.4.3 Forest Habitat: Socio-Ecological Connectivity and Implications

Across scale, the structure and composition of landscapes mirror socio-ecological connectivity and the legacies of past management impacting and impacted by it (Egerer et al. 2020; Pauleit et al. 2021). The over 30-times smaller mean core forest patch size than the national average (187 ha) (Reddy et al. 2013) could be attributed to faster forest habitat loss and fragmentation in urban than rural areas in the past. Given 150 ha as the threshold area for ecological functioning (Marulli and Mallarach 2005), more than 95% of core patches will be considered ecologically nonfunctional patches. Likewise, with mere three patches measuring >100 ha, the existing forests may not be effective in conserving insectivore and frugivore birds (Bregman et al. 2014). Changes in ecological connectivity (enlarged cores but not bridges connecting them) apparently were unintended outcomes of efforts to bridge the gap between existing and desired aggregate forest cover (one-third area of the country and two-thirds area of mountains set in National Forest Policy 1952).

Forest cover expansion served the purpose of carbon sequestration-mediated climate change mitigation but not of biodiversity conservation or socioeconomic upliftment of the poor migrants. Setbacks from silvicultural practices favoring tree species with high survival, fast growth, and low planting costs and industrial demands may be encountered in the long term (Semwal et al. 2013). Thus, exotic *Prosopis juliflora* turned invasive and extirpated native tree species like *Prosopis cineraria* serving the interests of the poor migrants (Ramakrishnan 1991; Ramakrishnan et al. 1998; Rao et al. 2022a, b). Similar setbacks followed large-scale planting of *Tectona grandis* or exotic *Eucalyptus camaldulensis* elsewhere in the Indian subcontinent (Saxena et al. 1993). *Mangifera indica* and *Syzygium cumini* yielding nutritious fruits attracted wildlife and also served the interests of the poor but remained infrequent taxa in forests because they are qualified as horticultural species and not forest species in the existing policies. Planting exotic *Eucalyptus camaldulensis* and *Jacaranda mimosifolia* or indigenous *Pterospermum acerifolium* did not serve a biodiversity conservation function in the absence of their natural regeneration. Recognized as an ecological keystone worldwide, *Ficus* (MacKay et al. 2018) might have favored avifauna diversity but, being a sacred species yielding perishable edible fruits, it served only limited income interests of people.

A twofold increase in carbon stock but extirpation of six tree species from conversion of open to dense forest underscores the criticality of striking trade-offs and inclusive spatial planning. Indeed, even the disconnect of forest cover expansion, biodiversity conservation, and equitable socioeconomic development observed in the present case, as also in neighboring mega-diverse China (Liu et al. 2019), is not as much detrimental as parallel forest habitat loss, fragmentation, and degradation in many other geographical regions (Wilkinson et al. 2018; Mengist et al. 2022).

The concepts of island biogeography, edge effect, nucleation, novel ecosystems, and integration within and between ecological and socioeconomic systems in improving water supply, checking populations of pathogens, pests, and depredators, and optimizing multiple landscape functions were ignored in the Master Plans (Apfelbeck et al. 2019; Rao et al. 2022b; Niu et al. 2023) possibly for three reasons. First, action research turning these concepts (Braaker et al. 2014; Ma et al. 2017; Lin et al. 2021) into tested practices was lacking. Second, governance lacked mechanisms fostering synergy between forest, agricultural, and socioeconomic development agencies (Rao et al. 2022b). Third, targets were set in terms of aggregate areas (proportion of total geographical area covered by forests or protected areas), physical activities (saplings produced, distributed, and planted), and social benefits (employment and income generated) and not in terms of social or ecological connectivity enhancing landscape multifunctionality. Participatory action research demonstrating the potential of adaptive, integrated co-management to promote synergy among multiple agencies and social–ecological connections enhancing landscape multifunctionality in rural landscapes needs to be extended to the urban ones (Mata et al. 2020; Rao et al. 2022b).

12.5 Conclusions

Peri-urban landscapes around Delhi portray an uncommon scenario of green space expansion despite population expansion and urbanization. This trend is a combined outcome of the age-old connectivity between forests and people engrossed in cultural heritage and a new kind of connectivity emerging from deterioration of human health due to air pollution, water shortage/pollution, and serene areas overcoming mental stress founded in rampant urbanization/industrialization in the past. Nodes in social and ecological networks would increase in the near future from the altogether new markets of carbon, organic food, and commodities coming out of sustainably-managed forests/other natural ecosystems and programs like corporate social responsibility (CSR). Thus, the challenge is to harmonize connectivity within and between social and ecological systems across scales falling in line with sustainable development goals.

The present study does suffer from several limitations. It assumes similar disruptions in connectivity between forest patches by all nonforest classes and insignificant changes in ecosystem structure and functions within open and dense forests over the 44-year period, samples only mature trees, and ignores the impact of land/resource property regimes on socio-ecological connectivity. Given these limitations, ecological losses from decline in spatial connectivity from reduction in "branch," "bridge," and "loop" coverage could have been outweighed by enlargement of "open-dense forest core mosaics." India being a megadiverse country partially covering four Biodiversity Hotspots (viz. Himalaya, Western Ghats and Sri Lanka, Indo-Burma, and Sundaland), migrants in urban areas are economically poor but are endowed with Indigenous knowledge and cultural values favoring forest/landscape multifunctionality. Forest cover expansion met the imperative of carbon sequestration mediated climate change mitigation but not of biodiversity conservation and socioeconomic upliftment of the urban poor. This deficiency can be overcome by (i) translating scientific concepts of island biogeography, edge effect, nucleation, novel ecosystems, and integrated landscape management underlying ecological connectivity into sustainable practices, (ii) improving coordination within and between government institutions and sociodemographic classes in the public, (iii) harmonizing genuine socioeconomic needs of the poor migrants with those of urban elites, and (iv) policy support for adaptive collaborative management.

References

Apfelbeck B, Jakoby C, Hanusch M, Steffani EB, Hauck TE, Weisser WW (2019) A conceptual framework for choosing target species for wildlife-inclusive urban design. Sustainability 11(24):6972

Bailey M, Cao R, Kuchler T, Stroebel J, Wong A (2018) Social connectedness: measurement, determinants, and effects. J Econ Perspect 32(3):259–280

Barton J, Hine R, Pretty J (2009) The health benefits of walking in greenspaces of high natural and heritage value. J Integr Environ Sci 6(4):261–278

Beacham AM, Vickers LH, Monaghan JM (2019) Vertical farming: a summary of approaches to growing skywards. J Hortic Sci Biotechnol 94(3):277–283

Beger M, Metaxas A, Balbar AC, McGowan JA, Daigle R, Kuempel CD, Treml EA, Possingham HP (2022) Demystifying ecological connectivity for actionable spatial conservation planning. Trends Ecol Evol 37(12):1079–1091

Braaker S, Ghazoul J, Obrist MK, Moretti M (2014) Habitat connectivity shapes urban arthropod communities: the key role of green roofs. Ecology, 95:1010-1021.https://doi.org/10.1890/13-0705.1

Bregman TP, Sekercioglu CH, Tobias JA (2014) Global patterns and predictors of bird species responses to forest fragmentation: implications for ecosystem function and conservation. Biol Conserv 169:372–383

Brockington D, Duffy R, Igoe J (2008) Nature unbound: conservation, capitalism and the future of protected areas. Routledge

Brown S (1997) Estimating biomass and biomass change of tropical forests: a Primer. UN FAO Forestry Paper 134, Food and Agriculture Organization, Rome

Buchel S, Frantzeskaki N (2015) Citizens' voice: a case study about perceived ecosystem services by urban park users in Rotterdam, The Netherlands. Ecosyst Serv 12:169–177

Budruk M, Thomas H, Tyrrell T (2009) Urban green spaces: a study of place attachment and environmental attitudes in India. Soc Nat Resour 22(9):824–839

Cairns MA, Brown S, Helmer EH, Baumgardner GA (1997) Root biomass allocation in the world's upland forests. Oecologia 111:1–11

Caryl FM, Thomson K, van der Ree R (2013) Permeability of the urban matrix to arboreal gliding mammals: Sugar gliders in Melbourne, Australia. Austral Ecol 38(6):609–616

Champion HG, Seth SK (1968) A revised survey of the forest types of India. Govt. Publication, New Delhi

Collins SL, Carpenter SR, Swinton SM, Orenstein DE, Childers DL, Gragson TL, Grimm NB, Grove JM, Harlan SL, Kaye JP, Knapp AK (2011) An integrated conceptual framework for long-term social-ecological research. Front Ecol Environ 9(6):351–357

D'Aloia CC, Naujokaitis-Lewis I, Blackford C, Chu C, Curtis JMR, Darling E, Guichard F, Leroux SJ, Martensen AC, Rayfield B, Sunday JM, Xuereb A, Fortin M (2019) Coupled networks of permanent protected areas and dynamic conservation areas for biodiversity conservation under climate change. Front Ecol Evol 7. https://doi.org/10.3389/fevo.2019.00027

Das R, Das G (2017) ARS-GIS based spatial analysis and assessment on the urban growth of Delhi National Capital Territory (NCT) from 1977 to 2014. IOSR J Human Soc Sci 22(6):103–123

De la Barrera F, Rubio P, Banzhaf E (2016) The value of vegetation cover for ecosystem services in the suburban context. Urban For Urban Green 16:110–122

Dutta D, Rahman A (2017) Assessing pattern of spatio-temporal change in NCT of Delhi and its peri-urban areas using geospatial techniques. In: Sharma P, Rajput S (eds) Sustainable smart cities in India: challenges and future perspectives. Springer International Publishing, pp 145–160

Egerer M, Fouch N, Anderson EC, Clarke M (2020) Socio-ecological connectivity differs in magnitude and direction across urban landscapes. Sci Rep 10:4252. https://doi.org/10.1038/s41598-020-61230-9

Elledge MF, McClatchey M (2013) India, urban sanitation, and the toilet challenge. RTI Press

Enssle F, Kabisch N (2020) Urban green spaces for the social interaction, health and well-being of older people—an integrated view of urban ecosystem services and socio-environmental justice. Environ Sci Pol 109:36–44

Fastré C, van Zeist WJ, Watson JEM, Visconti P (2021) Integrated spatial planning for biodiversity conservation and food production. One Earth 4(11):1635–1644

Forest Survey of India (2021) India State of Forest report 2021. Ministry of Environment, Forest and Climate Change, Dehradun

Forman RTT, Godron M (1986) Landscape ecology. Wiley, New York

Frantz CM, Mayer FS (2014) The importance of connection to nature in assessing environmental education programs. Stud Educ Eval 41:85–89

Google Earth Pro (2024) South Delhi, Delhi, India, viewed 3 March 2024

Higgins KF, Oldemeyer JL, Jenkins KJ, Clambey GK, Harlow RF (1996) Vegetation sampling and measurement. Res Manag Tech Wildlife Habit 5:567–591

Hodgetts T (2018) Connectivity as a multiple: in, with and as "nature". Area 50(1):83–90

Hooykaas MJ, Schilthuizen M, Aten C, Hemelaar EM, Albers CJ, Smeets I (2019) Identification skills in biodiversity professionals and laypeople: a gap in species literacy. Biol Conserv 238:108202

Intergovernmental Science-Policy Platform on Biodiversity and Ecosystem Services (IPBES) (2019) Global assessment report on biodiversity and ecosystem services of the Intergovernmental

Science-Policy Platform on Biodiversity and Ecosystem Services. In: Brondízio ES, Settele J, Díaz S, Ngo HT (eds) IPBES secretariat, Bonn

Jimenez JN, Lindemann-Matthies P (2015) Public knowledge and perception of toads and frogs in three areas of subtropical Southeast China. Soc Anim 23(2):166–192

Kar R, Reddy GO, Kumar N, Singh SK (2018) Monitoring spatio-temporal dynamics of urban and peri-urban landscape using remote sensing and GIS–A case study from Central India. Egypt J Remote Sens Space Sci 21(3):401–411

Kastner T, Chaudhary A, Gingrich S, Marques A, Persson UM, Bidoglio G, Le Provost G, Schwarzmüller F (2021) Global agricultural trade and land system sustainability: implications for ecosystem carbon storage, biodiversity, and human nutrition. One Earth 4(10):1425–1443

Keeley AT, Beier P, Jenness JS (2021) Connectivity metrics for conservation planning and monitoring. Biol Conserv 255:109008

Kirk H, Soanes K, Amati M, Bekessy S, Harrison L, Parris K, Ramalho C, van der Ree R, Threlfall C (2023) Ecological connectivity as a planning tool for the conservation of wildlife in cities. MethodsX 10:101989

Kishwan J, Pandey R, Dadhwal VK (2009) India's forest and tree cover: contribution as a carbon sink. Technical Paper by Indian Council of Forestry Research and Education

Kowarik I (2011) Novel urban ecosystems, biodiversity, and conservation. Environ Pollut 159(8–9):1974–1983

Labuschagne P, Zulch B (2016) Green rooftop systems: a South African perspective. Energy Procedia 96:710–716

Laishram J, Saxena KG, Maikhuri RK, Rao KS (2012) Soil quality and soil health: a review. Int J Ecol Environ Sci 38:19–37

LaPoint S, Balkenhol N, Hale J, Sadler J, van der Ree R (2015) Ecological connectivity research in urban areas. Funct Ecol 29(7):868–878

Lin J, Huang C, Wen Y, Liu X (2021) An assessment framework for improving protected areas based on morphological spatial pattern analysis and graph-based indicators. Ecol Indic 130:108138

Liu J, Coomes DA, Gibson L, Hu G, Liu J, Luo Y, Wu C, Yu M (2019) Forest fragmentation in China and its effect on biodiversity. Biol Rev 94(5):1636–1657

Lookingbill TR, Minor ES, Mullis CS, Nunez-Mir GC, Johnson P (2022) Connectivity in the urban landscape (2015–2020): who? Where? What? When? Why? and How? Curr Landsc Ecol Rep 7(1):1–14

Lyytimäki J, Petersen LK, Normander B, Bezák P (2008) Nature as a nuisance? Ecosystem services and disservices to urban lifestyle. Environ Sci 5(3):161–172

Ma L, Shen C, Lou D, Fu S, Guan D (2017) Ecosystem carbon storage in forest fragments of differing patch size. Sci Rep 7(1):13173

Maas J, Verheij RA, Groenewegen PP, De Vries S, Spreeuwenberg P (2006) Green space, urbanity, and health: how strong is the relation? J Epidemiol Community Health 60(7):587–592

MacArthur RH, Wilson EO (1967) The theory of Island biogeography, vol 1. Princeton University Press, Princeton

Maciejewski K, Currie P, O'Farrell P (2021) Social-ecological connectivity in Global South cities. In: Shackleton CM, Cilliers SS, Davoren E, du Toit MJ (eds) Urban ecology in the Global South. Springer Nature Switzerland, pp 347–364

Mackay KD, Gross CL, Rossetto M (2018) Small populations of fig trees offer a keystone food resource and conservation benefits for declining insectivorous birds. Global Ecol Conserv 14:e00403

Manavvi S, Rajasekar E (2020) Semantics of outdoor thermal comfort in religious squares of composite climate: New Delhi, India. Int J Biometeorol 64:253–264

Marulli J, Mallarach JM (2005) A GIS methodology for assessing ecological connectivity: application to the Barcelona Metropolitan Area. Landsc Urban Plan 71(2–4):243–262

Master Plan for Delhi 1962, Delhi Development Authority, viewed 6 March 2024. Retrieved from https://dda.gov.in/master plan 1962

Master Plan for Delhi 2001, Delhi Development Authority, viewed 6 March 2024. Retrieved from https://dda.gov.in/master-plan-2001

Master Plan for Delhi 2021, Delhi Development Authority, viewed 6 March 2024. Retrieved from https://dda.gov.in/master-plan-2021

Mata L, Ramalho CE, Kennedy J, Parris KM, Valentine L, Miller M, Bekessy S, Hurley S, Cumpston Z (2020) Bringing nature back into cities. People Nat 2(2):350–368

Mengist W, Soromessa T, Feyisa L (2022) Forest fragmentation in a forest biosphere reserve: implications for the sustainability of natural habitats and forest management policy in Ethiopia. Resour Environ Sustain 8:100058

Mennis J (2006) Socioeconomic-vegetation relationships in urban, residential land. Photogramm Eng Remote Sens 72(8):911–921

Niu H, Rehling F, Chen Z, Yue X, Zhao H, Wang X, Zhang H, Schabo DG, Farwig N (2023) Regeneration of urban forests as influenced by fragmentation, seed dispersal mode and the legacy effect of reforestation interventions. Landsc Urban Plan 233:104712

Nor ANM, Corstanje R, Harris JA, Grafius DR, Siriwardena GM (2017) Ecological connectivity networks in rapidly expanding cities. Heliyon 3(6):e00325

Paniagua A (2002) Urban-rural migration, tourism entrepreneurs and rural restructuring in Spain. Tour Geogr 4(4):349–371

Paul S, Nagendra H (2017) Factors influencing perceptions and use of urban nature: surveys of park visitors in Delhi. Land 6(2):27

Paul S, Saxena KG, Nagendra H, Lele N (2021) Tracing land use and land cover change in peri-urban Delhi, India, over 1973–2017 period. Environ Monit Assess 193(2):52

Pauleit S, Vasquéz A, Maruthaveeran S, Liu L, Cilliers SS (2021) Urban green infrastructure in the Global South. In: Shackleton CM, Cilliers SS, Davoren E, du Toit MJ (eds) Urban ecology in the Global South. Springer Nature Switzerland, pp 107–143

Penglee N, Christiana RW, Battista RA, Rosenberg E (2019) Smartphone use and physical activity among college students in health science-related majors in the United States and Thailand. Int J Environ Res Public Health 16(8):1315

Popescu C (2020) 'Back to the village': the model of urban outmigration in post-communist Romania. Eur Plan Stud 28(6):1200–1218

Prist PR, ReverberiTambosi L, Filipe Mucci L, Pinter A, Pereira de Souza R, de Lara MR, Rhodes JR, Comin CH, Fontoura Costa L, D'Agostini TL, Telles de Deus J, Pavao M, Port-Carvalho M, Del Castillo SL, Mureb Sallum MA, Fernandes Spinola RM, Metzger JP (2022) Roads and forest edges facilitate yellow fever virus dispersion. J Appl Ecol 59(1):4–17

Rae RA, Simon G, Braden J (2011) Public reactions to new street tree planting. Cities Environ 3(1):Article 10. https://digitalcommons.lmu.edu/cate/vol3/iss1/10

Ramakrishnan PS (ed) (1991) Ecology of biological invasion. International Scientific Publications, Delhi

Ramakrishnan PS, Saxena KG, Chandrashekara UM (eds) (1998) Conserving the sacred: for biodiversity management. Science Publishers, Oxford and IBH Publ, New Delhi

Randler C, Heil F (2021) Determinants of bird species literacy—activity/interest and specialization are more important than socio-demographic variables. Animals 11(6):1595

Rani A, Jaisoorya TS, Menon PG, Nair BS, Jeevan CR, Radhakrishnan KS, Jayarajan D, Adhin B, Thennarasu K, Benegal V (2019) Harm from other people's drinking among college students in India. Drug Alcohol Rev 38(7):774–780

Rao KS, Saxena KG, Tiwari BK (2016) Biodiversity, climate change and socio-economic development in the Indian Himalaya: an overview. Bishen Singh Mahendra Pal Singh, Dehradun

Rao KS, Maikhuri RK, Semwal RL, Saxena KG (2022a) The concept of sacred in natural resource management in India. Man India 102(1–2):41–65

Rao KS, Semwal RL, Ghoshal S, Maikhuri RK, Nautiyal S, Saxena KG (2022b) Participatory active restoration of communal forests in temperate Himalaya, India. Restor Ecol 30(1):e13486

Reddy CS, Sreelekshmi S, Jha CS, Dadhwal VK (2013) National assessment of forest fragmentation in India: landscape indices as measures of the effects of fragmentation and forest cover change. Ecol Eng 60:453–464

Saito O, Hashimoto S, Managi S, Aiba M, Yamakita T, DasGupta R, Takeuchi K (2019) Future scenarios for socio-ecological production landscape and seascape. Sustain Sci 14:1–4

Saura S, Torné J (2009) ConeforSensinode 2.2: a software package for quantifying the importance of habitat patches for landscape connectivity. Environ Model Softw 24(1):135–139

Saxena KG (1986) Forest cover changes between proposed Rajaji National Park and Corbett National Park during the period 1972-83 for identifying elephant corridors. In: Kamat DS, Panwar HS (eds) Proceedings seminar cum workshop on wildlife habitat evaluation using remote sensing techniques. Indian Institute of Remote Sensing and Wildlife Institute of India, Dehradun, pp 229–237

Saxena KG, Rao KS (eds) (2016) Soil biodiversity: inventory, functions and management. Bishen Singh Mahendra Pal Singh, Dehradun

Saxena KG, Tiwari AK, Porwal MC, Menon ARR (1992) Vegetation maps, mapping needs and scope of digital processing of Landsat Thematic Mapper data in tropical region of south-west India. Int J Remote Sens 13(11):2017–2037

Saxena KG, Rao KS, Purohit AN (1993) Sustainable forestry-prospects in India. J Sustain For 1(2):69–95

Semwal RL, Nautiyal S, Maikhuri RK, Rao KS, Saxena KG (2013) Growth and carbon stocks of multipurpose tree species plantations in degraded lands in Central Himalaya, India. For Ecol Manag 310:450–459

Sen S, Guchhait SK (2021) Urban green space in India: perception of cultural ecosystem services and psychology of situatedness and connectedness. Ecol Indic 123:107338

Shannon CE, Weaver W (1963) The mathematical theory of communication. University of Illinois Press, Urbana, p 19

Sinha GN (ed) (2014) An introduction to the Delhi Ridge. Department of Forests and Wildlife, Govt. of NCT of Delhi, New Delhi, India, p Xxiv, 154

Soille P, Vogt P (2022) Morphological spatial pattern analysis: OpenSource release. Int Arch Photogramm Remote Sens Spat Inf Sci 48:427–433

Sugiyama T, Thompson CW (2007) Older people's health, outdoor activity and supportiveness of neighbourhood environments. Landsc Urban Plan 83(2–3):168–175

Takahashi Y, Yoshida T, Shigeto S, Kubota H, Johnson BA, Yamagata Y (2024) Spatial exploration of rural capital contributing to quality of life and urban-to-rural migration decisions: a case study of Hokuto City, Japan. Sustain Sci 19:489–506

Taylor PD, Fahrig L, Henein K, Merriam G (1993) Connectivity is a vital element of landscape structure. Oikos 68(3):571–573

Tian Y, Yin K, Lu D, Hua L, Zhao Q, Wen M (2014) Examining land use and land cover spatiotemporal change and driving forces in Beijing from 1978 to 2010. Remote Sens 6(11):10593–10611

United Nations (2018) The world's cities in 2018: data booklet. Population Division, Department of Economic and Social Affairs

VanderWeele TJ (2017) On the promotion of human flourishing. Proc Natl Acad Sci 114(31):8148–8156

Vij S, Narain V, Karpouzoglou T, Mishra P (2018) From the core to the periphery: conflicts and cooperation over land and water in periurban Gurgaon, India. Land Use Policy 76:382–390

Voumard J, Derron MH, Jaboyedoff M (2018) Natural hazard events affecting transportation networks in Switzerland from 2012 to 2016. Nat Hazards Earth Syst Sci 18(8):2093–2109

Webster MS, Colton NA, Darling ES, Armstrong J, Pinsky ML, Knowlton S, Schindler DE (2017) Who should pick the winners of climate change? Trends Ecol Evol 32:167–173

Wiersum KF (2004) Forest gardens as an 'intermediate' land-use system in the nature-culture continuum: characteristics and future potential. In: Nair PKR, Rao MR, Buck LE (eds) New vistas in agroforestry: a compendium for 1st world congress of agroforestry. Springer, Netherlands, pp 123–134

Wilkinson DA, Marshall JC, French NP, Hayman DT (2018) Habitat fragmentation, biodiversity loss and the risk of novel infectious disease emergence. J R Soc Interface 15(49):20180403

Chapter 13
Empowering Youth in Stone Weirs Restoration for Ensuring Ecological Connectivity of Penghu Seascapes (SEPLS)

Chen-Yang Lee, Chi-Fang Chiu, Kuei-Lin Fu, Chen-Fa Wu, Szu-Hung Chen, Luu Van Thong Trac, Jui-Min Chu, Cheng Hsiu Tsai, Chuang-Chin Cheng, Chien-Hung Tsai, Hao-Yun Chuang, Chun-Hsien Lai, Ming-Ju Lee, Chen-Ju Wang, Chun-Chie Hu, Hsi-Chun Chen, Hao-Wei Hsu, and Ting-Yu Lin

Abstract The Penghu Islands possess rich marine life and a distinctive cultural heritage in the Taiwan Strait. Local governments and organizations are collaborating to restore sustainable practices in the area's socio-ecological landscapes and seascapes, preserving traditions like stone weir fishing. This restoration is crucial for biodiversity, ecosystem services, and cultural heritage, with youth involvement ensuring its long-term success. Challenges like declining fish populations, silt accumulation, and coral reef degradation hinder the fishing function of the stone weirs. However, Penghu authorities recognize and promote the stone weirs' diverse values, fostering ecological conservation and attracting young residents back to their hometowns. Initiatives promoting fish-eating culture, education, and tourism while preserving stone weirs have been implemented across communities like Hudong, Hongluo, and Nanliao. Stone weirs provide vital ecosystem services and support fishing villages along Penghu's coasts. The Youth Guidance Policy encourages young people to participate in environmental protection and village preservation,

C.-Y. Lee · H.-Y. Chuang
Agency of Rural Development and Soil and Water Conservation, Nantou City, Taiwan

C.-F. Chiu · J.-M. Chu · C. H. Tsai · C.-C. Cheng · C.-H. Tsai
Tainan Branch, Agency of Rural Development and Soil and Water Conservation, MOA, Tainan City, Taiwan

K.-L. Fu
Nantou Branch, Agency of Rural Development and Soil and Water Conservation, MOA, Nantou City, Nantou County, Taiwan

C.-F. Wu (✉) · L. V. T. Trac
Department of Horticulture, National Chung Hsing University, Taichung City, Taiwan
e-mail: cfwu@dragon.nchu.edu.tw

M. Nishi et al. (eds.), *Ensuring Ecological Connectivity in Socio-Ecological Production Landscapes and Seascapes (SEPLS)*, Satoyama Initiative Thematic Review, https://doi.org/10.1007/978-981-95-1474-8_13

which is crucial for the intertidal zone's ecological, economic, and social landscape. A youth cooperation platform raises awareness and accelerates intertidal zone conservation. In contrast, cross-sectoral cooperation ensures a sustainable future for Penghu's intertidal zone and stone weirs from ecological, economic, social, and cultural perspectives.

Keywords Oceanic island · Fishery resources · Ecological connection · Coastal ecosystem · Sustainable development

13.1 Introduction

For over 250 years, the Penghu area has relied on stone weirs for fishing in challenging ocean conditions. This traditional method, known as stone weir fishing, has not only endured but flourished. The number of stone weirs was only 69 in 1915 and increased by 592 in 2008, highlighting their resilience and importance for the community (Hong 1999; Lee 2008). However, pieces of information in a book entitled *Stone Weirs of Penghu* (Hong 1999) also documented that the quantity of stone weirs in Penghu was fluctuating over the years. For example, only 83 weirs were found in Huxi Township and 29 in surrounding waters, showing that many of them have succumbed to damage or destruction. As of 2021, only 60 weirs were recorded in Huxi Township by the Stoneweir Integrated Information Platform, with 23 weirs near our research area. Beyond their original purpose as fishing tools, stone weirs have evolved to provide various ecosystem services, including the preservation of cultural heritage, ecological conservation, opportunities for ecotourism, and scenic sightseeing.

S.-H. Chen (✉)
International Master Program of Agriculture, National Chung Hsing University, Taichung City, Taiwan
e-mail: vickey@dragon.nchu.edu.tw

C.-H. Lai · H.-W. Hsu · T.-Y. Lin
Taiwan Landscape Environment Association, Kaohsiung City, Taiwan

M.-J. Lee
Department of Tourist and Leisure, National Penghu University of Science and Technology, Magong City, Penghu County, Taiwan

C.-J. Wang
Huxi Township, Penghu County, Taiwan

C.-C. Hu
Department of Marine Sports and Recreation, National Penghu University of Science and Technology, Magong City, Penghu County, Taiwan

H.-C. Chen
Department of Soil and Water Conservation, National Chung Hsing University, Taichung City, Taiwan

Fig. 13.1 Different types of stone weirs in Penghu. (Source: Authors). (**a**) An arc stone weir. (**b**) A single-pool stone weir. (**c**) A double-pool stone weir. (**d**) A multi-pool stone weir

The stone weirs present diverse shapes influenced by factors like terrain, direction, and tide. These variations categorize them into three types: arc stone weirs, single-pool stone weirs, and double- or multiple-pool stone weirs (Fig. 13.1). Li's field investigation (2008) suggested that tidal traps (also known as arc stone weirs) accounted for the highest proportion at approximately 49% in Penghu. Single-pool weirs accounted for about 32% of the total, and double/multiple-pool stone weirs accounted for only 3%. Additionally, about 16% of the original double-pool stone weirs have been damaged with only remnants of their structure remaining, making it difficult to identify their original form.

Stone weirs are known for their species catch diversity (Hong 1999). Among the species caught are yellowtail (*Seriola dumerili* (Risso)), Pearly moray eel (*Gymnothorax margaritophorus*), Oriental bonito (*Sarda orientalis),* grouper (*Cromileptes altivelis),* clove fish (*Spratelloides gracilis*), neritic squid, green top shell (*Tectus pyramis*), crowned turban shell (*Lunella coronata*), and horseshoe crab (Limulidae). Additionally, stone weirs sometimes trap other larger-size fish unexpectedly, such as the longfin grouper (*Epinephelus quoyanus),* stingray, and shark.

In 1952, the Penghu County Government introduced a "revolving fund" initiative to facilitate the construction of motor-powered fishing vessels, which were subsequently lent to local fishermen. By 1957, the increased deployment of these motorized fishing boats significantly accelerated the development of the fishing industry, contributing to a substantial rise in the number of fishermen in Penghu, which exceeded 60,000 by 1964. However, as the industrial structure evolved and a processing export zone was established—attracting labour from Penghu's outlying islands—the region's fishery resources, heavily dependent on fishing, began to experiencing continuous decline in fish population (Lee 2008). The depletion of marine resources in Penghu can be attributed to two primary factors. First, the growing demand for seafood, both domestically and abroad, has led to large-scale, unregulated fishing practices, posing a serious threat to the survival of coastal marine species, including female fish, sea cucumbers, and sea urchins. Second, advancements in logistics and transportation have facilitated more efficient goods delivery, thereby increasing consumer demand and intensifying fishing activities, in turn further exacerbating the depletion of fish populations in the region.

Meanwhile, the decline of fishing stone weirs can be attributed to several factors. These include typhoons and climate change, which have caused rising sea temperatures and environmental pollution. These factors have resulted in the death of coral reefs and accumulation of coral reef debris in stone weirs, reducing their capacity to serve as a wildlife habitat. This has led to the declining fish population in these areas.

Fishing rights for stone weirs were historically divided and passed down from generation to generation in equal shares, but lack of transparent management led to confusion about the rights and responsibilities among generations. At their peak, there were up to 609 counts, but a subsequent survey conducted by Lin in 2011–2012 revealed that only approximately 15% of these stone weirs in Penghu were usable (personal communication).

As depicted above, stone weirs hold significant importance as traditional fishing sites for Penghu's fishermen. Nonetheless, economic limitations, property rights, technical skills, technological constraints, raw material availability, tidal fluctuations, and other factors have rendered the repair of damaged stone weirs increasingly challenging. Consequently, only about 18.7% of existing stone weirs maintain their fishing and habitat functionality. In light of the difficulties in swiftly restoring stone weirs, it is imperative to explore alternative functions within the intertidal zone and contemplate potential transformations for sustainability.

Furthermore, aside from biophysical conditions, social aspects such as an aging population, low birth rate, and low youth involvement increase the challenge of stone weirs' restoration. The rise in individuals aged 20–59 is primarily attributed to the migration of students and professors from the Penghu University of Science and Technology. Therefore, it is imperative to revitalize the fishing township industry and enhance living conditions to enable young and middle-aged people to reside in the township, care for elders, bear children, and sustain its growth.

13.2 Study Area and Description of Activities

13.2.1 Socioeconomic and Environmental Characteristics

The Penghu or Pescadores Islands, situated in the Taiwan Strait, span 126.864 sq km and boast rich volcanic landscapes (Fig. 13.2). Siliceous and alkaline basalt rocks, remnants of historical volcanic activity, dominate its geological makeup. Siliceous basalt, centered in the main eruption area, contrasts with the surrounding

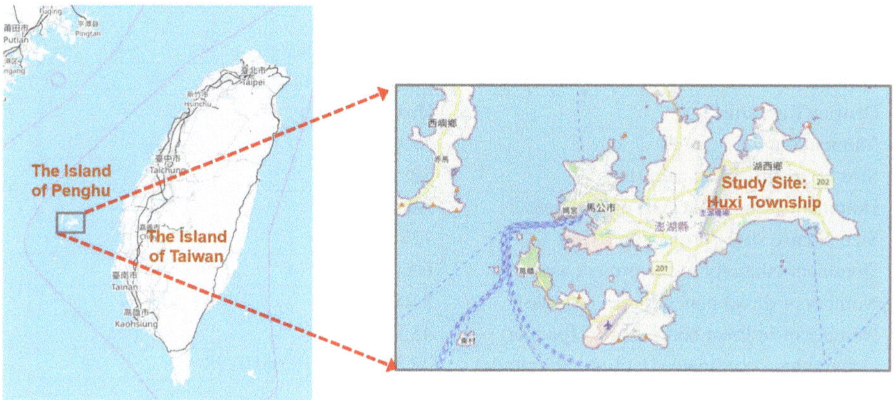

Fig. 13.2 Map of the case study area. Left: Relative location of the island of Penghu; Right: Location of Huxi Township. (Source: Created by authors using map data from OpenStreetMap; Map data source: openstreetmap.org/copyright)

distribution of alkaline basalt. The island's soil, primarily weathered from basalt, consists of red anorthite and pyroxene, which are low in organic matter and often saline due to minimal rainfall. The flat terrain, devoid of mountains, limits water capture and agricultural potential, relying on oceanic influences for temperature regulation and rain formation. This lack of rivers necessitates irrigation via groundwater or ponds. Despite agricultural challenges, Penghu's exposed basalt rocks are valuable construction materials for building walls and paving stones, integral to local life.

The coastal areas of Penghu showcase a diverse range of volcanic geological and coastal landscapes shaped by waves, currents, and tides. Erosive sea cliffs, arches, caves, and more emerge due to the sea's power. With a 164.1 sq km intertidal zone and a 3 m tidal range, basaltic magma forms a flat platform. The gravel-rich environment supports high biodiversity, hosting various organisms. This habitat variation fosters distinct biological communities, including shellfish, crustaceans, corals, and fish. Penghu boasts around 120 coral species, providing crucial biological resources. Coral remnants serve as natural building materials for constructing stone walls and homes by local residents (Wu 2021; Penghu County Government 2023).

The study area encompasses the villages and townships of Nanliao, Hudong, Huxi, and Hongluo, as well as the Qingluo Wetland in Huxi Township, Penghu County, covering approximately 7.63 sq km. The terrain varies, with higher elevations in the south tapering to lower levels in the north, displaying a mild elevation gradient—the average slope measures only around 12%, with an average elevation of 30 m. Huxi Township boasts the highest population, totaling 1379 residents, while Hudong Township has a smaller population of 400 but showing a slight population growth trend over the past decade. Demographically, the population distribution consists of approximately 13.36% aged 0–19, 29.78% aged 20–39, 35.5% aged 40–65, and 19.84% over 65 (Penghu County Government 2023) (Table 13.1).

Table 13.1 Basic information of the study area

Country/Region	Taiwan, China
Province	Penghu County
District/Township	Huxi Township
Municipality	N/A
Size of geographical area (ha)	728.8
Dominant ethnicity(ies), if appropriate	Han Chinese
Size of case study/project area (ha)	728.8
Dominant ethnicity in the project area	Han Chinese
Number of direct beneficiaries (people)	3406
Number of indirect beneficiaries (people)	45,312
Geographic coordinates (latitude, longitude)	23.578178, 119.6610198

Source: Authors

13.2.2 Protection of Stone Weirs as a National Important Wetland Species Habitat in the Intertidal Zone

Hongluo Village lies within Penghu's largest mangrove restoration area, the Qingluo Wetland, boasting vast expanses. This wetland serves as a crucial bird habitat and wintering spot, hosting thousands of migratory birds annually. Rich in both natural and human ecological resources, the area includes Hongluo Bay, a significant breeding ground for juveniles, and an intertidal zone teeming with fishery resources. Designated as a nationally important wetland by the Penghu County Government on January 28, 2014, under the Wetland Conservation Law, Hongluo Village encompasses approximately 250 ha. The Conservation and Utilization Plan for Qingluo Wetland aims to preserve its ecological integrity, foster sustainable environmental development, and bolster conservation and restoration efforts. The conservation strategy used in this project was to protect stone weirs as a comprehensive approach for habitat and species restoration in the project area.

The principles of conservation and wise use for managing the wetland's natural resources and ecological functions are as follows:

1. Priority should be given to protecting natural wetlands and maintaining their water resource systems.
2. The conservation of animal and plant resources in wetlands should be strengthened.
3. Sufficient planning and strategic management are essential for wetlands with ecological network significance, as well as for the surrounding environment and SEPLS (socio-ecological production landscapes and seascapes).
4. The overall planning of the wetland system should be promoted in conjunction with wetland restoration, flood control and detention, water purification, water resource conservation and utilization, and landscape and recreation. Semiartificial wetlands may be built in suitable locations and in appropriate ways when necessary.

13.2.3 Reviving Fishing Villages Through Eco-Conscious Investment in Stone Weirs to Enhance Ecological Connectivity and Functions

The primary responsibilities of the Agency of Rural Development and Soil and Water Conservation (ARDSWC) include:

1. Promoting the rejuvenation and revitalization of agricultural and fishing villages
2. Enhancing their self-sufficiency
3. Attracting youth to stay or return to rural communities
4. Encouraging their participation in regenerating these villages across different fields

The challenges encountered by these villages are varied and encompass a necessity for original creative concepts, limited capacity, and inadequate human resources. To tackle these obstacles, it is crucial to begin at the educational level and offer avenues for youth involvement in village development. The government agency has initiated various programs, such as a "Farm Long Stay for College Students," a "Rural Up! Competition," and a "Rural Up! Plus Action Plan." These are aimed at engaging students from non-agricultural backgrounds in rural development. The initiatives leverage students' creativity and diverse expertise to support agricultural and fishing communities in their redevelopment efforts. In Penghu, noteworthy initiatives include a rural migration competition for college students and an innovative research and demonstration plan for youth returning to rural areas.

The Rural Regeneration Act, enacted in 2000, aims to foster sustainable development in farming and fishing villages while revitalizing rural areas. ARDSWC is designated as the governing body for this endeavor. Key objectives include enhancing primary production conditions, preserving rural ecology and culture, improving quality of life, and fostering a prosperous new countryside.

13.2.4 Cross-Sectoral Partnerships for Stone Weir Preservation and Revitalization

Since 2017, the ARDSWC has encouraged college students and young people to conduct surveys and revitalize stone weirs. This effort includes youth surveys of stone weirs, stone weir-themed website development, stone weir craftsman training, ecological experiences, marine education, and more. Furthermore, the Oceanic Commission's "Salute to the Sea" policy promotes activities for Taiwanese people to understand the ocean, develop closeness to the ocean, engage with the ocean, and maintain the ocean. Stone weirs are also required for people to learn about the ocean's intertidal zone (Fig. 13.3).

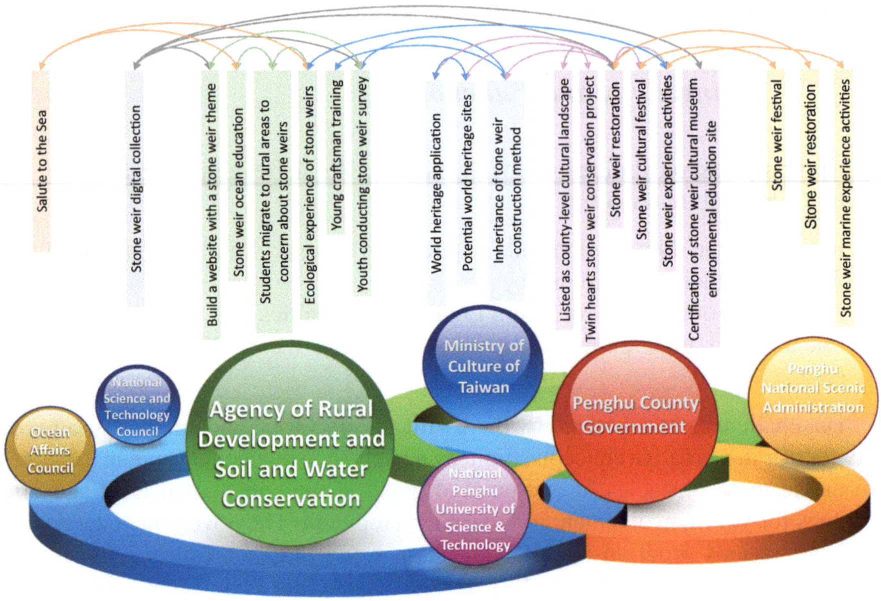

Fig. 13.3 The stone weir cross-sectoral partnership diagram. (Source: Authors)

13.3 Results

13.3.1 Applying Primary Data for Stone Weir Preservation

A local researcher, Dr. Ming-Ju Lee, received a grant to conduct field investigations and digitize authorized documents on stone weir sites. This effort resulted in developing a comprehensive digital database containing 592 records, including photographs, fundamental site information, satellite positioning data, shovel measurements, stone weir classifications, structural details, fishing methods, and current usage status. The research team developed 2D technical drawings, a 3D animation-based flight simulation system, and various documentary videos to enhance visualization and analysis.

Yang (2021), with funding from the ARDSWC's Innovative Research and Demonstration Project for Youth Returning to the Countryside, launched 39 aerial flight routes between 2020 and 2021 to capture high-resolution aerial photographs of Penghu. These images were utilized to identify and map the existing stone weirs (Fig. 13.4), enabling researchers to document their quantity, spatial distribution, morphological changes, and distinctive structural features. As of 2021, 612 stone weirs have been reported, including 338 newly recorded structures. In addition to mapping efforts, an integrated information platform was developed to consolidate knowledge of stone weir terminology, structural composition, construction techniques, and historical development. Expanding upon the foundational work initiated

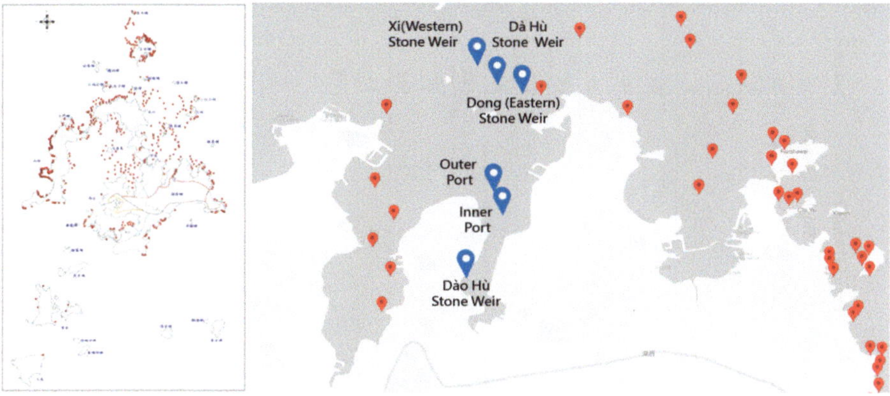

Fig. 13.4 Left: Distribution map of stone weirs (red dots) in Penghu County; Right: Distribution map of stone weirs in the Hongluo community. (Sources: provided by one of authors, Dr. Ming-Ju Lee)

by Dr. Lee, this initiative has compiled extensive databases to create a comprehensive knowledge repository encompassing construction methods, archival records, craftsman directories, training programs, educational materials, and ecotourism applications. This effort contributes significantly to the preservation and sustainable management of Penghu's stone weir heritage by bridging gaps in previous research and enhancing data integration (Table 13.2).

13.3.2 Enhancing Ecological Connectivity and Transforming Functions of Stone Weirs

We have identified 65 bird species across 22 families in the study area. Sandpiperidae, notably sandpipers, leads with 14 species, followed by Heronidae and Rallidae, each with nine species. Noteworthy conservation species include the black-faced spoonbill, swallow-tailed plover, tang egret, red-bellied hawk, gray-faced vulture, kestrel, osprey, red tern, lesser tern, gray tern, and red-tailed shrike. The coastal waters are renowned as clovefish fishing grounds, attracting many terns during early summer for foraging. Additionally, little terns breed in the dune terrain along the coastal region. A Penghu Marine Biology Research Center survey identified 159 fish species across 59 families. Dominant families include Gobiidae (10 species, 6.3%), Trevalidae and Snapperidae (8 species each, 5.0%), Exocoetidae (7 species each, 4.4%), and Filiaceae (6 species, 3.8%). Key species such as black-tailed sardines, short-nose stalked dogfish, mullets, and Aonai drill mouth comprise 72.8% of the total species. In a 2011 survey by the National Penghu University of Science and Technology at the Qingluo Wetland, 62 decapod species across 16 families were identified, showcasing rich biodiversity, including Ocypodidae, Portunidae,

Table 13.2 Lengths and locations of stone weirs in Penghu

City/Township	Magong City			Huxi Township			Baisha Township			Xiyu Township			Wangan Township			Qimei Township			Subtotal
Type	Arc-shaped	Single pool	Double pool	Arc-shaped	Single pool	Double pool	Arc-shaped	Single pool	Double pool	Arc-shaped	Single pool	Double pool	Arc-shaped	Single pool	Double pool	Arc-shaped	Single pool	Double pool	
>1000 m		0	0	0	1	1	0	1	0										3
900–1000 m		0	0	0	0	0	0	2	0										2
800–900 m		0	1	0	0	1	0	0	2	0									3
700–800 m		0	3	0	0	3	0	0	13	4									20
600–700 m	0	1	0	0	0	1	0	6	2				0	1	0				10
500–600 m	0	1	3	0	6	2	0	12	1	0	1	0							26
400–500 m	1	3	1	0	4	0	3	29	0	1	1	0							43
300–400 m	4	4	2	0	9	4	2	40	3	2	4	0	0	1	0				75
200–300 m	7	2	0	0	15	1	18	37	0	3	13	1	0	1	0	0	0	1	99
100–200 m	19	5	0	5	9	0	34	15	0	52	4	0	1	0	0	0	0	0	144
<100 m	18	0	0	11	0	0	15	3	0	43	0	0	0	0	0	0	0	0	90
Subtotal	49	16	6	16	48	9	72	160	10	101	23	1	1	2	0	0	0	1	515
No Data		7			10			9			17			0			0		43
Total	78			83			251			142			3			1			558

Sources: Made by authors based on the publication of Lee in 2008

Fig. 13.5 Embankment of a stone weir shows up at low tides. (Source: Photographed by an author, Dr. Chen-Fa Wu)

Grapsidae, Xanthidae, and Diogenidae. A 2018–2019 shellfish survey in Qingluo Wetland recorded 231 species across 84 families within three classes: Polyplacophora, Gastropoda, and Bivalvia. Dominant species include *Batillaria sordida, B. zonalis, Clypeomorus batillariaeformis, Isognomon nucleus, Littoraria undulata, Lunella coronata, Monodonta ladio, Nerita albicilla, N. costata*, and *N. plicata* (National Penghu University of Science and Technology 2016; Penghu County Government 2023) (Fig. 13.5).

The fishing function of stone weirs has waned due to declining fish populations, siltation of stone lakes, coral reef degradation, and inadequate maintenance. Nonetheless, various entities, including the Penghu National Scenic Area Management Office and Penghu County Government, have recognized and promoted the multifaceted value of stone weirs. Initiatives promoting ecological conservation, environmental education, and experiential learning have gained traction in fishing villages. Revitalization efforts have spurred young people to return to their hometowns. Concurrently, endeavors have been undertaken to promote fish-eating culture, education, and tourism while safeguarding stone weir and wetland ecosystems. For instance, the Hudong community focuses on freshwater wetland restoration and intertidal ecotourism, while the Hongluo community emphasizes stone weir and wetland conservation. In contrast, the Nanliao community prioritizes cultural preservation and sustainable tourism.

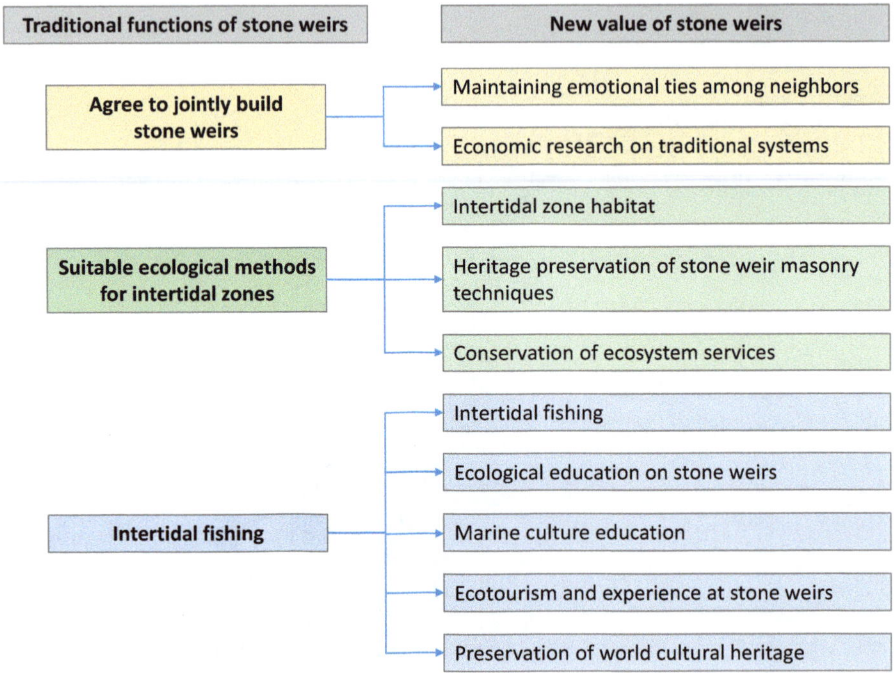

Fig. 13.6 Transformation of traditional functions to new values for stone weirs. (Source: Authors)

As of December 2022, 64.15% of the total 4232 rural communities (2715) have participated in the Grassroots Cultivation Project for rural residents. Additionally, 1000 communities (23.63%) have submitted community development plans following personnel training. These plans prioritize community environmental improvement, industrial activation, cultural preservation, and ecological conservation, aligning with local planning and development goals. The specific achievements over the years in the study area include: (1) improvement of the rural SEPLS and living quality; (2) creating highlights of rural and regional development (Figs. 13.6 and 13.7).

13.3.3 Cross-Sectoral Partnerships for Stone Weir Preservation and Revitalization by Establishing a Youth Cooperation Platform

In 2022, the Penghu Agro-Island-Ocean Youth Revitalization Association was established through collaboration between the Tainan Branch of ARDSWC, the Bureau of Agriculture and Fisheries of the Penghu County Government, and the Bureau of Culture. They initiated the "Penghu Partnership Collaborative Creation

Fig. 13.7 Ecological education and ecotourism in intertidal zones, observation, and identifying different marine organisms. (Source: Photographed by an author, Dr. Szu-Hung Chen)

Experimental Base" to facilitate cross-domain co-creation among young adults over 20, primarily in the hotel industry and those residing in Penghu's farming and fishing villages. The objectives include:

1. Conducting exchange planning and operational management guidance for youth in Penghu County
2. Encouraging the establishment and growth of youth industries, education and exchange platforms, cross-border collaboration, and support for innovation and entrepreneurship
3. Promoting youth entrepreneurship in Penghu and contributing to social welfare
4. Nurturing youth entrepreneurship and enhancing the integration of education with society

This collaboration offers an opportunity to implement initiatives in Penghu County, including construction using ecological enginnering methods, preservation of traditional culture, creation of unique village spaces, and promotion of food and agriculture education. It aims to identify suitable business models through cross-sectoral collaboration and generate mutual benefits. With community engagement, it will enhance sustainable tourism in agricultural and fishing villages, supporting Penghu's youth through business opportunities, tourism education, and ocean management.

With the restoration of stone weirs, a youth named Fu-Ci Yang organized a workshop on stone weir repair and collaborated with craftsmen, stone weir owners, and other relevant stakeholders, with the assistance of the 2016 "College Student

Migration Project" and the 2017 "Youth Returning Project" of the ARDSWC to discuss appropriate restoration points. The plan involved recruiting 17 people, including five locally trained young artisans, to restore two stone weirs within 4 weeks, achieving restoration levels of 60% and 80%, respectively. Additionally, participants partnered with residents to organize activities such as guided tours, working holidays, and events on repairing stone weirs. As a result of the restoration efforts, various species, such as red Hukui clams, cuttlefish, horseshoe crabs, lionfish, and other creatures, entered the pool one after another, indicating a restoration of biodiversity.

In addition, a growing generational gap in Penghu is leading to the gradual erosion of fish-eating culture and local traditions. Tourists visiting Penghu also lack adequate knowledge about sustainable fish consumption. The youth group is actively promoting education on fish consumption to enlighten people about the journey of seafood from sea to table. Through this initiative, the local youth group aims to deepen people's connection to the sea by educating them about fishing processes, techniques, fishermen's stories, and Penghu's cultural heritage. To further this cause, they have established a bookstore to promote fish-eating culture, created hands-on seafood classrooms to promote sustainable seafood concepts, invited community elders to cook traditional dishes, organized fish market tours and cooking demonstrations, and introduced a variety of engaging fish-related activities. These endeavors seek to bridge the knowledge gap in fish-eating culture between locals and tourists while preserving Penghu's cultural and culinary heritage for future generations. Additionally, they encourage consumers to opt for ocean-friendly seafood choices to sustain marine fishery resources. A key promotional strategy involves utilizing bycatch to enhance the Penghu seafood brand and boost business revenue.

13.4 Discussion

13.4.1 Stone Weir Landscape Protection, Restoration, and Revitalization System in Penghu

The number of stone weirs in Penghu has fluctuated, reflecting shifts in environmental conditions, socioeconomic factors, and conservation efforts. Historically, the construction of stone weirs required collective effort due to their large scale. Individuals could not undertake such projects alone; instead, they formed cooperative ventures with relatives or close associates, pooling resources for investment and labor. These ventures were traditionally passed down through generations based on familial or geographical ties, with ownership regarded as common private property. Each village established its regulations governing the catch's construction, maintenance, and distribution. In Penghu County, stone weirs have been categorized into four ownership types: (1) jointly owned by small community groups, (2) managed by the local temple, (3) held by small clan groups, or (4) owned by individual families (Lee 2008). Descendants continue to patrol and fish in these weirs based on inherited agreements, preserving a traditional system of catch sharing. The construction and management of stone weirs remain a compelling subject of study, offering valuable insights into the communal organization, economic strategies, and cultural heritage of fishing villages in Penghu.

Historical records indicate 69 stone weirs in 1915, increasing to 314 in 1917, 284 in 1938, 192 in 1957, 558 in 1999 (Hong 1999), and 592 in 2008 (Lee 2008). The 1988 publication *Stone Weirs of Penghu* reported 83 stone weirs in Huxi Township, with 29 located in its surrounding waters. However, many of these structures have since been damaged or destroyed. As of 2021, the Penghu Stoneweir Integrated Information Platform documented only 60 stone weirs in Huxi Township, with 23 situated near the study area. This decline underscores the evolving role of stone weirs, which have transitioned from being solely used for fishing to serving broader functions, including cultural preservation, ecological conservation, and ecotourism. This shift highlights the need for continued documentation, preservation, and sustainable management of these historical fishing structures.

13.4.2 Enhancement of Ecological Connectivity Functions of Stone Weirs

Stone weirs are a traditional form of ecological engineering commonly utilized in intertidal zones to enhance fishing efficiency and adapt to environmental conditions. Their construction requires careful material selection, with the outer layer typically composed of durable basalt or boulders for structural stability, while the inner layer consists of coral reef limestone to provide a more porous environment conducive to marine life. The tidal range is a critical factor in stone weir effectiveness, as a wider range facilitates fish entry during high tide and their retention during low tide. To endure the dynamic forces of wind, waves, and tidal currents, stone weirs are often built with arc-shaped walls, typically reaching heights of 1.5–2 m, with spiraled ends designed to stabilize the structure and enhance fish capture. Additional architectural features, such as fish-collecting houses and multiple fish-trapping pools, contribute to the complexity and functionality of these traditional structures. Various design elements, including entry openings, left and right dikes, and narrow passageways, are tailored to meet specific fishing needs and optimize capture efficiency (Hong 1999).

Beyond their historical role in traditional fishing practices, stone weirs also serve as critical ecological habitats, supporting marine biodiversity and fostering sustainable fishery resources. The intertidal pools created by stone weirs provide essential shelter, breeding grounds, and feeding areas for various aquatic species. In Hongluo Bay, primary fish species caught within the weirs include Japanese water skies, black-tailed small sardines, short-nose stalked dogfish, mullet, and Aonai drill mouth, collectively accounting for 72.8% of the total catch. These species predominantly exist in their juvenile stages, indicating that the stone weirs are vital nurturing habitats for fish population development (National Penghu University of Science and Technology 2016).

Ecological connectivity in seascapes refers to the movement and interaction of marine species across different habitats, ensuring healthy ecosystem functioning and biodiversity conservation. Coastal and intertidal zones are critical transition areas linking marine, estuarine, and terrestrial ecosystems. However, habitat fragmentation, climate change, and human activities have disrupted these natural connections, threatening marine biodiversity and ecosystem resilience. Restoring traditional stone weirs can significantly enhance ecological connectivity by re-establishing vital linkages between intertidal, nearshore, and offshore environments. Through their structural and ecological functions, restored stone weirs can facilitate species movement, improve water circulation, and support interconnected marine habitats. Stone weirs function as semipermeable corridors within the intertidal zone, creating a mosaic of habitats that support marine biodiversity. Their restoration can enhance ecological connectivity, such as habitat linkage between tidal zones, pathways for larval dispersal, breeding and spawning grounds for key species, and so on.

13.4.3 Cross-Sectoral Partnerships for Stone Weir Preservation and Revitalization Advance Youth Gathering and Cross-Sectoral Cooperation in Penghu

In Penghu, young individuals have established a youth cooperation platform aimed at safeguarding the marine environment and preserving local cultural heritage. The formation of this group follows two distinct paths. First, driven by the necessity to pursue higher education outside Penghu due to limited local academic opportunities, many young Penghu residents study in other areas or abroad. Despite their departure, efforts are made to encourage their return through programs like the College Student Migration Project, organized by ARDSWC. This initiative encourages students to explore local characteristics and envision development strategies, potentially leading to their return or continued involvement in Penghu. ARDSWC remains committed to engaging young people in activities promoting stone weir protection, intertidal zone conservation, fish-eating education, marine and domestic waste recycling, and other environmental conservation efforts in Penghu.

Another path for fostering the youth cooperation platform in Penghu involves individuals who were not originally from Penghu but relocated there for reasons such as education, marriage to a resident, or an appreciation for the environment. By joining the youth cooperation platform, they can swiftly assimilate into local life and culture, connect with peers of similar age, and engage in idea sharing and creativity, potentially leading to entrepreneurial ventures, family business succession, or participation in innovative research projects for returning rural youth, either

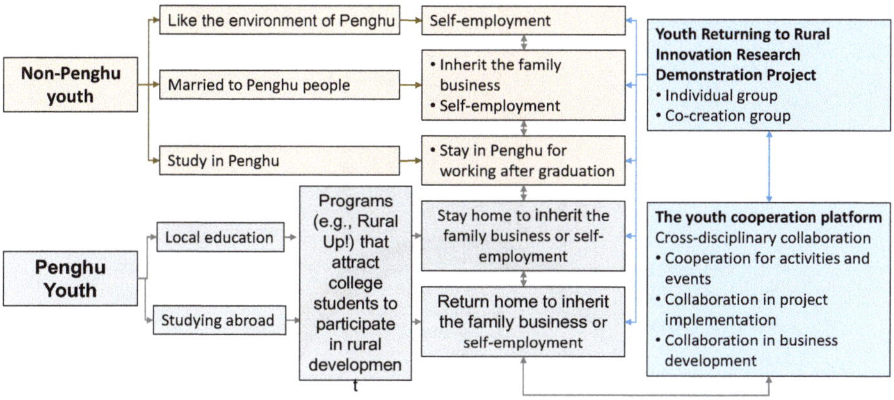

Fig. 13.8 The process of how young people gather or migrate to Penghu. (Source: Authors)

independently or as part of a collaborative endeavor. This avenue offers access to stable financial resources during the initial stages of entrepreneurship and the opportunity to develop distinctive industries. The allure of Penghu's environment and the cooperative atmosphere of the youth platform serve as significant draws for young individuals. The ARDSWC's Youth Returning to Rural Innovation Research Demonstration Project, accessible to individuals and collaborative groups, acts as a crucial catalyst for entrepreneurial endeavors and involvement in marine environmental conservation and settlement culture preservation. The outlying islands' lack of educational resources remains a primary reason for young people leaving Penghu (Figs. 13.8 and 13.9).

13.4.4 Synergetic Benefits of Youth and Community Development: A Cooperative, Sustainable Development Model

The linking of youth and community elders is driving the connection between the ocean and islands through food and agriculture education while spearheading the development of fishing village industries. Liu Jianrong and the Hudong community have established the Limited Liability Huxi Township Hudong, Penghu County Community Cooperative under the brand "Happy Mountain and Sea Flavor." This cooperative integrates traditional agriculture, processing technology, and new product development to bolster the agricultural and fishery sectors, proposing channel planning and establishing a network integration platform for comprehensive marketing. Prioritizing cultural heritage, the cooperative has devised a learning-oriented economic model encompassing cultural guides, heritage education, ecological experiences, and environmental education. They offer diverse itineraries focusing on fishing village environments, living culture, and intertidal ecological expertise.

Fig. 13.9 Demonstration of diverse values of stone weirs. (Source: Authors)

The collaboration between the community's youth and elders leverages their respective strengths. While elders concentrate on traditional crop and seafood production, fishing, and basic processing, youth contribute expertise in legal seafood processing methods, packaging design, brand establishment, and sales promotion. They manage advertising design, video production, and online publicity to promote the agricultural and fishery industries, processed products, and ecotourism domestically and internationally. This includes establishing physical stores, online shopping platforms, and participation in rural markets. The online presence facilitates the promotion of outlying islands' agricultural and fishery products, enhancing both the community's and young people's economic income and fostering a mutually beneficial outlying island business model (Fig. 13.10).

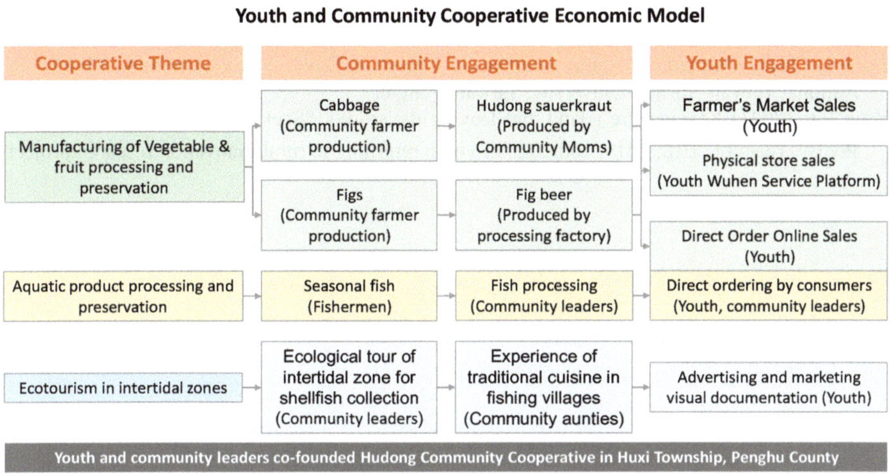

Fig. 13.10 Fishing village community development and youth cooperative economic model. (Source: Authors)

13.5 Conclusions: Key Messages

Penghu's stone weirs are vital components of the coastal ecosystem, supporting the livelihoods of fishing villages and fostering a balanced relationship between humans and nature. In this study, we shared the experiences and insights regarding trade-offs and synergies among efforts to enhance ecological connectivity in managing SEPLS. The ARDSWC's policies for youth effectively encourage young islanders to engage in environmental protection, village preservation, and tourism development, ensuring the holistic well-being of the intertidal zone. The youth cooperation platform further promotes awareness of marine environmental conservation and cultural heritage preservation, with interdisciplinary collaboration enhancing efforts to safeguard the intertidal zone and stone weirs. Preserving Penghu's intertidal zone demands a comprehensive approach considering ecological, economic, social, and cultural aspects. By establishing cross-sectoral cooperation mechanisms, sustainability initiatives can be advanced to preserve Penghu's intertidal zone and stone weirs.

References

Hong G-X (1999) Stone Weirs of Penghu. National Central Library, viewed 7 March 2023. Retrieved from https://tm.ncl.edu.tw/

Lee M-J (2008) Digital Archives - The Forms and Culture of Stone Tidal Weir in Penghu, viewed 7 March 2024. Retrieved from http://www.daweir.npu.edu.tw/

National Penghu University of Science and Technology (2016) Formulation of the report on the results of the Qingluo Wetland conservation and utilization plan

Penghu County Government (2023) Introduction, viewed 7 March 2024. Retrieved from https://www.penghu.gov.tw/en/home.jsp?id=6

Wu P-Q (2021) Caspian Spirit and connotation analysis of Stone Weir Community, Jibei Island. National Taiwan Normal University, Taiwan, Penghu

Yang F-C (2021) Leaving the island and leaving the studio. "Seeking Sea People Looking for Pools". Years of 2018, 2019, and 2020 of youth returning to rural innovation research projects

Chapter 14
Synthesis: Ecological Connectivity in the Context of Socio-Ecological Production Landscapes and Seascapes (SEPLS)

Maiko Nishi, Suneetha M. Subramanian, Philip Varghese, and Juliano Sènanmi Hermann Houndonougbo

Abstract This chapter provides the synthesis of the findings from the 12 case studies presented in this volume. It addresses the following questions: (1) how ecologi-

---

Contributing Authors: Vincent Awotwe-Pratt, Yu-Chun Chan, Mei-Yin Chen, Szu-Hung Chen, Emilio Rafael Díaz-Varela, Daniel Enti, Guido Gualandi, Stefan Hotes, Alice Jing-Juan Hsu, Gerald Jetony, Paulina G. Karim, Uma Khumairoh, Yuan-Jyun Wang, Hamilda Francisca Majit, Alessandra Markos, Yoji Natori, Daniel Osorio Dominguez, Mordecai Owidi Ogada, Somajita Paul, Andrés Quintero Angel, Gladys Chepkoech Rutto, Jeeranuch Sakkhamduang, Krishna Gopal Saxena, Chih-Chin Shih, Abdulai Sulemana, Wataru Suzuki, Luu Van Thong Trac, Leonor Valenzuela, Chemuku Wekesa, and Chen-Fa Wu.

The original version of the chapter has been revised. A correction to this chapter can be found at https://doi.org/10.1007/978-981-95-1474-8_15

M. Nishi (✉)
United Nations University Institute for the Advanced Study of Sustainability (UNU-IAS), Tokyo, Japan

Department of Global Coexistence Studies, College of Bioresource Sciences, Nihon University, Fujisawa, Japan
e-mail: nishi.maiko@nihon-u.ac.jp

S. M. Subramanian
United Nations University Institute for the Advanced Study of Sustainability (UNU-IAS), Tokyo, Japan

P. Varghese
Akita International University, Akita, Japan

United Nations University Institute for the Advanced Study of Sustainability (UNU-IAS), Tokyo, Japan

CHRIST University, Bangalore, India

J. S. H. Houndonougbo
Integrated Science and Engineering for Sustainable Societies, Chuo University, Tokyo, Japan

United Nations University Institute for the Advanced Study of Sustainability (UNU-IAS), Tokyo, Japan

M. Nishi et al. (eds.), *Ensuring Ecological Connectivity in Socio-Ecological Production Landscapes and Seascapes (SEPLS)*, Satoyama Initiative Thematic Review, https://doi.org/10.1007/978-981-95-1474-8_14

cal connectivity is conceptualized in the context of managing socio-ecological production landscapes and seascapes (SEPLS); (2) how to measure and evaluate ecological connectivity and monitor its level and progress for benefiting people and nature through managing SEPLS; and (3) how challenges are addressed and opportunities are seized to ensure and enhance ecological connectivity through managing SEPLS for biodiversity, ecosystems, and human well-being. Building on the synthesis of the case study findings, this chapter also offers policy recommendations that could support favoring ecological connectivity to ensure ecologically and socially sound outcomes. These recommendations illustrate ways to help simultaneously achieve multiple global goals and targets for sustainability and biodiversity, including the sustainable development goals (SDGs) (e.g., SDGs 1, 2, 3, 11, 13, 14, 15, and 17) and the goals and targets of the Kunming-Montreal Global Biodiversity Framework (e.g., Goal A, Targets 1, 2, 3, and 11).

Keywords Socio-ecological production landscapes and seascapes · Landscape approaches · Ecological connectivity · Biodiversity · Methodologies · Sustainable Development

14.1 Conceptualizing Ecological Connectivity in SEPLS

Ecological connectivity, as highlighted in the Introduction chapter, is the connection between different ecosystems within a landscape or seascape and the various types of land and sea uses, which may involve different governance regimes and management systems. For instance, ecological connectivity underpins key ecological processes, preserving ecosystem functionality at different levels such as habitats, biomes, and space (Lemieux et al. 2022). Thus, ecological connectivity ensures the unimpeded movement of species and the flow of natural processes that sustain life on Earth (CMS 2020, p.1). In the SEPLS context, it would ensure the thriving of biological resources; the conservation of critical ecological functions between the different mosaics of a land/seascape; and the continuity and connectedness of cultural trails, practices, livelihoods, and identities. It also contributes to climate change mitigation and adaptation by enabling the persistence and recovery of flora and fauna through the dispersal and movement of organisms and materials in response to environmental change (Berkström et al. 2022; Lemieux et al. 2022).

The preceding chapters have illustrated the need for ecological connectivity in SEPLS and challenges and efforts undertaken to enhance it in different contexts. In this chapter, we seek to synthesize the key messages and lessons learned from these experiences, gleaned both from manuscripts and from the sharing of experiences at an authors' workshop held in Tokyo, Japan, on 23–25 June 2024. In this section, we highlight how the concept of ecological connectivity operates in the context of

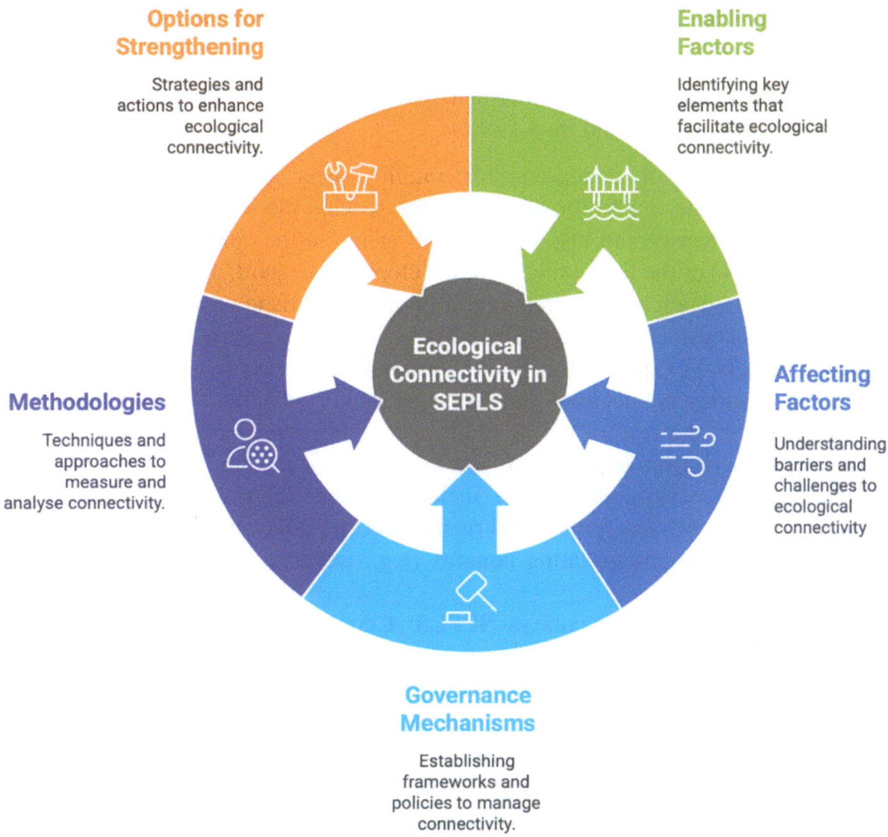

Fig. 14.1 Overview of key aspects relating to ecological connectivity in SEPLS, as detailed in this chapter. (Source: Authors)

SEPLS, what factors enable and affect connectivity, and what governance mechanisms are put in place. The next section highlights various methodologies used to operationalize ecological connectivity within SEPLS. Finally, the last section highlights the various options identified for ensuring and strengthening ecological connectivity in SEPLS. Figure 14.1 illustrates the structure of and relationships between the key aspects of this chapter.

Ecological Connectivity in SEPLS

Some unique aspects related to ecological connectivity in the context of SEPLS include the fact that connectivity is provided in multifunctional landscapes or seascapes, which involve a mosaic of activities covering production, conservation, and cultural dimensions. The emphasis is on local contexts and is determined by the local geographies, species, and human activities. This is highlighted through the illustrations in the following subsections.

14.1.1 Key SEPLS Elements to Ensure Connectivity

The areas between different types of land and sea uses can be connected via multiple means, both tangible and intangible:

- Different species of plants, animals, fungi, insects, and microbes: These also include keystone, umbrella, and indicator species that are critical to ensuring ecological integrity and also confer high value on the landscape and seascape (e.g., the case of the long-tailed "goral," elephant, leopard, and Asian golden cat in Thailand, which require greater conservation efforts to ensure their long-term survival as keystone species—Chap. 5). At the same time, they could also allow faster dispersal of species that could harm native biodiversity, although the benefits of ensuring connectivity between species far outweigh the risks of adverse impacts (e.g., the case of biological invasions favored by certain kinds of connectivity; Berkström et al. 2022).
- Habitats and ecosystems: These include patches, corridors, watersheds and waterways, coastlines, forests, and rivers that can ensure an uninterrupted flow of different ecosystem-regulating benefits (e.g., water regulation, pollination, air quality).
- Nutrient-water-energy nexus in SEPLS: Certain types of movement in key SEPLS components, notably nutrient, water, and wind flows and/or energy transfer between habitats, need to be taken into account to ensure good ecological connectivity (e.g., the case of fertilizer inputs and nutrient run-off [N and P] at a landscape-seascape interface). In many instances, water can serve as a good connector between key elements of the landscape or seascape, while riparian vegetation can function as a wildlife habitat. A good example of this is connectivity between the landscape-seascape elements at a watershed scale within the Xiuguluan River and the Coastal Mountain Range conservation corridors in Hualien County of Chinese Taipei—Chap. 10).
- People and their production activities: This links to the diverse ways in which people connect to different parts of the landscape or seascape for their various needs (e.g., food, healthcare, building materials, and cultural and spiritual practices), and connected parts of the landscape enable them to access and secure their well-being in a manner aligned with their preferences. Structures like roads and other infrastructures could enable such connectivity, while they could also serve to disrupt the connectedness between different ecosystems within a landscape or seascape.
- Traditional ecological knowledge (TEK) and related cultural practices: These can serve as socio-ecological connectors. Memories, language, sense of place, and identity further strengthen connections to a given landscape or seascape across generations. Integrating diverse values, modern knowledge, and TEK is very important for promoting connectivity in SEPLS. Some examples include the following: (1) local people's perceptions of the close relationship between the promotion of eco-friendly paddy rice farming and the maintenance of healthy freshwater habitats for the recovery of "*hala*" (monk goby, *Sicyopterus japoni-*

cus)—an endemic migratory fish species of the Turtle River in Fengnan SEPL in Hualien County of Chinese Taipei—Chap. 10; (2) the major role played by cultural traditions and beliefs in the preservation of iconic species like *banyan* trees (*Ficus benghalensis*), cows, peacocks, and owls in India over time—Chap. 12; (3) the significant contribution of cultural values to the conservation of agrobiodiversity in the Rabai Kaya forest landscape in Kenya—Chap. 3; and (4) the capacity building for local farmers and beekeepers in the use of innovative techniques/methods in the construction of traditional stone enclosures for effective protection of beehives against bear attacks in Northwestern Spain—Chap. 6.

14.1.2 Types of Ecological Connectivity

Overall, the way in which ecological connectivity is nurtured often determines the outcomes (positive or negative). Specifically, positive outcomes might require improved access to various ecosystem services such as "provisioning"—food, clean water, and medicines; "regulating"—climate regulation, pollination, and temperature regulation; "cultural"—customary rituals, spiritual practices, traditional art, ecotourism, and place-based education; and "supporting"—soil fertility and ecosystem resilience. Conversely, negative outcomes could include impaired water regulation, droughts, disease for nature and people, and inequity, essentially in the event of disruption.

These links are influenced by direct drivers such as human activities like types of land use and introduction of invasive alien species, as well as climate change, pollution, natural processes such as hydrological flows, micro and macro-climate regimes, and natural calamities like earthquakes. They are also influenced by indirect drivers such as market demands, demographic pressures, political regimes, and conflicts that determine the use and management of landscapes or seascapes. It is essential to examine where the disturbance occurred and identify the "disconnection" (e.g., loss of species) and what existed before. A similar approach has been successfully used to restore water quantity and quality in the landscapes and seascapes of Hualien County, Chinese Taipei (Chap. 10).

Furthermore, connectivity may be multidimensional. This is the case with "water flow," which has four dimensions: connectivity between earth's surface and atmosphere, connectivity between upstream and downstream, connectivity between water and land area, and vertical connectivity between groundwater and surface water. It is therefore essential to examine the different dimensions of water flow. For instance, the uncontrolled use of chemical fertilizers and pesticides in rice cultivation in Chinese Taipei between the 1990s and 2000s has affected ecological quality over time. Accordingly, the decline in raptor populations, driven by pesticide use, water pollution, and landscape fragmentation, disrupted ecological connectivity during that period. These challenges subsequently led to the promotion of eco-agriculture and organic farming in the Wufeng District in recent decades (Chap. 11).

The extent and quality of ecological connectivity also depend on the ways people perceive what it is. Having lived in SEPLS for a prolonged period of time, local communities understand the spatial-temporal changes in the area and can potentially provide site-specific adaptive co-management strategies. This is expressed in several traditional norms, rules, and practices, for example, regulating which resources can be used and how, or which cannot be used. Some examples include the following: (1) Local people following the community protocol that spells out rules on activities in forests in the area between Kinabalu and Crocker Range Parks in Sabah, Malaysia (Chap. 8); and (2) Indigenous and local knowledge being integrated with expert knowledge and modern science along with robust multi-stakeholder consultation processes while cross-sectoral partnerships enabled the restoration of declining populations of local freshwater shrimp varieties in Xinshe SEPLS, Hualien County of Chinese Taipei- (Chap. 10).

14.1.3 Current Challenges and Options for Ensuring Connectivity in SEPLS

- Understanding landscape is a mosaic: Local communities often see connectivity as an intrinsic part of their lived environment, perceiving it as one integral whole and noting disconnections (e.g., the population decline of local freshwater shrimp varieties, known as an indicator of the landscape-seascape ecological connectivity in the Xinshe SEPLS, was mainly attributed to increasing anthropogenic pressures on local habitats/ecosystems in Hualien County, Chinese Taipei—Chap. 10). Communities are sometimes unable to adapt their local activities to maintain ecological connectivity because it affects their livelihoods or because they are more vulnerable to certain social and environmental changes. For example, the benefits of tourism in a region sometimes do not reach the local communities, who receive little money from tourists or whose local products are not sold. Thus, an integrative ecotourism platform involving multiple stakeholders (i.e., community members and tourists) was proposed to create added value within local communities by simultaneously achieving the goals of sustainable agriculture, ecotourism, and alternative livelihoods, as implemented at the Kinabalu Ecolinc project sites in Malaysian Borneo. As a result, this project has contributed to the identification of some key areas for ecological connectivity via promotion of the establishment of community-conserved areas (CCAs), sustainable agricultural interventions, and community-based ecotourism (Chap. 8).
- Raising awareness of the close relationship between the key elements of SEPLS (tangible and intangible) for all stakeholders: Connectivity requires an open mind toward the multiple meanings of connectivity—ecological, sociocultural, spatial, temporal, and others. This can support development and appropriate and lasting co-management arrangements and conservation strategies. For example, the transformation of agricultural landscapes over a few decades (1990–2000), following heavy use of chemical fertilizers and pesticides, has led to a strong push toward promotion of organic farming and eco-agriculture in the Xinshe

SEPLS in Hualien County, Chinese Taipei. This change in farming practices by local farmers has paved the way for the establishment of an integrated landscape-seascape approach in the management of the Xinshe SEPLS, ensuring ecological connectivity (Chap. 10).

- Fostering culture and intergenerational transfer of knowledge: Ensuring the sustenance of traditional knowledge and cultural practices is challenging in the modern context due to changes in population size and other factors (e.g., increasing anthropogenic pressures and environmental degradation), but they can be combined with modern scientific research and tools. Knowledge co-production and weaving is very important, and appropriate mechanisms need to be identified in different contexts.
- Attending to scale of activities: It is important to take into account not only distances—short vs. long—between areas that connect, but also the temporal and spatial scales or extent of connectivity and what is managed. This will help to more effectively plan, regulate, and monitor the effectiveness of actions. Also, at a cross-SEPLS scale, immediate connectivity synergies are more likely to occur between SEPLS in close proximity or with similar geoecological, sociocultural, and economic contexts. This is the case of the four SEPLS (Fengnan, Xinshe, Gangkou, and Luoshan) located in the Taiwan Ecological Network (TEN) and described as shaping the two conservation corridors (between the Xiuguluan River and the Coastal Mountain Range) in Hualien County of Chinese Taipei (Chap. 10).

In sum, these gaps in perception, knowledge, capacity, and implementation that hamper efforts to ensure and enhance ecological connectivity need to be addressed. In the following sections, we highlight a range of methodological approaches and solutions being actively explored to facilitate desirable actions.

14.2 Measuring, Evaluating, and Monitoring Ecological Connectivity in the Context of SEPLS

Ecological connectivity, given its nuances and challenges in the context of SEPLS, requires effective assessment strategies. Effective assessment strategies help in making informed decisions for ensuring ecological connectivity to adapt to climate change, support ecosystem services, sustain biodiversity, and maintain the resilience of both ecosystems and human societies (Hilty et al. 2020; McRae et al. 2012). This volume specifically takes a holistic approach and focuses on local initiatives carried out in partnership with various stakeholders that are aimed at ensuring and strengthening ecological connectivity.

Connectivity in general is evaluated on the basis of the flow of species, materials, and information. It is thus important to know the flow, the disruptions to the flow, and then the value of the flow to understand the impact of connection, disconnection, or even of fragmentation. The SEPLS perspective reiterates connectivity as a multidimensional process that integrates the ecological, social, cultural, and economic dimensions of changes within landscapes and seascapes. The vantage points highlighted are community rights (Chap. 2), biocultural heritage (Chap. 3), traditional wisdom (Chaps. 4 and 8), community forest (Chap. 5), integrated SEPLS management or multi-stakeholder platforms (Chaps. 6, 7, 8, 9 and 10), ecological networks (Chaps. 10 and 13), sustainable agricultural practices (Chap. 11), and peri-urban habitat connectivity (Chap. 12). This section further expands on the benefits of measuring connectivity, indicators to determine the effects, and tools/methods for measuring the outcomes of connectivity.

14.2.1 What Are the Benefits of Measuring Ecological Connectivity?

Knowledge of habitats, habitat linkages, multifunctional effects, movement of species, and how ecological processes operate across any given landscape or seascape is critical for maintaining and enhancing the benefits and reducing the detriments that result from degradation, habitat loss, and fragmentation. The following are some of the benefits of measuring ecological connectivity and outcomes of initiatives aimed at enhancing ecological connectivity.

Multidimensional benefits
As discussed in the previous section, ecological connectivity is multidimensional, and thus it is crucial to carefully consider its potential impacts, particularly the negative ones. The complexity of the process has led practitioners and experts to widely accept the interdisciplinary nature of connectivity, recognizing the many competing factors involved. Connectivity encompasses more than just ecological structure and function; it also includes aspects such as policy, climatic conditions, governance, economics, politics, and social and cultural dimensions.

Tangible and intangible benefits
The most tangible benefit is habitat preservation and enhancement of biodiversity. Intangible benefits include the human connection to the environment, traditional knowledge, and pride in contributing to ecological connectivity, mutual understanding, and knowledge co-production. These benefits through ecosystem services like pollination, clean water, and climate regulation directly contribute to healthier ecosystems and as a result impact human well-being.

Enhancement of ecosystem services
Ecosystem services are commonly grouped into four categories—provisioning, regulating, supporting, and cultural services. Localized interventions and indicators help to quantify these services and assess their sustainability and accessibility. For example, the promotion of green belts or forests with low-flammable species surrounding settlements may favor connectivity for wildlife, enhance the accessibility

to ecosystem services (e.g., chestnut production) of local communities, and be a barrier against wildfires (Díaz-Varela et al. 2025).

Sociocultural and economic benefits (e.g., income, tourism)
Enhanced ecosystem services through enhanced connectivity provide opportunities for consistent income and resources. For example, the Kaya forests in Kenya's coastal region (Chap. 3), valued for their spiritual, cultural, and historical significance, also support the livelihood of the Mijikenda community. Apart from hosting rare plants and diverse fauna, the forests provide an ecosystem that is crucial to support livelihoods through the supply of timber, medicine, fruits, water, and other essential resources for the community. Similarly, Chap. 9 discusses the integrated management of natural resources of Kinabatangan landscapes in Sabah, Malaysia, by using the platform of UNESCO's Man and the Biosphere Programme. Not only aimed at minimizing fragmentation and pursuing ecosystem restoration, these efforts also made the area an ecotourism hotspot with its rich biodiversity and unique ecosystem to support the local community.

Environmental benefits (biodiversity)
Examples of biodiversity benefits range from protection of species and forests, creation of corridors, and pollination to environmental education activities. These benefits are also ensured by identifying, taking stock of, understanding, and mitigating ecosystem disservices, such as invasive species, allergens, wildfires, and decreased productive capacity, as well as human-wildlife conflicts, including the increased cases of predator species attacking livestock. Measuring helps with knowing what ecosystem services and disservices are, why and how they occur, and what impacts they have on ecosystem connectivity in SEPLS.

Noticeable cascading effects (e.g., water contamination affecting fish life, human health, and nutrient cycling)
Connectivity through ecological corridors may have counterproductive or unintended consequences or trade-offs. For example, the corridor that was rebuilt between Cote d'Ivoire and Ghana ended up becoming a funnel for poachers as it harbored many animals that were used as bush meat (Asare et al. 2014). The connectivity could also facilitate the spread of pests, diseases, and invasive alien species, resulting in the growth of pathogens and allergens. The propensity for unintended or unforeseen consequences regardless of negative or positive is something to be viewed cautiously. It is also seen that establishing connectivity through protected forests at times makes the forest resources inaccessible to the local community. If access to the forest is lost, then the related traditional ecological knowledge is also lost, as it can no longer be used. For details, see the Thailand case where the law is strict on human interventions (Chap. 5).

14.2.2 What Are the Indicators That Determine the Positive or Negative Outcomes of Ecological Connectivity?

Indicators of ecological connectivity have lately become a central topic in the study of conservation science and ecology (Rudnick et al. 2012). Landscape connectivity, species movement corridors, and habitat patch size, shape, isolation, and distance are commonly understood as the key indicators of ecological connectivity. These kinds of indicators are crucial for assessing the impact of development, designing restoration projects, and chalking out effective conservation strategies. The evidence-based approach holds the key to producing the most accurate monitoring outcomes. The points below could serve as checkpoints to align the indicators better with overall goals.

Use of mixed approaches
As each approach has its own advantages and disadvantages, it is deemed better to combine quantitative and qualitative approaches. Population size, spatial scale, frequency, density, human perception, and collaboration activities are some of the indicators that are commonly used in measurement. In addition, visual mapping of various human activities—such as commercial operations, activity diversity, vegetation patterns, and the spatial scales of land use and landscape features—is integrated with perception-based studies in measuring ecological connectivity. In particular, the transdisciplinary approach helps synthesize sociocultural as well as biophysical knowledge. For example, when an environmental scientist studies the soil health and crop yield, farmers could contribute with their insights on traditional farming practices, whereas a sociologist could investigate the cultural values that shaped the farmers' decisions. This transdisciplinary approach ensures that the solutions are scientifically sound, economically viable, socially acceptable, and tailored to the specific context of the community involved for co-creating solutions to the challenges. Knowledge co-production approaches with engagement of Indigenous and local knowledge and modern science can lead to more comprehensive monitoring outcomes and reduce uncertainty.

Mapping and assessing the natural capital
The framework of natural capital takes into account the diverse values of nature as assets. While at times deemed impractical and controversial as some argue that nature should not be reduced to a monetary figure, monetary valuation of nature may be necessary to ensure nature is considered when arriving at decisions for resilience and sustainability (Natural Capital Coalition 2021). The key components for mapping and assessing ecosystem services and biodiversity health include the following:

- Ecosystem Services: Identifying and assessing the services provided by ecosystems, including provisioning, regulating, cultural, and supporting services.

- Health Index: Using illustrations in maps of consequences such as images of dead fish, dead birds, and loss of biodiversity, which indicate declining ecosystem health.
- Gene Pool: Evaluating return or increase of native species that suggest improvements in the gene pool with enhanced biodiversity.
- Population Size: Assessing changes in population size of native and even non-native species that indicate changes in ecosystem health.
- Space, Habitat Distributions, and Species Movement: Analyzing the land altered by natural events (like wildfire clearing land for pinecones) that impact habitat distributions and species movement. At times, species have adapted to human activities, illustrating that humans are integral to their habitat (e.g., managed rice paddy landscapes, managed secondary forests) while human-managed landscapes have also seen their effects on habitat space and species distribution.
- Resilience or adaptability of species: Identifying the degree of recovery or adaptation of species in human-modified habitats that indicates the role of ecological connectivity in ensuring species can thrive across different landscapes.
- Temporal Variation: Assessing temporal changes in ecosystems as species adapt to disturbances or human activity over time.

Community involvement and localized indicators
Community involvement and capacity development are critical indicators in the context of sustainability, particularly in SEPLS. The role and participation of community encapsulate the essence of local knowledge, traditional practices, and collaborative governance, which are essential for sustainable environmental management. It is imperative to ensure the greater involvement of Indigenous peoples and local communities (IPLCs), women, youth, and other marginalized groups in documenting local knowledge and monitoring the status of ecological connectivity based on localized indicators. Below are some of the useful community-level (localized) indicators:

- Human perception and traditional knowledge: Understanding local perceptions and traditional knowledge of native species can enhance biodiversity conservation efforts. Communities often possess valuable insights into local ecosystems, which can guide sustainable practices and enhance resilience. For example, in the Hualien region of Chinese Taipei, engaging local knowledge immensely helped inform strategies for optimizing ecological connectivity (Chap. 10). In the case study of Kenya, local perceptions and traditional knowledge were shown to enhance the cultivation of traditional crop varieties and native tree species on farms for food and medicinal purposes, contributing to agrobiodiversity conservation (Chap. 3).
- Activities for awareness building: Raising awareness about the benefits of diverse agricultural practices—such as mono-mixed, homogenous, and heterogeneous farms—encourages local stakeholders to recognize the interconnections within their landscapes. Horizontal integration among components like home gardens and agroforestry, along with vertical connections that link different landscape

layers (e.g., watershed management), can enhance ecosystem services and resilience (Chaps. 11 and 13).

- Participation in decision-making: Involving local communities in policy formulation ensures that development plans reflect the diversity of their environments, experiences, and needs. The participatory approach helps in informing key decision-making that impacts their ecosystems, fostering a sense of ownership and responsibility.
- Communication channels and discussion platforms: Creating platforms for community-based discussions with multiple stakeholders facilitates communication and the development of a common language among stakeholders. Incorporating social, qualitative, and policy elements into indicators ensures that the diverse needs and perspectives of local communities are reflected in environmental assessments (Chaps. 9 and 10). The Turtle River case exemplifies the importance of incorporating local voices in the development of indicators (Chap. 10). Thus, actively listening to community concerns and experiences fosters trust and enhances the effectiveness of conservation efforts.

14.2.3 Tools and Methods Used to Identify the Effectiveness of Ecological Connectivity

Actualization of the benefits of ecological connectivity would require a context-specific and purposeful selection of relevant tools and methods. The case studies in this volume highlight the use of various tools by practitioners and field experts for realizing the full potential of measuring and evaluating connectivity. For the best results, the tools and indicators need to be adjusted to the local context in the SEPLS.

Technology-based tools

At the species level, camera traps, GPS tracking, and genomic technologies for destructive (e.g., hair or feces analysis) and non-destructive (e.g., tissue or blood analysis) genomic analysis may be used. At the habitat level, mapping, satellite or drone images, and soil and water analysis may be used. Both levels together contribute to obtaining a holistic evaluation of ecological connectivity and understanding of the ecosystem.

Some useful software packages and add-on modules to GIS and image processing software that are widely used to assess habitat fragmentation and landscape connectivity through computation of various landscape metrics are Fragstats (McGarigal and Marks 1995), Morphological Spatial Pattern Analysis (MSPA) tool (Soille and Vogt 2022), Patch Analyst in ArcGIS (Lin 2007), Arc_LIND in ArcGIS (Yu et al. 2019), and LecoS in QGIS (Jung 2016). GIS analysis with models such as corridor modeling, least-cost path estimation, habitat suitability, circuit theory, graph theory, and agent- or individual-based models (Keeley et al. 2016; Correa Ayram et al. 2016; Unnithan Kumar and Cushman 2022) are used to study connectivity.

Resilience assessment tools and methods
Resilience assessment workshops based on the 20 indicators of resilience in SEPLS possess a high potential as a community-based tool for monitoring ecological connectivity in SEPLS (Dunbar et al. 2020). They can help to understand the precise meaning of ecological connectivity in each SEPLS context (e.g., as the integrity of all landscape elements or as the presence of an indicator migratory species) and elicit priority issues and challenges at each stage of the adaptive co-management process (Chap. 10). Resilience assessment workshops, for example, were applied as the main tool for identification of priority connectivity issues in Fengnan, Xinshe, Gangkou, and Luoshan SEPLS, Hualien County, Chinese Taipei (Chap. 10). This also integrates the use of traditional methods such as community tracking tools and models suggested by the Intergovernmental Science-Policy Platform on Biodiversity and Ecosystem Services (IPBES) such as the scenario model (IPBES 2016). Likewise, a coarse-filter approach, in contrast to a fine-filter approach, as a conservation strategy that focuses on broad habitat types and ecosystem-level processes rather than focusing on individual species (Haufler et al. 1996), can be useful for habitat-based conservation planning.

Multi-stakeholder approach
This approach is adopted with the aim of bringing diverse groups and individuals to chalk out strategies to address complex environmental, social, and governance issues. By involving various stakeholders—local communities, governments, NGOs, scientists, and private sector actors—this approach ensures that multiple perspectives and knowledge systems contribute to sustainable solutions. Through participatory approaches to land-use planning, such as Public Participation Geographic Information System (PPGIS), and participatory mapping, local communities can engage in the decision-making process. Citizen science and citizen patrolling of forest, river, coastal, and marine areas involving bird watching and coral reef checks (with trainings and licenses) can lead to empowerment of local stewardship. Finally, this multi-stakeholder approach emphasizes stakeholder engagement and cross-sectoral collaboration, ensuring all interests are aligned toward a common goal.

Economics and governance of ecosystem services
Economics and governance play a vital role in enhancing ecological connectivity through the management of ecosystem services. By understanding the economic value of the services, we can create frameworks that promote sustainability while ensuring that local communities benefit from their natural resources. Below are key pointers that are used to identify the effectiveness of measures adopted to foster ecological connectivity.

- Implementing Payments for Ecosystem Services (PES) programs: This incentivizes landowners and communities to maintain or restore ecosystems by providing financial compensation for the services they offer, such as carbon sequestration, water purification, and biodiversity conservation. For instance, in the Komodo Islands, PES initiatives encouraged local communities to protect their ecosystems while simultaneously benefiting economically (Yu et al. 2020).

- Exploring alternative services and products: Those services and practices that arise from sustainable practices can enhance economic viability. This can involve supporting local industries that rely on natural resources such as fisheries, agriculture, and ecotourism, creating diverse income sources that reinforce the importance of maintaining ecological connectivity.
- Examining the dependency on tourism: The dependency on tourism is a crucial aspect as it is often assumed to drive economic growth in regions rich in biodiversity. However, tourism must be managed sustainably to ensure that it does not lead to habitat degradation or ecological disruption. Strategies should include community-based tourism initiatives that empower locals and prioritize conservation by assessments through tourism carrying capacity studies.
- Establishing certification schemes and public recognition: Such programs can enhance market access for sustainably produced goods. Green labelling provides consumers with information about environmentally friendly practices, encouraging them to support businesses that contribute to ecosystem conservation.
- Addressing harmful subsidies: The subsidies that incentivize environmentally damaging practices such as converting wetlands into fruit lands should be discouraged. Governments must reassess their investment and taxation schemes to ensure that they do not encourage practices that undermine ecological connectivity.
- Promoting nature-positive subsidies: Subsidies can guide economic activities toward sustainable practices. These subsidies can support reforestation, wetland restoration, and sustainable agriculture, fostering economic growth while enhancing ecosystem health.
- Implementing tax measures: Taxes on activities that harm ecosystems, such as converting wetlands, can serve as a deterrent and encourage landholders to engage in sustainable practices. The revenue generated can be reinvested into conservation initiatives, creating a positive feedback loop for ecological connectivity.

Communication, Education, Participation, and Awareness (CEPA)

To tailor communications effectively, information and data should be shared in formats that resonate with different stakeholder groups. Key points with regard to CEPA are as follows:

- Compensating for resource scarcity and bridging gaps: One of the key challenges in communication is the resource crunch for updating data and information. This highlights the need for a freely accessible platform where data can be shared across sectors, ensuring transparency and collaboration. Further, facilitators or boundary brokers play an important role in bridging gaps between various stakeholder groups, ensuring effective and lasting connectivity processes.
- Long-term monitoring: Capacity development and knowledge transfer are essential for long-term ecological monitoring. For instance, local communities can be

trained in community-based biodiversity monitoring and patrolling of their environments, fostering local stewardship and ongoing engagement. Similarly, capacity development, cross-site learning, and knowledge exchange among stakeholders are critical to expanding the scope and effectiveness of these partnerships for sustainable outcomes. For instance, the long-term monitoring of the brown bear population in Northwestern Spain over several decades has enabled the successful detection of changes in its demographic structure, growth performance, viability, and dispersion, allowing for formulation of effective conservation measures in the context of local disappearance or even decline of the species in Southwestern Europe (Chap. 6).
- Intergenerational knowledge exchange: Since knowledge is held in specific vernacular languages depending on the context, the dissemination of ideas can fail if the language used is not adapted to the local context (e.g., training chestnut growers in Spain using communication methods entirely adapted to the local context—Chap. 6). It is vital for raising awareness, as seen in programs like mentorships or intensive communication and knowledge sharing between older and younger generations with careful consideration of language. An example includes the Indigenous Bunun practices in the Nan'an Tribe (Chap. 4), where traditional knowledge is passed down, enhancing understanding of ecosystem management over time.

14.3 Ways Forward: How to Ensure Ecological Connectivity in the Contexts of SEPLS

14.3.1 Interconnected Nature of Efforts for Ecological Connectivity

The efforts to maintain, enhance, or restore ecological connectivity largely take the form of either or both changes in land and sea uses and/or changes in management practices. A prime example of the former is setting aside some farmland for a conservation corridor. The cases of the latter include the shift from conventional agricultural practices to organic farming and its expansion (Chap. 11) and the promotion of agrobiodiversity conservation for enhanced ecological connectivity between forests and surrounding farmlands (Chap. 3). Despite the initial intentions, these efforts may give rise to unintended consequences (e.g., a spread of invasive species, an increase in human-wildlife conflicts) as SEPLS are complex social-ecological systems inherently involving nonlinear dynamics and uncertainties.

Effects of interventions for ecological connectivity, either positive or negative, are often intimately and immediately felt on site. Some of them are directly related to land and sea uses (e.g., loss of agricultural income generation opportunities for locals), but others are also associated with land tenure, access rights to natural resources, and various types of capital. For instance, on the one hand, despite the intention to promote sustainable and equitable production, the state-led sustainability certification schemes for palm oil in Indonesia appeared to be less effective for the smallholders who had neither clear land legality nor organizational capacities in the face of several challenges such as additional administrative procedures, stricter restrictions for farm management, and higher upfront and recurring costs (De Vos et al. 2023). On the other hand, biocultural enterprises contributed to an increase in incomes for locals and reduced pressures on Kaya forests in Rabai, Kenya (Chap. 3). Importantly, these effects can extend to broader spatial and temporal scales. For instance, fertilizer use in an inland area can impact downstream and marine and coastal areas, whereas plastic wastes running into the ocean can damage fisheries and human and ecosystem health. The shift from conventional to eco-friendly farming may also incur short-term losses but may pay off in the long run. Furthermore, the values of a certain species may not be recognized immediately but could be noticed and appreciated once it is lost.

As mentioned above, these effects are often conflicting or contradictory in terms of cost, benefits, interests, and needs among different stakeholders. Such incompatible results can arise from unintended consequences attributed to the dynamic complexity of human-nature interactions (e.g., less yields from high-cost organic production). Yet, certain efforts may result in anticipated effects but can entail conflicts in values and interests among stakeholders across different spatial and temporal scales. For instance, the enhanced connectivity for growing populations of brown bear in NW Spain can threaten beekeepers and other stakeholders due to an increased possibility of bear attacks on livestock and humans (Chap. 6). Thus, such incompatibility, almost by design, results from diverse human standpoints and attitudes—including different responsibilities, mandates and jurisdictions for interventions, different levels of awareness, diverse conceptualization and perspectives, and varied capitals and capacities of stakeholders to address ecological connectivity.

14.3.2 Key Aspects of Landscape Approaches for Ecological Connectivity

SEPLS management rests on landscape approaches, which render long-term collaborative processes for diverse stakeholders to negotiate and deliberate to balance multiple objectives in a given geographical area. These approaches help to align and coordinate the varied human standpoints and attitudes for more concerted efforts for quality ecological connectivity on a landscape or seascape scale and beyond. In particular, the following features of landscape approaches can facilitate synergistic efforts and systemic changes in landscapes and seascapes for socially and ecologically sound outcomes.

Place-based schemes

A place-based scheme allows for placing local actors at the center of decision-making and emphasizing the role of Indigenous and local knowledge in the efforts for ecological connectivity. For instance, selection of species or areas important for conservation could be made in consultation primarily with local stakeholders (e.g., farmers who use them for their livelihoods). This allows for turning certain ecological problem(s) into opportunities when looking to solutions in combination with other social, economic, environmental, and cultural issues relevant to the area. In this connection, it is also crucial to consider the broader historical, normative, and legal contexts of the place, including land tenure (e.g., tensions between local communities and the government on types of land conservation as in Chap. 8) and land and sea stewardship (e.g., local commitments to sustainable management practices as in Chap. 3). Such considerations help local actors to better engage in the project for ecological connectivity with a stronger sense of ownership.

A "place" should be considered not static but dynamic as discussed below for adaptive co-management strategies. As a first tangible step, however, the place-based approach helps to set an appropriate boundary of an intervention attending to manageability or feasibility and representation of the area, while identifying specific measures on a certain scale where a relatively full cycle of ecological processes manifests along with a few key components connected. Nevertheless, it is also important to contemplate social-ecological processes outside of the boundary that may affect the phenomena inside the boundary, whereas the boundary may need to be adjusted over time to reflect emerging needs and changes (e.g., species movement under climate change).

Multi-stakeholder processes

Building on the place-based approach, a multi-stakeholder process enables multiple stakeholders across different sectors and jurisdictions to complement and coordinate with each other through sharing their views, knowledge and expertise, and tasks and practices in the efforts for ecological connectivity. Key stakeholders may include Indigenous peoples and local communities (IPLCs)—including women, youth, and local champions (who are passionate about collective benefits and committed to proactive engagement in sustainable development), government agencies (including agricultural extension officers), private sector, academia, NGOs, and religious organizations.

Among these stakeholders, facilitators or boundary brokers (e.g., NGOs as in Chap. 6, academic institutions as in Chap. 10) serve as key intermediaries to ensure local participation and promote interactions between different stakeholders. Qualified as being honest, dedicated, and flexible, they skillfully enact transdisciplinary approaches to ecological connectivity, ensuring and enhancing cross-sectoral commitments, sectoral coherence, policy alignment across various levels, and adaptability of the efforts where new issues or stakeholders emerge from time to time. Furthermore, not only individuals or organizations but also certain mechanisms enable multi-stakeholder collaboration and coordination. For instance, although a governmental structure in many cases does not easily allow for seamless communications and effective cross-sectoral collaboration, integrated policies or governance structures sometimes serve as a facilitation mechanism to foster synergistic efforts at a multi-stakeholder interface (e.g., Taiwan Ecological Network

(TEN) including the landscape-seascape interfaces as in Chap. 10, and the Asunafo-Asutifi Landscape governance structure at the local level as in Chap. 2).

Adaptive co-management strategies
Throughout the long-term collaborative processes of implementing landscape approaches, an adaptive co-management approach helps to reflect on short- and mid-term effects toward a long-term vision, address risks and uncertainties associated with the interventions, incorporate lessons learned into next steps, and scale up the initiatives for broader impacts to meet the higher-level agenda for sustainability. In these processes, monitoring and evaluating the interventions is imperative so that the results from one cycle of a project could be reflected in its next cycle for changes or adjustments (e.g., SEPLS boundaries for the project), whereas new phenomena, new ideas, or unintended consequences (e.g., human-wildlife conflicts as in Chap. 6) should also be sensitized by those involved in the project. Also, the adaptive co-management approach allows for cautious and deliberative processes of upscaling the place-based good practices. Rather than by simply replicating such practices from one site to somewhere else as a one-size-fits-all solution, this could be done through communication and iterative adaptation (e.g., a mixture of bottom-up and top-down approaches as in Chap. 13) to reflect the place and scale contexts and deliver tailored solutions in accordance with such specificities.

14.3.3 Operationalizing Landscape Approaches in Spatial Planning and Management Practices

The process of operationalizing landscape approaches in spatial planning and related management activities for ecological connectivity may involve the following iterative and overlapping steps: (1) setting the scene for spatial planning, (2) establishing incentive and financing mechanisms, and (3) building partnerships for integrated spatial planning and management practices.

Setting the scene for spatial planning
The first step is to have local stakeholders recognize and understand the issues to be addressed, easing any tensions and building trust between different stakeholders. It is necessary to foster a common vision among them to pursue subsequent efforts collectively. At this stage, facilitators or boundary brokers (e.g., local champions or outsiders who have long worked with local communities) can serve critical roles, for instance, in raising awareness of local stakeholders for collective action (e.g., an NGO intervention for awareness raising of farmers leading to their concessions of land for corridor development as in Chap. 7). In such a process, a *common platform* helps to align different views and needs to build a common vision (e.g., Chaps. 9 and 10). Here, finding a *common language* across different stakeholders is a key to facilitate mutual understanding and knowledge sharing. Additionally, in this process, it is essential to secure the rights of local stakeholders and empower them for sound decision-making and action.

Establishing incentive and financing mechanisms
Appropriate incentives and financial mechanisms should be devised and established for continued stakeholder engagement. Developing such mechanisms requires consideration of incentives suited for different stakeholders' contexts. In this regard, short-term needs should be addressed in many cases to realize long-term benefits. For instance, fixed-term subsidies can be provided for farmers who encounter certain livelihood loss from the change in farming practices until their produce can be profitable, though it is often challenging to set an appropriate timeline for such finance. At the same time, long-term considerations are needed for financial sustainability and continuity of the efforts for ecological connectivity. For instance, youth are expected to be continuously involved for long-term engagement rather than as volunteers on a temporary or ad hoc basis to support the initiatives (e.g., Chap. 13). In this connection, job markets for the youth, which could even be competitive with those in cities, are key to securing their livelihoods and attracting them to engage in the initiatives.

Potentially useful mechanisms are available to promote and enhance the financial sustainability of interventions, but they require caution as they sometimes bring about challenges for their development and implementation. For instance, biodiversity offsets offer opportunities for financing the creation or enhancement of ecological connectivity with economic incentives for certain stakeholders such as the business sector and landowners. This requires the provision of sufficient and accurate information to all stakeholders so that they understand what is proposed to be done. It is also important to maintain a biodiversity credit market with the participation of quality buyers and sellers for proper implementation (Conservation International 2024). Another example is a tax system, through which, for instance, all the property owners (e.g., private companies and citizens) in many of the regions in Italy have been supporting the maintenance of river corridors and wetland areas since the mid-1970s. However, the enforcement of this relatively top-down approach built on the historical legal evolution since the late nineteenth century was possible in the 1970s and 1980s with favorable political will, but may be impossible to do so under the current political climate. To develop and implement these mechanisms, results-based step-by-step approaches may work to better engage stakeholders in the initiatives and deliver sound conservation outcomes (e.g., payments or reward schemes for the local communities evidently contributing to the good condition of ecosystems).

Building partnerships for integrated spatial planning and management practices
Related to the abovementioned mechanisms, strategic partnership building is crucial in the process of planning and implementation. Among various stakeholders, private sector entities are expected to be actively brought in for economic opportunities and financial sustainability. They can be one of the major funders of the initiatives, along with other contributors such as governments and multilateral donors (e.g., international banks). The private sector involvement may allow for not only a one-time investment or initial fund provision but also long-term engagement, for instance, in sustainable business activities that also help other stakeholders to remain in the process. To ensure the quality of the initiatives, however, it is

important to avoid greenwashing (e.g., just naming business-as-usual practices as sustainable practices) and to accord with and even go beyond corporate social responsibility (CSR), for example, habitat conservation for endemic species by Toyota as a CSR initiative in Thailand.[1] Furthermore, private sector engagement helps with devising practically effective strategies as in the case of Italy where truffles associations took care of some corridors for ecosystem conservation in exchange for their exploitation of truffles and associated resources that benefited from conservation (Gualandi 2025).

Even with the strategic involvement of the private sector as a key actor, multi-stakeholder partnerships—including local communities, NGOs, and public and private sectors—should be established and strengthened for mutually enabling contexts in the process of planning and implementation. A series of multi-stakeholder meetings underpinned by landscape approaches allow for not only sharing and aligning different interests and objectives for common commitments but also drawing on a wide range of knowledge systems and technologies (e.g., traditional knowledge, AI for monitoring) for innovative solutions such as stone wires restoration for multiple purposes (Chap. 13). Furthermore, certain guidelines or certification schemes may help to ensure that different stakeholders comply with protocols, standards, or criteria required for ecological connectivity. This may further facilitate complementing each other to address the lack of technical expertise and financial resources. In this connection, ensuring international links of local and regional initiatives (e.g., disclosure recommendations and guidance by the Taskforce on Nature-related Financial Disclosures (TNFD), Kunming-Montreal Global Biodiversity Framework, Other Effective area-based Conservation Measures (OECMs), Natural Capital Protocol, and Verified Carbon Standard) should be important for concerted efforts across different governance levels.

As a result, integrated plans for spatial arrangements and management practices should be developed and implemented. Such a plan consists of at least three major components. First, scale-specific concrete actions should be laid out as short-term and long-term strategies for adaptive and anticipatory management actions. In this regard, a clear implementation plan is fundamental besides action items. For instance, the cocoa planting design was developed along with the provision of certified sourcing materials, financial investments, and awareness raising of farmers in Ghana, ensuring the feasibility of the plan (Chap. 2). Second, strategies for extending or upscaling of landscape- or seascape-scale collaboration to different or broader areas should be explored to address diffusive and potentially uncontrollable effects of the interventions for ecological connectivity. The challenge of collaborative action among a large number of actors has to be overcome with sufficient information sharing among them. Finally, the components of monitoring and evaluation should be incorporated in the plans, where small steps should be taken and adjusted. Although more funds for data collection and analyses may be required for this

[1] The example of Biodiversity & Sustainability Learning Center launched by Toyota to complement its Biotope project in Thailand in 2016 can be found at https://global.toyota/en/detail/12564657

process, some existing frameworks and tools may be useful, for example, IUCN's management effectiveness tracking tool (Stolton et al. 2021).

14.4 Conclusion

Landscape approaches can facilitate the optimization of multiple tangible and intangible benefits from ecological connectivity in SEPLS, while minimizing negative trade-offs arising from the efforts in ensuring and enhancing ecological connectivity. A place-based scheme, which underpins landscape approaches, centers local actors in the decision-making processes, rendering Indigenous and local knowledge for locally adequate and adaptable interventions. It also helps to set the scene for spatial planning by better understanding and sharing local issues and building on them to foster a common vision among different stakeholders. Additionally, it aids in setting an appropriate spatial boundary of an initial intervention, whereas social-ecological processes beyond the boundary can be explored and contemplated essentially from the local perspectives.

Also, multi-stakeholder processes of landscape approaches enable effective partnership building for ecological connectivity among different stakeholders, ranging from IPLCs, government agencies, NGOs, and academia to the private sector. Among others, qualified facilitators or boundary brokers could serve as key intermediaries to capitalize on the place-based scheme with local participation and promote transdisciplinary approaches to complementary cross-sectoral collaboration and policy coherence. Strategic partnerships in which the private sector is actively brought along with other stakeholders would also help to develop effective incentive and financing mechanisms. Such mechanisms can be designed to address both short-term needs and long-term benefits for all the stakeholders so as to ensure sustainability of the interventions. Other mechanisms or tools, such as a common platform, guidelines, and certification schemes, may also be leveraged to enhance and promote quality and competent partnership building for concerted efforts among different stakeholders.

Finally, building on landscape approaches, adaptive co-management strategies for ecological connectivity help to address risks and uncertainties associated with the interventions in a timely manner by managing short- and mid-term effects to achieve the long-term vision for socially and ecologically sound outcomes. Here, monitoring and evaluation play a crucial role to facilitate adaptive strategies, wherein the stakeholders may choose, adapt, and apply certain methods among a diverse set of indicators, methodological approaches, and tools earlier introduced, in accordance with the local and regional contexts. If place-based good practices are successfully upscaled through adaptive strategies, local solutions would effectively contribute to simultaneously achieving the sustainable development goals (SDGs) (e.g., SDGs 1, 2, 3, 11, 13, 14, 15, and 17) and the goals and targets of the Kunming-Montreal Global Biodiversity Framework (e.g., Goal A, Targets 1, 2, 3, and 11).

References

Asare R, Afari-Sefa V, Osei-Owusu Y, Pabi O (2014) Cocoa agroforestry for increasing forest connectivity in a fragmented landscape in Ghana. Agrofor Syst 88:1143–1156

Berkström C, Wennerström L, Bergström U (2022) Ecological connectivity of the marine protected area network in the Baltic Sea, Kattegat and Skagerrak: current knowledge and management needs. Ambio 51(6):1485–1503

Conservation International (2024) Nature Positive: Conservation International's Position on Nature Credit Markets. May 2024. https://cicloud.s3.amazonaws.com/docs/default-source/s3-library/publication-pdfs/conservation-international-nature-credit-position-statement.pdf

Convention on Migratory Species (CMS) (2020) Improving ways of addressing connectivity in the conservation of migratory species, resolution 12.26 (REV.COP13), Gandhinagar, India (17–22 February 2020). Available at https://www.cms.int/sites/default/files/document/cms_cop13_res.12.26_rev.cop13_e.pdf

Correa Ayram CA, Mendoza ME, Etter A, Salicrup DRP (2016) Habitat connectivity in biodiversity conservation: a review of recent studies and applications. Prog Phys Geograph Earth Environ 40(1):7–37. https://doi.org/10.1177/0309133315598713

De Vos RE, Suwarno A, Slingerland M, Van Der Meer PJ, Lucey JM (2023) Pre-certification conditions of independent oil palm smallholders in Indonesia. Assessing prospects for RSPO certification. Land Use Policy 130:106660

Díaz-Varela ER, Álvarez-Álvarez P, Pérez-Girón JC, Díaz-Verla RA, Ferreira Golpe MA, García-Arias AI (2025) Chestnut production-related businesses in the Courel Mountains of Galicia, NW Spain: an opportunity for biodiversity conservation, ecosystem restoration and rural development. In: Nishi M, Subramanian SM, Varghese P (eds) Business and biodiversity: reciprocal connections in the context of Socio-Ecological Production Landscapes and Seascapes (SEPLS). Satoyama Initiative Thematic review, vol 9. Springer, Singapore, pp 31–49

Dunbar W, Subramanian SM, Matsumoto I, Natori Y, Dublin D, Bergamini N, Mijatovic D, Alvarez AG, Yiu E, Ichikawa K, Morimoto Y, Halewood M, Maundu P, Salvemini D, Tschenscher T, Mock G (2020) Lessons learned from application of the "Indicators of Resilience in Socio-ecological Production Landscapes and Seascapes (SEPLS)" under the Satoyama Initiative. In: Saito O, Subramanian SM, Hashimoto S, Takeuchi K (eds) Managing socio-ecological production landscapes and seascapes for sustainable communities in Asia. Science for Sustainable Societies. Springer, Singapore, pp 93–116. https://doi.org/10.1007/978-981-15-1133-2_6

Gualandi G (2025) Traditional environmental knowledge and trees conservation: the example of the White Truffle (*Tuber Magnatum Pico*). In: Nishi M, Subramanian SM (eds) Business and biodiversity: reciprocal connections in the context of Socio-Ecological Production Landscapes and Seascapes (SEPLS). Satoyama Initiative Thematic review, vol 9. Springer, Singapore, pp 51–70

Haufler JB, Mehl CA, Roloff GJ (1996) Using a coarse-filter approach with species assessment for ecosystem management. Wildl Soc Bull 24(2):200–208

Hilty J, Worboys GL, Keeley A, Woodley S, Lausche B, Locke H, Carr M, Pulsford I, Pittock J, White JW, Theobald DM, Levine J, Reuling M, Watson JEM, Ament R, Tabor GM (2020) Guidelines for conserving connectivity through ecological networks and corridors. Best Practice Protected Area Guidelines Series No. 30. IUCN, Gland, Switzerland

IPBES (2016) Methodological assessment report on scenarios and models of biodiversity and ecosystem services

Jung M (2016) LecoS — a python plugin for automated landscape ecology analysis. Eco Inform 31:18–21. https://doi.org/10.1016/j.ecoinf.2015.11.006

Keeley ATH, Beier P, Gagnon JW (2016) Estimating landscape resistance from habitat suitability: effects of data source and nonlinearities. Landsc Ecol 31:2151–2162

Lemieux CJ, Beazley KF, MacKinnon D, Wright P, Kraus D, Pither R, Crawford L, Jacob AL, Hilty J (2022) Transformational changes for achieving the Post-2020 Global Biodiversity Framework ecological connectivity goals. FACETS 7(1):1008–1027

Lin JP (2007) Availability of patch calculator, an ArcGIS v. 9 tool for the analysis of landscape patches

McGarigal K, Marks BJ (1995) FRAGSTATS: spatial pattern analysis program for quantifying landscape structure. Gen. Tech. Rep. PNW-GTR-351. U.S. Department of Agriculture, Forest Service, Pacific Northwest Research Station, Portland, OR, p 122

McRae BH, Hall SA, Beier P, Theobald DM (2012) Where to restore ecological connectivity? Detecting barriers and quantifying restoration benefits. PLoS One 7(12):e52604. https://doi.org/10.1371/journal.pone.0052604

Natural Capital Coalition (2021) Natural Capital Protocol. https://capitalscoalition.org/capitals-approach/natural-capitalprotocol/?fwp_filter_tabs=guide_supplement

Rudnick D, Ryan SJ, Beier P, Cushman SA, Dieffenbach F, Epps CW, Gerber L, Hartter J, Jenness J, Kintsch J, Merenlender AM, Perkle RM, Preziosi DV, Ryan SJ, Trombulak SC (2012) The role of landscape connectivity in planning and implementing conservation and restoration priorities. Issues Ecol 16:1–23

Soille P, Vogt P (2022) Morphological spatial pattern analysis: open-source release. Int Arch Photogramm Remote Sens Spatial Inf Sci XLVIII-4/W1-2022:427–433. https://doi.org/10.5194/isprs-archives-XLVIII-4-W1-2022-427-2022

Stolton S, Dudley N, Hockings M (2021) METT handbook: a guide to using the Management Effectiveness Tracking Tool (METT). Second edition guidance for using METT-4. WWF, Gland

Unnithan Kumar S, Cushman SA (2022) Connectivity modelling in conservation science: a comparative evaluation. Sci Rep 12:16680. https://doi.org/10.1038/s41598-022-20370-w

Yu M, Huang Y, Cheng X, Tian J (2019) An ArcMap plug-in for calculating landscape metrics of vector data. Eco Inform 50:207–219

Yu H, Xie W, Yang L, Du A, Almeida CMVB, Wang Y (2020) From payments for ecosystem services to eco-compensation: conceptual change or paradigm shift? Sci Total Environ 700:134627. https://doi.org/10.1016/j.scitotenv.2019.134627

Correction to: Ensuring Ecological Connectivity in Socio-Ecological Production Landscapes and Seascapes (SEPLS)

Maiko Nishi, Suneetha M. Subramanian, Philip Varghese, and Juliano Sènanmi Hermann Houndonougbo

Correction to: Chapters 1 and 14 in : M. Nishi et al. (eds.), *Ensuring Ecological Connectivity in Socio-Ecological Production Landscapes and Seascapes (SEPLS)*, Satoyama Initiative Thematic Review, https://doi.org/10.1007/978-981-95-1474-8

The original version of the book was inadvertently published with errors. The following corrections have been made after the original publication.

Chapter 1:

- Part of caption for Fig. 1.1 "(Map template: Geospatial Information, United Nations). Note: Details of the case study locations, including geographic coordinates, are described in each chapter." was placed separately in the original version, whereas it has now been included within the figure caption.

Chapter 14:

- The source for caption for Fig. 14.1 was not published in the original version, and it has now been updated as "Fig. 14.1 Overview of key aspects relating to ecological connectivity in SEPLS, as detailed in this chapter (Source: Authors)."
- Third affiliation for editor Philip Varghese, "CHRIST University, Bangalore, India" was included which was missing in the original version.

The updated versions of these chapters can be found at
https://doi.org/10.1007/978-981-95-1474-8_1
https://doi.org/10.1007/978-981-95-1474-8_14

M. Nishi et al. (eds.), *Ensuring Ecological Connectivity in Socio-Ecological Production Landscapes and Seascapes (SEPLS)*, Satoyama Initiative Thematic Review, https://doi.org/10.1007/978-981-95-1474-8_15

The opinions expressed in this chapter are those of the author(s) and do not necessarily reflect the views of the United Nations University.

MIX
Papier aus verantwortungsvollen Quellen
Paper from responsible sources
FSC® C105338

If you have any concerns about our products,
you can contact us on
ProductSafety@springernature.com

In case Publisher is established outside the EU,
the EU authorized representative is:
Springer Nature Customer Service Center GmbH
Europaplatz 3, 69115 Heidelberg, Germany

Printed by Libri Plureos GmbH
in Hamburg, Germany